PRESERVING THE LEGACY

Preserving the Legacy

Site Characterization

Sampling and Analysis

HMTRI

NATIONAL SCIENCE FOUNDATION

IN-TELE-COM®
INTELLIGENT TELECOMMUNICATIONS

COMMUNITY COLLEGES ★ GOVERNMENT ★ INDUSTRY

JOHN WILEY & SONS, INC.

Technical Illustration: Richard J. Washichek, Graphic Dimensions, Inc.
Photo Research: Beatrice Hohenegger

This material is based on work supported by the National Science Foundation under Grant No. DUE94-54521. Any opinions, findings, and conclusions or recommendations expressed in this material are those of the author(s) and do not necessarily reflect those of the National Science Foundation.

This book is printed on acid-free paper. ♾

Published by John Wiley & Sons, Inc., Hoboken, New Jersey
Published simultaneously in Canada.

Library of Congress Cataloging-in-Publication Data:

ISBN: 0-471-29270-2

Printed in the United States of America

20 19 18 17 16 15 14 13

DEDICATION

To the wonderful blending of knowledge and skills characteristic of today's technician.

Table of Contents

To the Student

The environmental technician is a valuable team member in the site characterization process. By effectively carrying out the intent of a sampling plan, the technician is critically important to the entire process of obtaining accurate and precise laboratory results. High quality lab data enable the data users to make decisions about a site – sometimes multi-million dollar decisions! The goal of *Site Characterization – Sampling and Analysis* is to provide you, the technician, with the knowledge and skills necessary to assist a site characterization project planner in the sampling and monitoring process. The text repeatedly emphasizes the expectation that the technician adhere to procedures set forth in a sampling plan; this should not imply, however, that the environmental technician performs tasks in a lockstep manner. Unexpected obstacles often prevent such an approach and require creative thinking and alternative solutions. In those instances, a rational decision on how to proceed can only be reached by applying critical thinking skills to an understanding of the sampling plan's objectives. Thus, this book is designed not only to inform the environmental sampler about the procedural steps to follow but also to explain much of the underlying philosophy of sampling, monitoring, and data interpretation.

To help you focus your studies, each chapter includes learning objectives, introductory concept statements, questions to check the understanding of each section, illustrative diagrams and photographs, and a summary of the main ideas. Within the text, regulatory information is consistently cited so that you may supplement your studies by reading pertinent passages from the U.S. Code of Federal Regulations. The chapters conclude with suggested activities to encourage you and your class to engage in more complex thinking. Special attention has been given to providing an extensive glossary and index as well as a comprehensive bibliography to guide further exploration of the topics.

Highlighted in special text boxes are many occupational applications of math, science, and technical concepts. The concepts are among those identified by the ATEEC Fellows in concert with industry practitioners. ATEEC, the Advanced Technology Environmental Education Center, has convened its annual Fellows Institute bringing together teachers from across the nation to study and advise upon the knowledge and skill base necessary for performing the work of various environmental technology occupations as well as to make curricular recommendations. INTELECOM has asked its authors to include concepts from the ATEEC knowledge and skill base within the **Preserving the Legacy** series textbooks. The efforts of INTELECOM and ATEEC together thus contribute to courses that are valid for industry needs. Of course, a knowledge and skill base is the beginning; comprehension of the subject matter in *Site Characterization – Sampling and Analysis* is further enhanced by associated coursework in chemistry, biology, environmental regulations, and college-level mathematics, which are usually prerequisites or corequisites to environmental sampling courses in two-year college degree programs.

The writing team and all the partners in this project acknowledged below wish you a safe and productive career in environmental technology and hope that *Site Characterization – Sampling and Analysis* will be a helpful learning tool and reference.

Acknowledgments

The following persons and institutions have provided invaluable support to our writing team:

The National Science Foundation (NSF) for its generous support of environmental technology education by funding the INTELECOM **Preserving the Legacy** project;

Intelligent Telecommunications Inc. (INTELECOM) and the Partnership for Environmental Technology Education (PETE), for the opportunity to contribute to a project they have worked so hard to realize;

The Advanced Technology Environmental Education Center (ATEEC) – which is supported by NSF and is a partnership of the Hazardous Materials Training and Research Institute (HMTRI), PETE, and the University of Northern Iowa – for the time provided me to lead the writing team;

The reviewers for the **Preserving the Legacy** National Academic Council (NAC), representing the six PETE regions, for studying drafts; especially Dr. Bill Engel, University of Florida TREEO Center for his skill in editing, and Howard Guyer, NAC team leader, for his hours of thoughtful reading, suggestions, and leadership;

Steve Fenton for his wise leadership as the HMTRI project manager for this book, and Doug Feil, the HMTRI representative on the NAC, for believing in us;

The team of Graphic Dimensions, Inc. for their skill in page and graphic design;

Our HMTRI-ATEEC instructional design colleagues for their constant encouragement and good humor during the project; also, our families for their inspiration, especially Linda Peterson who shared her ideas and expertise as a chemistry laboratory analyst.

Most particularly, Beatrice Hohenegger, our INTELECOM editor, who not only demanded but also demonstrated excellence. Thank you for your guidance and long hours, Beatrice.

Diane Dickens Gere, Instructional Designer
HMTRI
ATEEC, 1997

The HMTRI Writing Team

The Hazardous Materials Training and Research Institute or HMTRI which is listed as the author of this book was established in 1987 by the Eastern Iowa Community College District (EICCD) headquartered in Davenport, Iowa and Kirkwood Community College in Cedar Rapids, Iowa. The writing team includes an instructional designer and three subject matter experts:

David Foss, Ph.D.

Dr. David Foss is a chemical engineer and environmental coordinator for the Rock Island Arsenal, a U.S. Army major manufacturing center. David is a liaison between the Arsenal and local, stateand federal government agencies. He earned his B.S. in chemistry from Wisconsin State College at River Falls and his M.S. and Ph.D. in Chemistry from Michigan Technological University. He is a member of the American Chemical Society.

Prior to joining the Rock Island Arsenal, David discovered his interest in environmental chemistry when he was a chemistry laboratory supervisor with the John Deere Company. David's experience includes working with the State of Iowa in developing test procedures for waste analysis as well as a training curriculum for asbestos remediation in Iowa schools. David has served since 1987 as an adjunct instructor and advisory committee member for the Environmental Compliance and Technology program (formerly the Hazardous Materials Technology program) of the Eastern Iowa Community College District.

Sally Gaines

Sally Gaines is lead instructor in the Environmental Compliance and Technology credit program at Eastern Iowa Community College District and participates in team teaching in HMTRI short-term training. Sally has a B.S. in Geology from Indiana University, an M.S. in Geology from the University of Kentucky, and is a registered geologist. Her memberships include the Association of Groundwater Scientists and Engineers, the American Geophysical Union, the Iowa Ground Water Association, and the American Society for Testing and Materials.

Sally teaches courses in EPA and OSHA Regulations, Characteristics of Hazardous Materials, Sampling and Monitoring Procedures, Introduction to Environmental Technology, Health Effects, and Contingency Planning. Prior to teaching, Sally worked as a project manager and hydrogeologist for environmental consulting firms located in the Iowa Quad Cities area. In this capacity, she gained experience in soil and groundwater investigations, remediation of petroleum release sites, underground storage tank (UST) investigations and closures, RCRA (Resource Conservation and Recovery Act) closure plans and investigations, site safety plans, air emissions inventories, landfill hydrogeologic studies, landfill monitoring, and investigations related to property transfers.

Diane Gere

Diane Gere is the coordinator and HMTRI editor for this textbook. She received her B.S. in English and Speech at Iowa State University and her M.A. in Instructional Design and Technology from The University of Iowa. She began her English and speech teaching career in 1970 and joined HMTRI as an instructional designer in 1990. In this capacity she is on the coordinating team and staff for the ATEEC Fellows Institute and its curricular projects, provides resources to two-year colleges and high schools that are starting or improving their environmental technology programs, and produces and maintains ATEEC's World Wide Web site at www.ateec.org, which is a dissemination tool for the work of the ATEEC Fellows Institute.

Among Diane's HMTRI publications are *Solvent Waste Reduction and Recycling – Practical Advice for Small Business*, written with David Foss for the Iowa Waste Reduction Center of the University of Northern Iowa, and *Living and Writing Responsibly – Environmental Readings for Classic and Technical Composition.*

Julie Plummer

Julie Plummer was a project manager and laboratory analyst for the environmental division of a metallurgical laboratory before establishing her own consulting firm, Riverview Environmental Services. Julie is experienced in management of hazardous waste site field investigation and remediation, environmental property assessment, industrial compliance permitting, cost estimation, field investigation planning, collection and analysis of environmental samples, field and laboratory quality control, interaction with regulatory agencies, and interpretation of regulations to clients.

Julie received her B.S. from Augustana College (IL) in Biology with Environmental Concentration. She is a Certified Emergency Medical Technician-A, a Certified Drinking Water Analyst for Total Coliform Bacteria & Nitrates, and a Certified Class K Industrial Wastewater Pretreatment Operator.

1

Site Investigation

Chapter Objectives

Upon completing this chapter, the student will be able to:

1. **Describe** the purposes and components of site characterization.
2. **Differentiate** among the following:
   — Environmental investigation
   — Regulatory inspection
   — Environmental audit
3. **Identify** the objectives associated with planning for a site investigation.
4. **Give** examples of federal site investigation requirements.
5. **Understand** the site investigation process used by EPA and by environmental consultants.
6. **Discuss** the technician's role in avoiding potential liability associated with site investigation.
7. **Identify** and discuss the components of a Phase I environmental site assessment for property transfer.

Chapter Sections

Figure 1-1: "The care of the Earth is our most ancient and, after all, our most pleasing responsibility. To cherish what remains of it, and to foster its renewal, is our only legitimate hope."

1-1 The Purpose of Site Characterization

Concepts

- Site characterization identifies the source, type, concentration and extent of contamination.
- Site characterization involves: 1) research into the history and use of the site, 2) field investigation, 3) laboratory analysis of environmental samples, and 4) evaluation of the laboratory data.
- It is necessary to move beyond routine sampling and toward a fully developed site characterization process when there are indications of "the unknown."

Site (sīt) n. [ME, OFr, Latin, *Situs*, place, position, situation] 1. a place or setting for something. 2. an area considered from the standpoint of its use for some specified purpose.

In the case of site characterization, the above definition may unfortunately be used in connection with contaminants that could pose a threat to our health or to our natural environment. A "place or setting" that is suspected of being contaminated becomes the subject of an investigation to determine the nature and extent of the problem. It is the safe, thorough and properly performed site investigation for contamination that is the topic of this book.

What then constitutes a site? A site may be a specific property having specific legal boundaries, such as the land owned by a business, an individual, a city or other governmental entity. Contamination, however, can affect multiple properties; thus a site could be all of those properties together. Furthermore, a single property may have more than one site to be investigated for contaminants; for example, a property may contain suspected contamination from an electroplating shop in one area and a leaking underground storage tank in a different area. Each area – that is, each "site" – could undergo its own separate investigation.

The following scenarios are examples of several different types of sites that would be subject to investigation.

For twelve years a battery-manufacturing facility has been vacant and inactive after forty years of producing lead-acid batteries. The once busy site is surrounded by a high chainlink fence and has become an eyesore of crumbling concrete and brick buildings in a community of 25,000 people. The property is near the outskirts of the community, not far from a main interstate highway. Commercial districts and residential neighborhoods are just across the roads surrounding the site. There are new business and industry developments in the community, but none of the future owners want to move onto the old battery-manufacturing facility because of the risk of contamination from lead, sulfuric acid and polychlorinated biphenyls (PCBs). Nearby property owners and residents are concerned about health issues and property values. The city is committed to rejuvenating the site and wants to find out just how "bad" the contamination really is.

ABC Dry Cleaner has operated on a busy street in a metropolitan area for decades. It has been a family-owned and operated business for four generations. ABC has a good reputation with its customers, but it has experienced increasing competition from a larger dry-cleaning operation with faster order turnaround. The great-grandson of ABC's founder wants to expand and modernize operations. The lender, whom the owner of ABC Dry Cleaner has approached for financing, is requiring that an environmental study be performed before financing is approved.

A vehicle-repair shop, located at the edge of a small town, services all kinds of vehicles, such as cars, trucks with gasoline or diesel engines, the school district's school buses, even farm tractors. A parts washer – purchased three years ago – is used to clean oily vehicle parts. Four times a year the parts washer's dirty solvent, containing 10% trichloroethylene, needs to be replaced. The shop owner had started using this solvent after hearing that it was "safer" and "better" than the previous one. For the last three years an employee has taken the parts washer to the edge of the concrete outside the shop's garage, dumped the used solvent onto the ground, and then rinsed the washer out with soapy water. Recently, the state's environmental regulatory agency inspected the facility and issued a penalty and citation for illegal disposal of the waste solvent. The citation requires the shop owner to perform an investigation determining the extent of contamination and providing a plan for cleanup of the spilled solvent.

Box 1-1 ■ Relationship of Environmental and Health/Safety Regulations to Site Investigation

Sampling and monitoring activities on **environmental investigation** sites are mandated by numerous laws. The wording in laws may not be specific as to how these activities should be carried out. For example, a law may simply require that the nature and extent of contamination be determined. Professional societies, such as American Society for Testing and Materials (ASTM), as well as governmental agencies, particularly the Environmental Protection Agency (EPA), have developed standards and methods for properly carrying out sampling and monitoring activities to meet the intent of the law. Some of the important laws associated with site characterization are listed below. Although these laws are beyond the scope of this book, the student should study their requirements.

Resource Conservation and Recovery Act (RCRA)

The 1976 Resource Conservation and Recovery Act charged EPA with regulating the handling of ongoing hazardous waste generation by industries. Generators of potentially hazardous waste must evaluate their waste streams for all of the following characteristics: flammability, corrosivity, reactivity and toxicity. The evaluation of these characteristics is accomplished by analyzing samples according to the lab methods specified by the EPA publication called **"Test Methods for Evaluating Solid Waste, Physical/Chemical Methods,"** also called **SW-846**.

Comprehensive Environmental Response, Compensation and Liability Act (CERCLA)

The 1980 Comprehensive Environmental Response, Compensation and Liability Act, known as **Superfund**, requires EPA to provide a structure for cleaning up old, dangerous and hazardous waste sites. Superfund sites undergo the RI (Remedial Investigation) CERCLA process, which includes characterization of the site.

Monitoring is conducted as a preliminary step of site characterization to provide indications of the presence of hazards. If hazards are indicated, subsequent sampling will be necessary to determine the specific substances and their concentrations. Monitoring is also part of the emergency response procedures and the confined space entry procedures that may be necessary at a hazardous waste cleanup site. Sampling is required at all Superfund sites whether or not positive monitoring data exist.

Superfund Amendments and Reauthorization Act (SARA) Title III

Title III of the 1986 Superfund Amendments and Reauthorization Act, a major amendment of CERCLA, contains the Emergency Response and Community Right-to-Know laws. Businesses that process or otherwise use reportable quantities of certain listed hazardous materials must comply with the Community Right-to-Know regulations that require notification of authorities as well as contingency planning and employee training.

SARA Title III specifies how release of hazardous materials, both planned and unplanned, must be reported. Monitoring and sampling are pertinent to determining the quantity of a release, whether it be air, water, or any other contaminant.

Hazardous Materials Transportation Act (HMTA)

The 1975 Hazardous Materials Transportation Act brought together the four transportation modes of air, water, rail and highway under the Department of Transportation (DOT). HMTA regulates the transport and packaging of hazardous substances and commodities that consist of hazardous materials. The sampling activities performed under RCRA bear a relationship to HMTA, because the Department of Transportation regulates the packaging and transport of hazardous waste and hazardous waste samples.

Monitoring is also pertinent to the HMTA because of the labeling and placarding requirements for materials that present radiation hazards. The three levels of radiation placarding/labeling (Radioactive White I and Radioactive Yellow II and III) are determined by using a radiation monitoring device.

Clean Water Act (CWA)

The 1972 Clean Water Act regulates **point source** and **nonpoint source** water pollution control. Industrial wastes discharged directly into surface waters are controlled through National Pollutant Discharge Elimination System (NPDES) permits. Generators of wastewater must perform monitoring and sampling to comply with their various permit requirements.

Clean Air Act (CAA) Amendments

The 1990 Clean Air Act Amendments provide sweeping additions to the permit process for industries that generate air pollution. Such contaminants as sulfur dioxide, nitrogen oxide, ozone, particulates and halogenated hydrocarbons have come under increasing regulation which, in some cases, has led to their phase-out. Monitoring and sampling for release of the contaminants are common practice not only in heavy industry, but also in service businesses such as heating, ventilation, air conditioning (HVAC) and automotive repair.

Occupational Safety and Health Act (OSHA)

Congress passed the 1970 Occupational Safety and Health Act to provide standards for workplace safety and health. Monitoring and sampling for potential air contaminant exposures play an important role in the OSHA regulations, especially in 29 CFR 1910.1000, the standard that describes the requirements for permissible exposure limits. Sampling and monitoring are also part of the Hazardous Waste Operations and Emergency Response (HAZWOPER) regulations of 29 CFR 1910.120. The HAZWOPER regulations specify methods of operating in a hazardous waste cleanup site and present the framework for emergency response incident management.

The above is by no means a complete listing of legislation in which sampling and monitoring may play a role. A few examples of the diversity of environmental connections in other federal administrations include:

—The monitoring of permissible dosages for workers potentially exposed to radiation;

—The Department of Housing and Urban Development's guidelines for lead levels in paint within HUD public housing;

—The Fish and Wildlife Service's monitoring of wildlife habitats;

—The monitoring of biohazards by the Centers for Disease Control.

According to the Chemical Abstracts Service (CAS) there are more than 16 million chemical substances either existing in nature or created by humans; new ones are constantly being added to the list at a rate of approximately one million per year. Some of these substances are considered hazardous to humans and to the environment. This makes it imperative to investigate their release into the environment. Both the cleanup of the contaminants and the reduction of the associated risks are challenges that face our society. Environmental author and educator Daniel Chiras reminds his readers that "Sir Edmund Hillary once wrote that environmental problems begin with people as the cause and end with people as the chief victims." Chiras's writings encourage us to "live lightly on the land" as a way of becoming a sustainable society.

The public's demand for attention to a deteriorating environment emerged in the early seventies; environmental technology subsequently developed as an occupational field in the eighties; and the cleanup efforts of the nineties will extend well into the coming millennium. Daniel Chiras begins his 1992 book, "Lessons from Nature," with the words of Wendell Berry: "The care of the Earth is our most ancient and most worthy and, after all, our most pleasing responsibility. To cherish what remains of it, and to foster its renewal, is our only legitimate hope." The environmental technician is and will remain a key force in the Earth's renewal.

Addressing the issue of contamination begins with the process known as site characterization, which is performed to identify the source, type, concentration and extent of the contamination. Site characterization is performed to identify the contaminants' pathways and their deleterious effects on the public and on the environment.

Thorough planning and conscientious execution are necessary to perform satisfactory site characterization. The planner may be an engineer, a scientist, an environmental management professional, or even an experienced environmental technician. The planner's job is to determine the best way to obtain such information as: How many samples need to be taken? How large should the samples be? When and where should they be collected and in what order? What barriers to sampling must be addressed? What budget is available? Are the results valid? How should the results be used in decision-making for the site?

Environmental sampling does not always involve a full-blown site characterization process. Situations requiring only routine sampling include:

—Wastewater samples taken at a POTW (Publicly Owned Treatment Works or at a private industrial treatment facility

—Drinking water samples from taps

—Water quality samples at recreational areas

—Soil samples to determine suitability for building construction

—Continuous air sampling of stack emissions

—Worker breathing-zone sampling

However, it is necessary to move beyond routine sampling and toward a fully developed site characterization process when there are indications of "the unknown" (an unexpected find or occurrence which might change the course of the investigation). For example, in the following scenario a routine leak investigation turns into a more detailed site characterization process when an environmental consulting firm stumbles upon the "unknown."

> A gas station owner discovers that an underground storage tank (UST), containing unleaded gasoline, is leaking. The owner notifies the regulatory agency and takes measures to stop the leak.
>
> The regulatory agency requires the station owner to investigate how much fuel has leaked and what the extent of the contamination is downward and outward from the source. An environmental consultant is hired to sample the soil and the groundwater for contamination levels.
>
> As the consultant's employees drill boreholes to install groundwater monitoring wells, they encounter a thick, black, sludgelike material in the soil. Since gasoline is pale yellow, it could not by itself have produced such a substance. Something else must have been released into the soil besides the gasoline. Uncertain of the type of material encountered and concerned about their safety, the consultant stops the routine procedure to evaluate the situation.
>
> This situation has moved from the predictable to the unknown; it requires the consultant to "begin at the beginning." In addition to knowledge about the present use of the site, past uses of the location must now be investigated.
>
> In checking the history of the property the consultant determines that until about 50 years ago a coal gasification plant had been located on the property for a period of 49 years. The plant had produced gas from coal which had, in turn, generated several waste products, including tar. The tar had been stored in a cistern or tar well and had evidently seeped from it. Years after the coal gasification plant had been torn down, a gas station had been built on the same property. In sampling for the gasoline leak, previously produced coal tar has been encountered.

In this case the initial investigation followed guidelines established in the federal regulations for USTs (40 CFR 280). But when the former property use came under suspicion of contaminating soil and groundwater, the federal CERCLA regulations came into play (40 CFR 300). Given the history of the site it is reasonable to suppose that other "surprises" could threaten the soil and the groundwater. A systematic site characterization then became the appropriate response in order to determine the location and the extent of contamination on the property.

Four Stages of Site Characterization

The full site characterization process includes four stages of activity:

- Background research on the history and use of the site
- Field investigation
- Laboratory analysis of environmental samples
- Evaluation of the laboratory data

During the first stage the environmental technician assists the project planner in studying the past and present uses of a property as well as the influence of the environment surrounding the site. Data gleaned from the research should help the project planner determine where contamination would most likely occur; this information provides the basis for

Figure 1-2: A technician packs a sampling jar with soil from a bucket auger.

Box 1-2 ■ Differentiating Monitoring and Sampling

Monitoring and sampling are essential procedures in determining the presence of hazardous substances. Although these terms are sometimes used interchangeably, we will distinguish between the two according to the following definitions:

Sampling will be defined in this textbook as 1) the collection of representative portions of gas, liquid or solid environmental **media** and 2) their preparation for transport to an analytical laboratory.

Monitoring will be defined as the use of devices that provide real-time information about the presence and relative concentration of hazardous substances. Among the monitoring devices that will be introduced are the oxygen meter, the combustible gas indicator and various toxicity monitors. A word of caution: although in our electronic society the word "real-time" usually suggests "at the time of use," monitoring devices seldom yield information instantaneously. The technician may wait seconds or minutes for an instrument reading, while the atmosphere potentially grows more dangerous. Those seconds or minutes can make a difference in explosive or toxic conditions. Thus, the technician must keep in mind that "**real-time**" does not necessarily mean "immediate response."

Some site activities tend to blur the lines between sampling and monitoring. For example, the performing of wet chemistry tests in the field can yield real-time data and, like a laboratory analysis, can help characterize a substance and determine concentration. Our textbook definitions of sampling and monitoring are not meant to be rigid, as the technician may encounter other ways of considering these words. They are rather meant to be used for convenience in categorizing site characterization technologies.

development of a site-sampling plan. Historical research also provides a prospective property buyer with a form of protection against Superfund liability. The topics of historical research and of liability are expanded later in this chapter.

During the second stage technicians go onsite to take environmental measurements and to gather observations as well as to collect, handle and store samples. If second stage research indicates that the onsite technicians could be exposed to hazardous substances, then air monitoring with detection instruments will be performed on an ongoing basis to warn of potentially dangerous air contamination.

The third stage is mainly performed in the lab by chemistry laboratory technicians. However, the environmental technician should have enough knowledge about lab processes to be able to interact with the lab personnel in an informed way and to determine that the lab has a quality control program in place.

During the fourth stage the project planner is most likely the data user (the person who will evaluate the lab data), although sometimes qualified environmental technicians may perform this task instead. The data user needs to be able to compare the analytical results with the applicable regulatory limits. For example, data for water samples collected from wastewater discharge points would be compared to pollutant limits published by EPA under their National Pollutant Discharge Elimination System (NPDES). In order for site decisions to be made within acceptable and reliable parameters, the data user must also be able to evaluate a lab report's quality control information.

Checking Your Understanding

1. List the purposes of site characterization.
2. List and briefly describe the four stages of site characterization.
3. Indicate how "sampling" may be differentiated from "monitoring."

1-2 Background and Historical Research

Concepts

- Photographs, maps, interviews, governmental records, as well as onsite observations, allow an investigator to reconstruct the history of previous activities on or around a site.
- Phase I assessments are intended to provide the property owner with the "innocent purchaser defense" to CERCLA liability.

Historical Research/Data Gathering

The site investigation begins with the use of off-site methods. These range from low-tech approaches, such as interviews of people involved with the site and research of public records, to high-tech approaches, such as aerial photography and **Geographic Information Systems (GIS)**. A GIS is a computer system that allows large volumes of spacially referenced data to be viewed together in different forms, such as a series of layered maps. This gives the user the ability to analyze separate types of information simultaneously and in relation to each other.

Aerial Photographs

Interpretation of aerial photography can yield information on site size, drainage patterns, burial sites, lagoons, **impoundments** and general conditions of a site. Aerial photography also can indicate locations and descriptions of tanks, drums, open storage areas, evidence of vegetation stress, onsite obstacles, structures, equipment and access routes. The comparison of photographs taken at different times may help an investigator reconstruct the history of previous activities on or around a site. Historic aerial photographs may be available in a public or educational institution's special collections library.

U.S. Department of Agriculture Farm Service Agency (USDA FSA) Photos – The USDA's Farm Service Agency, formerly called the Agricultural Stabilization and Conservation Service, produces aerial photographs and makes them available to the public, either free of charge or for a nominal fee. Since 1987 the National Aerial Photography Program has provided color infrared aerial photographs in which one full frame covers approximately 32 square miles of terrain. Until 1987 the National High Altitude Program provided infrared aerial photographs which covered approximately 72 square miles. The FSA National Aerial Photography Field Office may be reached at 801 975-3500.

Environmental Photographic Interpretation Center (EPIC) Photos – This EPA center provides a wide range of remote sensing and aerial photographic analyses. Among EPIC's technical capabilities are: 1) aerial photographic flights; 2) photogrammetric techniques, in which maps are developed from aerial photographs; 3) photogeologic methods, which identify rock types and fracture zones in bedrock that may provide a path for the movement of pollutants; 4) thermal infrared imagery that can detect heat sources, silty water and stressed vegetation; and 5) the already mentioned Geographic Information Systems (GIS) which uses data to develop computer models of landforms, vegetation and land use.

Department of Transportation Photos – Each state's Department of Transportation (DOT) makes aerial photographs of state roads and highways available to the public for a fee. Many properties adjacent to the roadways are also captured on these aerial photographs, making DOT photographs often useful for site investigation.

Maps

Sanborn Fire Insurance Maps – These are detailed city plans, usually drawn at scales of 50 or 100 feet to an inch, which provide insurance companies with information for risk assessment on commercial buildings, factories and dwellings. They show construction details of buildings, property features, and even uses of the property, such as industrial processes and chemical storage tanks. **Sanborn Fire Insurance maps** are produced primarily for major cities, and by contract for smaller cities, by the New York-based Sanborn Map Company which has been in business since the 1800s.

Figure 1-3: Aerial photograph of Kearns, Utah.

A technician will most likely find Sanborn maps in city government planning, zoning, public works, engineering or redevelopment offices. Many public libraries contain collections of the local community's Sanborn maps. They may also be directly purchased from the company.

United States Geological Survey (USGS) Maps – Maps available from the USGS include topographic, hydrologic and hydrogeologic maps, as well as bedrock geology, surficial geology and land use maps. Topographic maps graphically represent the surface features of a region. Hydrologic or hydrogeologic maps detail surface water and groundwater features. Bedrock geology maps show the bedrock present beneath the surficial deposits which are depicted on a surficial geology map. Land use maps specify agricultural, residential, industrial, commercial and recreational areas as well as those that are undeveloped.

Farm Service Agency Agricultural Stabilization and Conservation Service (ASCS) Maps – Maps available from the ASCS include soil maps, which detail the types of soils in an area; they also show some land uses.

Other Maps – Specialized maps are available for identification of features such as pipelines, transportation routes, and sewer systems. Sources for

Figure 1-4: Students analyzing an aerial photograph and a USGS map.

specialized maps vary from private companies (e.g., pipeline distribution companies and utility companies) to city, county and state governmental units (e.g., the Department of Transportation and the Department of Natural Resources).

Records

Useful site information is found in a variety of publicly available documents, such as county or city building permits; county or city assessor's records; county data about property abstracts; and state or federal agency documents, such as permits and correspondence. A list of these sources follows.

Property Abstracts – The abstract or title search for a property will identify property owners and the period of ownership. Often the abstract will include covenants for use of the property such as the sale of petroleum products or rights-of-way.

Assessor's Records – Each county has assessor's records which pertain to property taxes. Tax assessments are based on size of and improvements to the property. Assessor's records typically indicate the use of a property, but they also summarize structures, surfacing materials and large chemical storage systems such as aboveground and underground storage tanks (USTs).

Building Records – Most communities require that building permits be obtained when work is being done to a structure. Inspections usually are done by the city building inspection department. This is to protect property owners and property values, as well as to ensure that work is done according to local building codes. Building records may contain details of buildings or other structures being built, remodeled or torn down.

City Zoning Records – Zoning records indicate what type of businesses are located in a given area and what general use the community has approved for the area.

Fire Department Records – Fire departments generally keep records pertaining to fire runs and bulk storage of chemicals or petroleum products. During a large fire, soil and groundwater can become contaminated from the spillage of chemicals stored at a property. Fire departments may be called to contain a release of chemicals at a facility, and would therefore have records of the operation.

City Directories – A number of companies, the R. J. Polk Company among them, publish city directories with cross-indexes of names, addresses, phone numbers and occupations. These directories contain information about the ownership and type of business present in a location, and are generally updated once a year. Libraries often keep copies of city directories in their collections. Depending on the community, city directories may be available back to the 1800s.

EPA or State Regulatory Agencies – The federal EPA and state regulatory agencies maintain databases pertaining to properties that have come under regulatory scrutiny, or that have an obligation to register activities with the appropriate agency. The federal EPA records for sites pertaining to the **National Priorities List (NPL)**; CERCLA; RCRA generators, transporters, or TSDFs (treatment, storage, disposal facilities); and ERNS (Emergency Response Notification System), a list of spills large enough to be reported to the National Response Center. State regulatory agencies maintain lists of registered USTs, Leaking USTs also called LUSTs, hazardous waste sites, landfills and solid waste disposal site lists. Many of these state and federal lists and databases are available either from companies that specialize in searching such lists, or by subscription to a service that updates the lists monthly. Detailed information pertaining to a site listed in any such database may be obtained from the regulatory agency by a Freedom of Information Act (FOIA) request. EPA and state regulatory agency databases should be checked not only for the property in question, but for the surrounding prop-

Technical Corner ■ Topographic Maps and Symbols

Topographic Maps

Topography is the technique of graphically representing the exact physical features of a place or region on a map.

Part of the preliminary study of a site involves obtaining one or more **topographic maps**. U.S. Department of the Interior Geological Survey (USGS) maps may be ordered from a state's Geological Survey Bureau or from the national USGS Earth Science Information Center at 1-800 USA-MAPS. Farm Service Agency (FSA) maps are usually obtained from the county.

To order a map or aerial photo from the state GS Bureau or from the FSA, the project planner needs to specify the township and section of the site undergoing preliminary investigation. A section is 640 acres or one square mile. A township is a public land surveying unit of 36 sections, or 36 square miles. Section and township information can be obtained from a city or county planning office

60 seconds	=	1 minute
60 minutes	=	1 degree
7-1/2 minutes	=	the distance across a quadrangle

The map delivered to the planner by the USGS or the FSA will usually be a **quadrangle map**, or some portion of a quadrangle. A quadrangle is the area of land shown on one atlas sheet charted by the USGS. The measurement of a quadrangle, which represents an area between two latitude lines and two longitude lines, is expressed as 7-1/2 minutes by 7-1/2 minutes. Therefore, a quadrangle map might be referred to as a 7-1/2 minute map. The minutes of latitude-longitude are, like clocks, composed of seconds. One minute on a topographic map equals 60 seconds. However, unlike the clock, 60 minutes on a topographic map equals one degree, NOT one hour.

Topographic Symbols

USGS topographic maps are developed from sets of aerial photographs of every part of the country. For each land area to be charted, two aerial photos taken from slightly different angles are viewed through an optical instrument, a type of stereoscope, which imparts a three-dimensional effect. The cartographer (and, these days, a computer) interprets the three-dimensional result and translates it into **contour lines**, which graphically depict the variations in the land surface.

A USGS brochure called "Topographic Map Symbols" describes contours as imaginary lines that follow the land surface or ocean bottom at a constant **elevation** (altitude) above or below sea level. To visualize the land as one looks at a topographic map, one must consider the contour lines relative to their elevation and their contour interval.

Elevation

Elevation is expressed in numeric values and stated periodically on the map in relationship to the contour lines. Different numeric values represent variations in altitude (feet above sea level) of the site. From one given point on the map to another, decreasing elevation values indicate a downward slope. Likewise, increasing elevation values indicate an upward slope.

Contour Interval

The **contour interval** is stated once on the map. It is defined as the elevation difference between adjacent contour lines. For example, if the contour interval of a map is specified as one foot, then whatever the spacing between the contour lines might look like on the map, those spaces represent a one-foot difference in elevation. Therefore, if contour lines are closely spaced, the slope is steep and if they are widely spaced, the slope is gradual.

The contour interval is chosen on the basis of the map scale and of the local variation in elevation. Large variation in elevation, such as hilly or mountainous terrain, is expressed by a large contour interval; for example, spacing between contour lines may represent 100 feet or more.

Scale

The scale bar is shown once on the map. **Map scale** is the relationship between distance on a map and the corresponding distance on the ground. Scale is expressed as a ratio, such as 1:24,000, and is shown on the map as a bar marked in feet and miles or in meters and kilometers.

Scales vary from one map to another depending on the size of the area to be charted. For example, a civil engineering site map of a property has a scale of one inch on the map corresponding to forty feet on the ground. In another example, a USGS map of the United States has a scale of 1:1,000,000. This is a small scale map expressed in centimeters: one centimeter on the map represents 1,000,000 centimeters on the ground, which converts to 10 kilometers (in English measurement the ratio would be one inch to 16 miles). On the other hand,

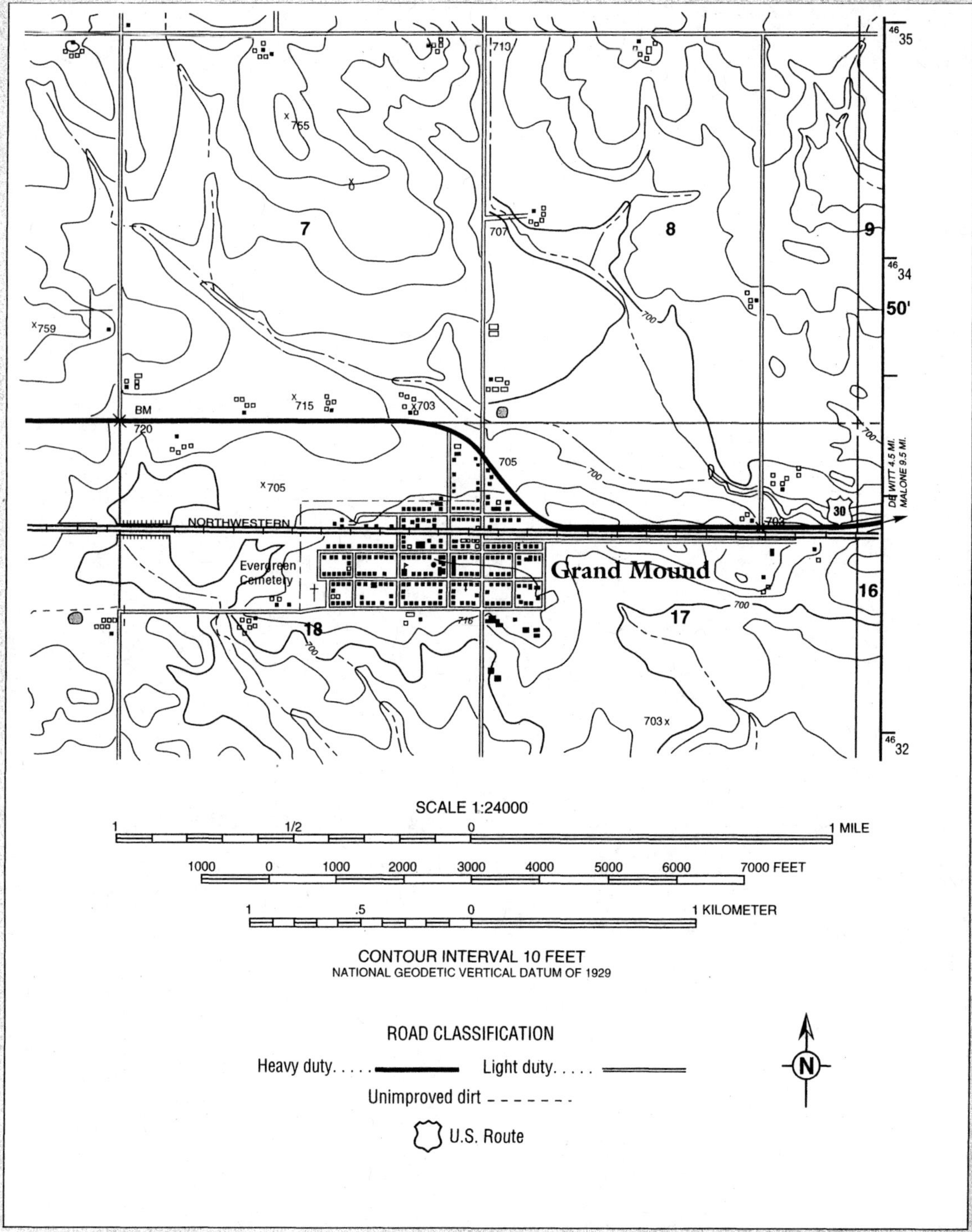

Courtesy of Department of the Interior U.S. Geological Survey National Mapping Division

a USGS map that has a scale of 1:24,000 is a larger scale map, which provides a more close-up depiction of an area.

A quadrangle map, or portion thereof, depicts a large area. This kind of topographic map is used to show the relationship of a site to its surrounding region. A more detailed topographic map on the 1-inch to 40-foot scale, such as a type developed by a civil engineer, is appropriate for identifying sampling points for a site.

Legend

The **legend** is shown once on a map and contains symbols that represent various natural and human-made features. For example, one site map of a property includes among its symbols the following: manhole, power pole, power pole with light, fire hydrant, storm water inlet, sanitary sewer, storm sewer, water main, gas main, deciduous tree, coniferous tree, fence and traffic signal standard.

Direction

The top of a map represents north. The north arrow is depicted once on the map, often next to the scale. Although orientation may seem to be an elementary concept, it is surprising how some people can have difficulty orienting themselves on a site in relationship to a map. The technician should include a compass in the field equipment to ensure accuracy and to aid in orienting the workers at a site.

Benchmark/Pin Set

Official, permanent **benchmarks** are markers placed by the USGS at the corner of each section (i.e., every mile) throughout the country. The top of the marker is a bronze medallion that contains the specific location of the benchmark, such as longitude, latitude and elevation. The benchmarks provide reference points or starting points for surveying sites. On a geologic map benchmarks are identified by the initials "BM."

A **pin set** is a permanent marker that is placed in one corner of a site being surveyed. The pin's size and material is specified as 5/8 inch steel, so that it may be picked up by a metal detector. Like the benchmark, the pin is used as a reference or starting point for surveying or measuring. A surveyor may use as a starting point either a USGS benchmark or a pin set placed during another survey.

If the technician must select an existing site feature – rather than a USGS benchmark or a pin set – as a starting point for measuring, then he or she will have to make sure that the chosen feature will be there for future reference. For example, a fire hydrant or a power pole may be a more permanent measurement starting point than a tree or a building.

Technicians need to know how to interpret maps and how to visualize land from a map's data. They must develop skills in analyzing map information that might affect field work, such as drainage patterns or potential barriers to sampling.

Series	Scale	1 inch represents approximately	1 centimeter represents	Size (latitude x longitude)	Area (square miles)
Puerto Rico 7.5-minute	1:20,000	1,667 feet	200 meters	7.5 x 7.5 min.	71
7.5-minute	1:24,000	2,000 feet (exact)	240 meters	7.5 x 7.5 min.	49 to 70
7.5-minute	1:25,000	2,083 feet	250 meters	7.5 x 7.5 min.	49 to 70
7.5 x 15-minute	1:25,000	2,083 feet	250 meters	7.5 x 15 min.	98 to 140
USGS/DMA 15-minute	1:50,000	4,166 feet	500 meters	15 x 15 min.	197 to 282
15-minute	1:62,500	1 mile	625 meters	15 x 15 min.	197 to 282
Alaska 1:63,360	1:63,360	1 mile (exact)	633.6 meters	15 x 20 to 36 min.	207 to 281
County 1:50,000	1:50,000	4,166 feet	500 meters	County area	Varies
County 1:100,000	1:100,000	1.6 miles	1 kilometer	County area	Varies
30 x 60-minute	1:100,000	1.6 miles	1 kilometer	30 x 60 min.	1,568 to 2,240
U.S. 1:250,000	1:250,000	4 miles	2.5 kilometers	1" x 2" or 3"	4,580 to 8,669
State maps	1:500,000	8 miles	5 kilometers	State area	Varies
U.S. 1:1,000,000	1:1,000,000	16 miles	10 kilometers	4" x 6"	73,734 to 102,759
U.S. Sectional	1:2,000,000	32 miles	20 kilometers	State groups	Varies
Antarctica 1:250,000	1:250,000	4 miles	2.5 kilometers	1" x 3" to 15"	4,089 to 8,336
Antarctica 1:500,000	1:500,000	8 miles	5 kilometers	2" x 7.5"	28,174 to 30,462

Courtesy of Department of the Interior U.S. Geological Survey National Mapping Division

erties as well. ASTM spells out recommended minimum distances from a property to conduct searches.

Local Utility Companies – Electric utility companies may maintain records concerning locations of, as well as service and damage to, electrical equipment containing PCBs.

Interviews

First-hand interviews with individuals acquainted with a site may provide insights into activities and situations related to possible contamination. The site investigator might interview current neighbors, the fire chief or fire marshal, a representative from the local health department, the city engineer, and present and past owners of the site. Interviews often provide additional leads and contacts to be pursued. Records might be available of employees of businesses formerly associated with the site. The employees might supply information about particular processes and practices that may have resulted in contamination.

Observations from Site

The power of personal observation is a valuable aspect of onsite data gathering. When first approaching the site, the technician should be alert to any unusual signs, such as:

—An absence of expected wild birds and animals

—A strange odor

—Conspicuously absent vegetation; dead or stressed vegetation

—Stained or discolored soil or other surfaces

—Mounded soil

—Discolored water or water with a "rainbow" sheen on the surface

—Drums or other containers in disorder

If performing a walk-through of the site, the technician should note onsite activities, structures, equipment, processes, natural features such as bodies of water and, most particularly, the chemical storage. The technician should also note off-site clues such as proximity to gasoline service stations, manufacturers, or agricultural areas that may affect the property.

Phase I Environmental Site Assessments for Property Transfer

All site investigations should begin with a historical study of a site's former and current uses. A particular type of historical study is the **Phase I Environmental Site Assessment**, sometimes also called ESA. Phase I assessments are usually driven by the banking or the real estate industries. In most cases, financial institutions and prospective buyers do not want to lend money or become a part of the chain of title on a property that may have environmental problems. The worst environmental classification for a property is to be declared a Superfund (CERCLA) site. CERCLA liability includes the costs associated with the investigation as well as with the cleanup of hazardous substances; these operations can easily exceed the $1 million mark. Phase I environmental

Figure 1-5: The oily staining and discoloration of the doors and concrete indicate there is reason to suspect that the doors, concrete, soil and groundwater may be contaminated.

Technical Corner ■ Freedom of Information Act (FOIA) Requests

The Freedom of Information Act (FOIA) was passed in 1966 by Congress to facilitate public access to government records. This includes any information about a property that may be held in government agency files.

EPA and other federal, state and local government agencies must accept written requests for information under the Freedom of Information Act. Many times an individual in the agency will be designated the "Freedom of Information Officer" and assigned the responsibility of fielding and filling these requests.

FOIA requests need to be concise, specific and complete. They should be identified as being FOIA requests (as opposed to any other type of request). Name, address and phone number of the person making the request must also be stated. The type of information being sought should be clearly indicated; the site name and address as well as the site's township, range and section must be included.

EPA or another government agency will respond to a request in writing and let the inquirer know when the information will be sent, or what additional information or fee the agency needs in order to comply with the request.

Although the government has an obligation to supply information, it is nonetheless allowed to charge a fee for copying. Sometimes files can be several inches thick and photocopying charges can be staggering at 10 to 25 cents or more per copy. It is therefore best to specify in the written request what charges one is willing to incur for the information. It is also possible to appear in person at the agency, review the files and copy only the necessary pages.

In some cases FOIA requests cannot be filled because the information in the agency's files falls into one of several categories exempted from public disclosure. These categories include information being used in certain legal proceedings and in trade secrets. If the agency cannot release information for this reason, it will send written notification and explanation of the matter.

The sample letter shown here is intended as a general guideline and may be modified to meet specific needs.

Sample Letter

(Date)

(Name of individual, if known)
Freedom of Information Officer
US EPA Region (or other governmental agency)
Street Address
City, State, Zip Code

Dear (name of individual or) FOIA Officer:

(Name of Company) is conducting a Phase I Environmental Site Assessment for Property Transfer at (name of site) located at (property street address including city, state and zip code). This property is located in (township, range, section and/or other location information, if known).

(Name of Company) would like to obtain information pertaining to the environmental history of the site (e.g., regulatory compliance history for past and current owners, reports, permits, inspection logs and violations).

We would appreciate your timely response to this request. Please send the requested information to:
(Your Name)
(Your Company's Name)
(Company Address)

If you have questions about this request, I can be reached at the address listed above or by phone at (phone number), by fax at (fax number) or by e-mail at (your e-mail address, if you have one). If charges for this request exceed $ (insert dollar amount), please contact me before filling this request.

Thank you for your assistance.

Sincerely,

(Signature)
(Your name, typewritten)
(Your title)

site assessments are intended to provide "all appropriate inquiry into the previous ownership and uses of the property....", they are an effort to provide the property owner with the "innocent purchaser defense" to CERCLA liability, as outlined in 42 U.S.C. 9601(35)(B) and 9607(b)(3). The innocent purchaser defense may be used against CERCLA liability suits if a property is later determined to have been contaminated from a past use. If this defense is successful, the property owners are not held liable for the contamination.

An environmental technician should be able to take responsibility for assisting an environmental professional such as an engineer, geologist or other manager in the research required for a Phase I environmental site assessment. Investigating a site's history is much like detective work as the technician follows up on leads, generates data, and formulates conclusions about present and past site activities.

Checking Your Understanding

1. Describe different historical data sources and list what types of information might be found in those materials.
2. List the types of observations that may be made concerning a site or the properties adjacent to it.
3. Identify and describe what defense a property owner has with respect to CERCLA liability.

1-3 Field Investigation Teams and Duties

Concepts

- The health and safety program for a field investigation must meet EPA and OSHA's requirements in 29 CFR 1910.120(b).
- EPA defines eight roles within a CERCLA field team. RCRA field teams cover the same roles, but often under different job titles.

The field investigation process involves development of a written sampling plan by either the engineer, the scientist or the environmental management professional. The plan includes components such as a project management plan, data quality objectives, a field sampling plan, a quality assurance project plan, a **site health and safety program** and a community relations plan. The first four components address the methodology to be used during sampling and analysis to ensure that the portions collected will be representative of the site as a whole, and that analysis results will be reliable. The community relations program is designed to inform the local community of site activities. The site health safety program is developed according to the levels of hazard present on the site and following specifications by EPA and OSHA (see Appendix I for an example of a health and safety program).

The Site Health and Safety Program

The health and safety program must meet the requirements of 29 CFR 1910.120(b), which is the Hazardous Waste Operations and Emergency Response (HAZWOPER) worker protection standard developed jointly by EPA and OSHA. The employer's general health and safety program will include a written health and safety plan for each site to be investigated. The following list, which represents minimum requirements for a written plan, is abbreviated from the complete text of 29 CFR 1910.120(b)(4)(ii). The elements of the written site-specific health and safety plan include:

—A safety and health risk or hazard analysis for each site task and operation found in the work plan.

—Adequate employee training according to job functions and responsibilities.

—**Personal protective equipment (PPE)** to be used by employees for each of the site tasks and operations being conducted.

—Medical surveillance requirements.

—Frequency and types of air monitoring, personnel monitoring, environmental sampling techniques and instrumentation to be used, including methods of maintenance and calibration the equipment.

—Site control measures.

—Decontamination procedures.

—An emergency response plan.

—Confined space entry procedures.

—A spill containment program.

The site-specific health and safety plan must provide for pre-entry briefings to be held prior to initiating any site activity. Inspections are conducted by the site health and safety supervisor or other designated, qualified person (referred to later as the safety officer). Any deficiencies in the effectiveness of the site health and safety plan must be corrected by the employer.

Organization of the Field Team

EPA defines eight roles that are typically included in a field team. According to EPA's publication "A Compendium of Superfund Field Operations Methods," the number of roles needed at each site is dictated by potential hazards and the specific needs of the site. Dual-role assignments may be acceptable if hazardous substances and physical conditions at a site are well documented and if the nature of the work is limited. Typical CERCLA Field Team titles may include:

—Site Manager

—Field Team Leader

—Site Safety Officer

—Personnel Decontamination Station Operator/ Equipment Specialist

—Communications Supervisor

—Initial Entry Party

—Work Party

—Emergency Response Team

Typical RCRA Field Team titles may include:

—Project Manager

—Project Engineer/Field Engineer

—Project Geologist/Field Geologist

—Field Technician

—Field Crew (drillers, heavy equipment operators, etc.)

The field teams have slightly different organization schemes as well as different titles. Although the descriptions below pertain to CERCLA field team titles, RCRA field teams allot those same functions among their members. A team member may fulfill multiple functions; for example, the field team leader or site manager (i.e., project manager) may also fulfill the function of the site safety officer. Experienced environmental technicians may also be required to fulfill multiple functions such as field team leader and work party member.

Following is a list of job titles and job descriptions for CERCLA team members. The corresponding RCRA job titles are included in each paragraph.

Site Manager – The site manager is generally a scientist, engineer, or environmental management professional whose responsibilities include the preparation and organization of project work as well as the overall management of personnel, final reporting and coordination with the lead agency. Under a RCRA investigation this role may be filled by a project manager. The project manager may also be the project engineer/field engineer or project geologist/field geologist.

Field Team Leader – The field team leader is responsible for the overall operation and safety of the field team. Duties include the execution of the site work plan; safety procedure compliance through coordination with the site safety officer; coordination with laboratories and subcontractors; community relations; site control; and compliance with the requirements. In many RCRA investigations the person fulfilling this role may have the title of project manager, field geologist or field engineer.

Site Safety Officer – The site safety officer is responsible for safety procedures and operations at the site. He or she reports to the safety administrator in the main organization rather than to the field team leader or to the site manager. This independent reporting system establishes a separate line of authority – in addition to the one in the field – which ensures that proper decisions on safety are made irrespective of scheduling considerations and investigation deadlines. In RCRA investigations the site safety officer duties may be fulfilled by the field geologist or field engineer.

Box 1-3 ■ Terminology Related to Site Investigation

In the environmental industry the technician is likely to encounter a variety of terms related to site investigation. In each case a site is certainly "investigated," but the terms may imply rather different investigative purposes. Technicians should be able to help reduce confusion by clarifying all these terms according to the local usage. Some examples follow.

— *Environmental investigation or remedial investigation*
Site characterization is often called "environmental investigation." **Remedial Investigation (RI)** is EPA's special terminology for a hazardous waste site characterization performed according to Superfund or CERCLA regulations. With respect to RCRA the site characterization process is referred to as a **RCRA Facility Investigation (RFI)**. According to EPA, RIs and RFIs consist of data-gathering activities undertaken to determine the presence, the degree and the extent of contamination at a site. These data are used in the identification, screening and evaluation of remedial alternatives. A CERCLA **Feasibility Study (FS)** or a RCRA **Corrective Measures Study (CMS)** both consist of developing and evaluating remedial alternatives, so that decisions may be made regarding the limitation of the spread of contaminants and the reduction of their toxicity.

— *Regulatory inspection*
A **regulatory inspection** is conducted by an agency inspector to determine compliance with federal, state or local requirements. For example, the hazardous waste management practices of a particular company may be subject to EPA inspection. Let's suppose that the company's business management cannot prove that waste streams have been sampled and analyzed for hazardous characteristics and that the hazardous wastes have been disposed of through sewers or solid waste landfills. The inspection findings are then backed by the force of law and penalties may be levied for noncompliance.

— *Environmental audit*
An **environmental audit** is an analysis of an industrial or commercial process to determine compliance with applicable environmental laws and regulations. The environmental audit can also be used to identify opportunities for waste reduction, energy conservation and pollution prevention.

— *Pollution Prevention Assessment*
A **pollution prevention assessment** determines opportunities for waste reduction, energy conservation and pollution prevention.

— *Phase I, II and III Environmental Site Assessment*
A **Phase I Environmental Site Assessment** is a historical study specifically requested by the banking and real estate industries to ensure against liability, should a site become involved in environmental litigation or corrective actions. Like the first stage of any site investigation, a Phase I Environmental Site Assessment entails researching the background of a property. Phase I Environmental Site Assessments for property transfers have not so far been regulated by the federal government. Professional organizations such as ASTM (American Society for Testing and Materials), the Federal Deposit Insurance Corporation (FDIC) and the National Resolution Trust Corporation (NRTC) have developed standards or guidelines for Phase I Site Assessments for property transfer. If a Phase I Environmental Site Assessment indicates that a site might be contaminated, then a **Phase II Environmental Site Assessment** may follow, which is a field investigation. The Phase II environmental site assessment involves collecting samples which are analyzed by a laboratory to determine if contamination is present. If necessary, a **Phase III Environmental Site Assessment** process can be completed. The Phase III process fully characterizes the vertical and lateral extent of soil and groundwater contamination; it identifies potential cleanup alternatives and their feasibility; and it implements appropriate cleanup activities.

The site safety officer may accompany the work party to the hazardous worksite and closely supervise the activities in person; he or she may also monitor the work party and site activities from the clean side of the exclusion area (this is the area of greatest contamination, where entry and exit is strictly limited to authorized personnel wearing protective equipment). The site safety officer is responsible for maintaining communications and line-of-sight contact with the work party; determining the level of personal protection required; adjusting the levels of protection to specific site needs if the site observations make it necessary; ensuring and monitoring compliance with safety and health regulations; interrupting activities to protect personal worker safety if it becomes necessary; coordinating response to safety and health emergencies; examining work party members for symptoms of exposure or stress; and providing emergency medical care and first aid as necessary at the site. Depending on the nature of the hazards onsite, the safety officer may perform additional duties such as photography, air monitoring, headspace analysis, or sample logging, if it does not interfere with the primary duty of maintaining safety.

Personnel Decontamination Station Operator/ Equipment Specialist – The personnel decontamination station operator/equipment specialist functions in two roles. The role of equipment specialist includes selecting and inspecting equipment before it is taken to the work site as well as determining that it is properly maintained and operational during the site activities. The role of personnel decontamination station operator includes designing and setting up the "decon" station; preparing the necessary decontamination solutions; managing the mechanics of removing contaminated clothing from the work party; and properly disposing of discarded contaminated clothing and decontamination solutions. In RCRA or other investigations, decontamination and equipment maintenance responsibilities may be assigned to several team members, such as the field engineer, field geologist, driller, field technician or heavy equipment operator. In many cases the team member who uses the equipment is also responsible for its maintenance and decontamination.

Communications Supervisor – The communications supervisor functions as the clearinghouse for communications. Among the responsibilities are notification of emergency support personnel as necessary; logging of communications and site activities, such as duration of work periods with respirators or personnel and equipment movement onto and off the site; and maintenance of good community relations in the absence of the field team leader. The communications supervisor may also be responsible for logging and packaging samples for transport, maintaining a weather watch, and providing security for the emergency response vehicle and other equipment. In RCRA investigations the communications supervision is a duty fulfilled by the project manager and the field engineer or field geologist.

Initial Entry Party – First access to a site is performed by the initial entry party. This team operates with specialized instrumentation to characterize site hazards. In order to become familiar with the conditions and dangers associated with the site, the field team leader should be a member of the initial entry party. The main purpose of this team is to measure existing hazards and to determine whether previously set levels of protection are sufficient or should be adjusted. The initial entry party may consist of as few as two people using the "buddy system". In many investigations, especially those carried out under RCRA, the initial entry party is the same as the work party. Even though many RCRA sites – often a gas station or a field adjacent to a manufacturing facility – may be seemingly harmless locations, caution and care should always be used when performing any type of environmental investigation. An initial evaluation of hazards present or potentially present should be performed for each site, and conditions should be re-evaluated periodically.

Work Party – The work party performs preliminary onsite tasks necessary to proceed with the investigation, such as obtaining samples or determining locations for monitoring wells. The work party consists of a minimum of two individuals. No single work party member should enter or exit the exclusion area alone. If a number of activities are taking place simultaneously in several areas at a distance from each other, the site safety officer may need to assign assistants to each of the smaller work teams. The work party in many RCRA site characterizations includes the field geologist or field engineer, the driller, the backhoe operator and the field technician. Often these individuals visit the site in phases; this means that they may not all be present on the site at the same time. One phase may require drilling boreholes and installing monitoring wells, the next phase might be the sampling of the monitoring wells after they have stabilized. Other phases may require digging test pits in soils or waste piles for sampling those materials. In all phases,

Figure 1-6: The decontamination process is demonstrated at a training event.

however, good communication is vital, both among the teams and with the site safety officer and the field team leader. The system of communication may include two-way radios, cellular phones, beepers, and designated times to call in to the team leader.

Emergency Response Team – Extensive and particularly hazardous assignments requiring long hours and large work parties may sometimes necessitate the use of a standby emergency response team. Members of this team – which often works on a rotation basis with the rest of the group – are "half-dressed" in personal protective equipment, so they can quickly enter the exclusion area in an emergency. The site safety plan specifies procedures to follow if an emergency arises. At many CERCLA sites, and in most cases for a RCRA investigation, there will not be a half-dressed emergency response team waiting nearby. Instead, the field team may need to leave a site if an emergency situation arises; or a call may need to be placed to an emergency response team better prepared to deal with the situation.

The Technician's Role in Avoiding Liability

As a member of a team performing an environmental investigation, a technician has a duty to act carefully and responsibly with regard to collecting and preserving data and samples obtained during the investigation. All sampling documentation, including field notes, photographs and chain of custody, are considered legal documentation. Sample collection must be carried out as specified in the sampling plan. Any deviations from the plan must be justified in writing and documented in the field notes. Collections of a sample in an inappropriate manner, omissions of any kind, or destruction of field notes for the purposes of covering up contamination, are all actions that may be considered a criminal offense. Civil or regulatory penalties may also be applied to such acts.

The samples, field notes and other data collected during an investigation are used to prepare the reports of the environmental investigations. These

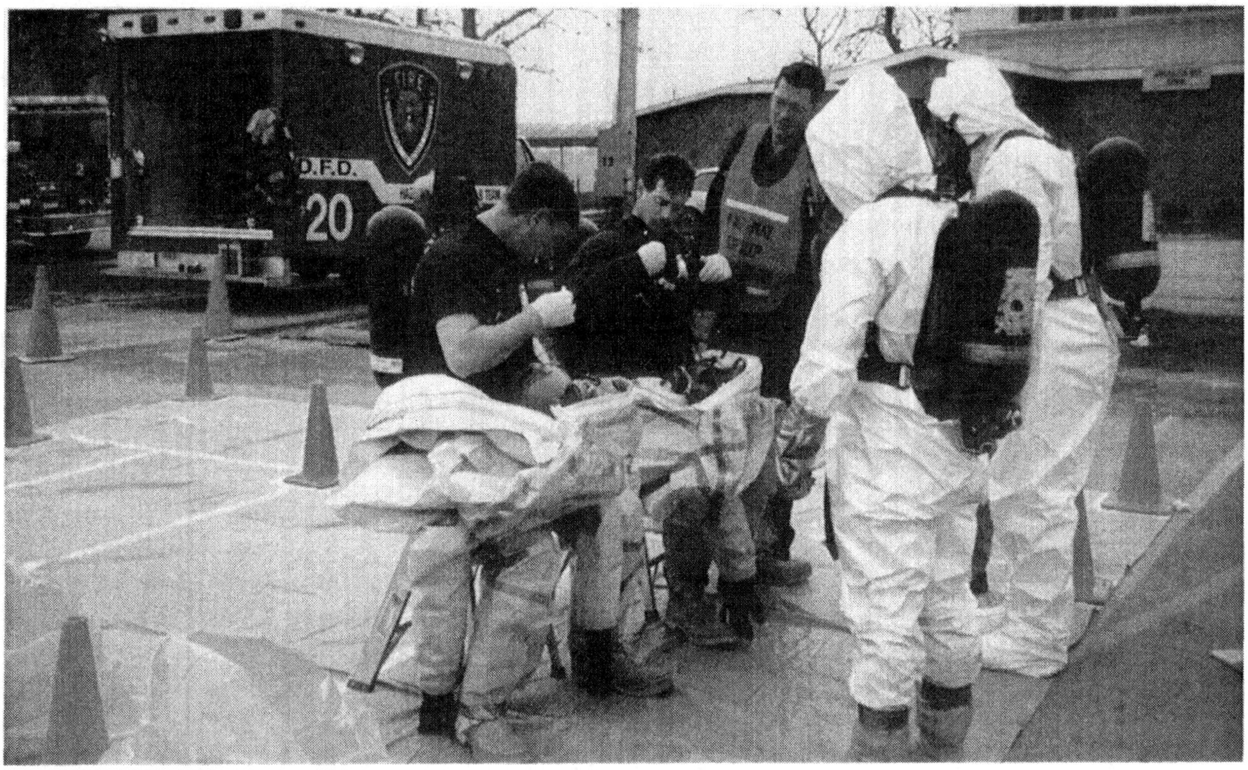

Figure 1-7: Students and local Hazmat team members practice donning personal protective equipment.

reports are used to make decisions on the regulatory status and the course of action to be taken for a site. The cost of a cleanup can run into millions of dollars. Litigation may result as responsible parties are identified. It is not unusual for one or more members of the environmental investigation team to be questioned during legal proceedings. Data collection methods and samples may come into question. In these situations the importance of thorough procedures and documentation becomes evident. Environmental consulting firms may be held liable for producing erroneous or inadequate reports. The field team and the environmental technician have a profound responsibility in the collection of representative and accurate data and samples.

Environmental investigations may draw the attention of the media, the neighbors and any other interested parties. During an investigation it is best to be careful in making statements which can be misunderstood, misinterpreted or later proven wrong as more data are collected. Technicians and field crews working on a project or at a site may encounter people who are curious and ask questions about what is happening. These situations must be handled carefully and with a minimum of interruption. It may be best to direct the inquiries of external parties to someone such as the project manager, site manager or site owner.

Checking Your Understanding

1. Why are the sampling methodologies specified in a written plan?
2. Describe the elements of a written site-specific health and safety plan.
3. List and describe eight typical roles that need to be fulfilled by members of a field investigation team.
4. Describe what might result if a sample were purposefully collected in a manner that conceals the existence of contamination.
5. List the types of materials generated by an environmental technician that are considered to be legal documents.

Summary

Site characterization involves investigating a site through a variety of strategies, particularly through sampling and analysis of its soil, water, air, **biota** (plants and animals), and human-made materials such as the contents of drums, tanks and other containers.

The process of site characterization is used to identify the source, type, concentration and extent of contamination on a property. The process involves four stages:

1. Research into the history and use of the site, employing several research methods;
2. Field investigation, including real-time monitoring as well as the collection of samples;
3. Laboratory analysis of samples;
4. Evaluation of the laboratory data and implementation of the data in decision-making.

An environmental manager or technician who plans a field investigation must include a written site-specific health and safety plan. The safety officer is a member of the field team, which includes a variety of management and technical personnel.

Whether or not to embark upon a full site characterization process is dependent upon factors such as EPA's RCRA requirements for a controlled hazardous waste operation or CERCLA requirements for an uncontrolled hazardous waste site; a bank or other lender's requirements in a real estate transaction to protect against potential liabilities; or a property owner's own initiative to avoid future liability.

The general process of sampling involves the following broad steps, most of which are discussed in greater depth in subsequent chapters.

1. Gather historical information from local governmental sources and from persons who have knowledge of present and past uses of the site, as described earlier in this chapter.
2. Contact and consult with the analytical laboratory. Chapter 8 discusses considerations for selecting and consulting with laboratories.
3. Prepare the sampling objectives and the plan, including the health and safety plan as well as the quality assurance project plan. Chapter 2 outlines planning for the collection of representative samples and implementation of quality assurance and quality control methods.
4. Organize the field team according to the site-specific requirements as described in this chapter.
5. Prepare to go onsite. Chapter 3 explains what the technician's considerations should be when preparing to go onsite. Chapter 5 lists additional preparatory steps applicable to any sampling and an example of an equipment checklist.
6. Conduct a preliminary investigation with real-time methods, including monitoring instruments and field analysis methods. Chapter 4 introduces the use of monitoring instruments and Chapter 8 introduces field analysis methods.
7. Conduct the sampling event. Chapters 4 through 7 detail the procedures for sampling air, soils and other geologic materials; surface water and groundwater; and tanks and other containers.
8. Ship the properly prepared samples to the laboratory. Chapter 3 discusses the preparation, storage and shipment of samples.
9. Use the analytical results for decision-making. Chapter 9 introduces data validation and interpretation.

Environmental technicians who participate in sampling programs need the technical skills introduced throughout this text, but they also need to cultivate several important personal qualities, such as the ability to pay attention to detail; to follow directions precisely; to be meticulous about using clean, well-maintained equipment; and to carefully document all aspects of a field assignment. Environmental technicians must above all think clearly, because there will be times when – no matter how carefully planned the field assignment is – they will need to make a decision or alter a course of action. Technicians who are not only trained to collect a sample, but also educated about the principles of sampling, can make informed judgments when the need arises. These technicians are an asset to their employers.

The best and most exacting laboratory in the country cannot possibly make up for haphazard planning, collection and handling of environmental samples. No matter how precise and accurate the lab's work is, the results are meaningless – and costly for the client – if based upon incorrect or careless approaches to designing and carrying out the sampling event itself.

Application and Critical Thinking Activities

1. Differentiate a Phase I environmental site assessment from other types of environmental investigations such as a remedial investigation.

2. Identify and define the site investigation terminology used in your area.

3. a. Collect records – Obtain assessor's records and building records for a property. Visit a public library and locate city directories and Sanborn maps for a specific property. Make a FOIA request to a state regulatory agency for records or information associated with a particular property or spill.

   b. Evaluate records – What type of information did each source provide? Does any of the information give you cause for concern that soil or groundwater might be contaminated or that hazardous substances might be present on the property? Why or why not?

4. Prepare a list of appropriate questions to ask if you were planning to interview someone in connection with a Phase I site assessment for property transfer. Formulate some open-ended questions to obtain more than a simple yes-or-no answer. How would you feel if you were asked questions regarding your property or business?

5. Obtain a copy of the Hazardous Waste Operations and Emergency Response Standard (29 CFR 1910.120). Based on this standard, what is the training required for an environmental technician who collects samples during an investigation at a suspected hazardous waste site? How long is the required training? How often must the training be repeated?

6. List questions you think you might be asked as an environmental technician working on a site. What would be your response to the various questions? Would your response vary depending on whom you encountered? Consider the following individuals/groups as you formulate your questions and responses:

   —the media

   —the site owner/operator

   —employees working at the site (but not working for your employer)

   —neighbors/concerned citizens

   —local or state governmental officials/employees (including environmental regulators)

   —utility employees

7. Obtain from an environmental consultant a copy of the site-specific health and safety plan for a sampling site. Study the various requirements (for example, regarding personal protective equipment and procedures).

2

The Sampling Plan

Chapter Objectives

Upon completing this chapter, the student will be able to:

1. **Identify** the sampling plan elements required by EPA.
2. **Distinguish** between biased and unbiased sampling techniques.
3. **Distinguish** between homogeneous and heterogeneous samples.
4. **Distinguish** between grab samples and composite samples and indicate what determines the choice between the two.
5. **Define** the following sampling techniques and differentiate among them. Indicate where and when each technique should be used.
   —Simple random sampling
   —Stratified random sampling
   —Systematic random sampling
6. **Given** a hypothetical situation, determine sampling points and frequency by applying the following statistical methods:
   —SW-846 formula
   —Random number chart
7. **Develop** a graph paper grid with sampling points and apply the grid in the field.
8. **Define** sampling precision, accuracy and completeness.
9. **Identify** and describe quality assurance and quality control procedures.

Chapter Sections

2-1 The Written Sampling Plan

Concepts

- Site investigation of EPA-enforced remedial sites requires formal planning, quality assurance/quality control (QA/QC) as well as representative sampling programs.
- The sampling plan is designed as a step-by-step guide for conducting the sampling activities while covering all possible contingencies.
- The two main elements of the written sampling plan that concern the technician's activities are the field sampling plan and the quality assurance project plan (QAPjP).

Sampling is the process of physically collecting a representative portion of environment. Just as the carpenter could not build a house without a set of plans, so the environmental technician could not perform sampling activities without a written sampling plan. The sampling plan details all the elements necessary for the job, such as historical data, sampling objectives, applicable regulations, sampling procedures, team composition, health and safety plan, and so on. Therefore, site characterization starts with the writing of a sampling plan (see Appendix 2 for an example of a written sampling plan).

In the previous chapter we briefly introduced one component of the sampling plan: the health and safety program. In this chapter we will describe the two main components that are essential to the activities of the field technician: the **field sampling plan** and the **quality assurance project plan**, often referred to as the **QAPjP**.

Purpose of the Written Sampling Plan

Sampling plans are a legal requirement only for EPA-regulated actions and must be approved by the site's **lead agency**. These actions may involve activities ranging from the gathering of baseline data for an **environmental impact statement (EIS)** to a remedial investigation of an abandoned waste site under CERCLA or a corrective action at an industrial site under RCRA. Although students may not anticipate being employed at an EPA-enforced remedial site, it will be nonetheless important to understand the concepts of planning, representative sampling and **quality assurance/quality control (QA/QC)**. All these concepts are applicable to routine sampling activities such as those associated with real estate assessments; soil analysis for nonregulatory reasons; or waste analysis to determine if a company is classified under EPA regulations as a hazardous waste generator. Written sampling plans are generally prepared prior to all these types of activities – even though EPA does not require them – to maximize efficiency and protect from future liability.

The general goal of the sampling plan is to obtain representative samples of potentially contaminated media. Its purpose is to detail a plan of action so as to meet this goal. Only after the following questions have been answered should samples be collected.

Figure 2-1: The environmental media air, water, soil and biota are the carriers of the contaminants.

—What are the objectives of collecting samples?

—What types of samples are needed?

—Where are the best sampling locations?

—How many samples are required and how often do we sample?

—How are samples to be collected and handled?

When answering these questions the following factors must be taken into consideration:

—The topographic, geologic and hydrologic characteristics of the site, i.e., surface, soil types and groundwater characteristics.

—Meteorologic conditions, such as temperature, air pressure, precipitation, wind velocity and seasonal variations.

—The flora and the fauna of the area, i.e. bio-accumulation and biotransformation in the plant and animal life, especially agricultural.

—Demographic and geographic information on and surrounding the site.

—Physical properties and hazardous characteristics of material involved.

The sampling plan presents in detail all activities, **sampling protocols**, and data gathering methods to be carried out during the sampling effort. Sampling protocols describe the procedures the technician must follow during each step of sampling including collecting, packaging, labeling, preserving, transporting, storing and documenting. The development of precise protocols reduces the potential for errors, and the correct implementation and documentation of precise protocols protects the sampling event from future liability. It is essential to understand that liability is an important issue in every remedial action, and that every sampling plan must be structured in a sound and legally defensible manner. This means that the technician will enter field notes in the log throughout the sampling activities, that the sampling protocol will be followed in all its details, and that the quality assurance program will be carefully implemented to ensure accuracy, precision and completeness of the data.

If the sampling is part of remedial activity associated with the cleanup of hazardous wastes, the sampling plan must be in accordance with the applicable agency regulations:

—RCRA regulations (40 CFR Part 264, 265) for remedial activity at a current industrial site;

Figure 2-2: Uncontrolled hazardous waste site simulated for training purposes.

Point of View ■ David Foss

David Foss is the Chief of the Environmental Division, Science and Engineering Directorate, Rock Island Arsenal. David also teaches in the Eastern Iowa Community College District's Environmental Compliance and Technology (ECAT) program. He co-authored this book's chapters on sampling plans and on the sampling of containers. David received a Ph.D. in chemistry from Michigan Technological University; he has worked as a chemist for twelve years, and in the environmental field for the past ten years. Being employed in the field gives David a real-life perspective to share with his students.

Q: *What is your involvement with sampling plans at the Rock Island Arsenal?*

A: I have been involved with developing sampling plans for several EPA cleanup efforts at the Rock Island Arsenal, which is a Department of Defense weapons manufacturing plant that dates back to the Civil War. In one effort, the EPA required us to conduct a RCRA cleanup of an old limestone building, which housed a metal plating shop that began use during the Civil War.

Q: *Do you have advice for your students about sampling plans?*

A: You could find yourself managing contractors whom your company hires to perform cleanup. Think of the contractors as your employees. In other words, while your contractor does the physical work, you as the employer are responsible for their work in the end. Now, this is where knowledge and experience in sampling come into play. Because no matter whom you hire to do the job, you still have to know the EPA requirements. As the authorized representative of the owner, you may be considered subject to personal liability for problems. So, to protect yourself as well as your company, everything done on the site must follow regulations, have proper intent and be fully documented.

Q: *What do you expect of technicians regarding the sampling plan?*

A: Before they go on the job, I run a training meeting with every last person who will work on the site – contractors, subcontractors, even the subs of subs. I expect the technicians to understand the sampling plan language just as the engineers do, and to pass a written test to demonstrate that they can comply with requirements. I sit around a table with groups of technicians to get a sense of their knowledge. I can tell when an individual is not up on the requirements during these discussions, which means that the person will not be eligible to work on the site. Another thing I do is get out on the site with them. If I have to grab a shovel to show them a procedure, I will. When I as an administrator work beside the technicians, it shows I have confidence that my work request is rational and within the realm of the law. I would never ask anyone to do anything I wouldn't do myself.

Q: *How do you make sure that contractors are doing what they are supposed to do?*

A: I observe the site work daily and read the field logs, making sure everything is up to code. The contractors submit periodic status reports. If anything at all occurs at the site that is unanticipated or not meeting the sampling plan, the contractor must let me know immediately. Sometimes a contractor asks permission to modify the sampling plan if, for example, another technique might save some money. I study that request to make sure it is legally wise. If the contractor's idea does not clearly satisfy the regulations, then I have to deny permission.

Q: *How big is your sampling plan for the RCRA cleanup of that old building?*

A: About four inches thick!

—CERCLA regulations (40 CFR Part 112) for remedial activity at an uncontrolled site, such as an abandoned landfill, a former business that mismanaged wastes, or a catastrophic accident that released large amounts of chemical.

Sampling Plan Elements

A written sampling program that reflects RCRA and CERCLA requirements may include the following components:

1. A project management plan including overall management approach, schedules, budget, personnel;
2. **Data quality objectives (DQOs)** to ensure that the appropriate data are obtained to meet the needs of a remedial investigation or remedial activity;
3. A field sampling plan, including sampling objectives, sampling protocols and procedures, sample types, locations and frequencies;
4. A quality assurance project plan (QAPjP);
5. A health and safety plan;
6. A community relations plan.

In this section we are going to describe the third and the fourth components of the list above, the field sampling plan and the QAPjP. These are the two elements of the written sampling plan that directly concern the work of the technician, since they involve field implementation of objectives and procedures already put in place by the project planner.

Element 1: The Field Sampling Plan

The field sampling plan provides detailed guidance for all field work by defining the methods to be used on a project. EPA specifies the following field sampling plan elements, which the project planner then develops:

1. Site background: Specific information on former activities at the site (see Chapter 1).
2. Sampling objectives: A statement – based on previous research and knowledge of the site – regarding what particular contaminants the sampling event is directed towards.
3. Sample location and frequency: Location and quantity of samples to be taken.
4. Sample designation: Specific naming system for labeling sample locations and containers.
5. Sampling equipment/procedures: What equipment is to be used and in what manner.
6. Sample handling and analysis: Specific designation of analytical procedures.

Because the sampling plan is a document intended for use by the agency as well as by samplers and other field workers, it must be written so that it can be clearly understood by both site employees and regulatory agency employees. The plan must be thorough in describing the purpose of the sampling activity at the remediation site. The sampling should cover all possible contingencies that could arise on a remediation site. The sampling procedures and protocols should be written in such a manner as to provide a step-by-step guide for the sampling team.

Element 2: The Quality Assurance Project Plan (QAPjP)

The QAPjP describes the organizational responsibilities as well as the quality assurance and quality control procedures necessary to achieve the DQOs. The QAPjP is designed to ensure that the data obtained from sampling and analysis are of high quality, scientifically valid and, above all else, legally acceptable for the regulatory agency in charge of the remediation project. The project planner develops the QAPjP portion of the sampling plan with input from the laboratory. EPA specifies the following QAPjP elements:

1. Title Page, Introduction and Table of Contents
2. Project Description – Defines goals; includes site description and history, schedule of activities, intended data usage, identification of sample matrices and parameters, sampling design and description.
3. Project Organization and Responsibilities – Identifies the roles and responsibilities of field team members, such as the site manager, field team leader, site safety officer and technicians.
4. Quality Assurance (QA) Objectives for Data Measurement – Lists qualitative and quantitative statements that outline the decision-making

process and that specify the data required to support those decisions.

5. Sampling Procedures – Describes general procedures for matrices to be sampled (the field sampling plan provides site-specific procedures for both investigative samples and quality control samples).

6. Sample and Document Custody Procedures – Provides a procedure for records such as sample identification tags, chain of custody, receipt-for-sample forms, field notebooks, airbills/bills of lading, photographic logs; includes procedures for the document control officer.

7. Calibration Procedures – Describes processes for checking and adjusting laboratory equipment.

8. Analytical Procedures – Describes protocols for various analyses.

9. Data Reduction, Validation and Reporting – Reviews laboratory analyses for errors; determines inconsistencies; compares data against quality control samples and background data; deletes unusable data and explains limitations of qualified data; reviews and summarizes detection limits for nondetectable results; reviews sampling design for dealing with media variability.

10. Internal Quality Control (QC) – Describes the quality control methods for the laboratory.

11. Performance and Systems Audits – Describes the internal and external system of checklists for verifying implementation of the quality assurance program.

12. Preventive Maintenance – Provides the schedule and the requirements for maintaining the equipment, such as analytical instrumentation.

13. Data Assessment Procedures – Reviews completeness of work plans and supporting documentation; assesses calibration and standardization procedures, measuring/monitoring equipment maintenance as well as repair records, personnel qualification records and documentation kept by the personnel, such as logbooks.

14. Corrective Actions – Investigates causes of significant or repetitious unsatisfactory conditions related to laboratory quality assurance procedures; establishes methods for implementing and documenting corrective actions in the laboratory.

15. Quality Assurance Reports – Issues periodic (e.g. monthly) QA reports to management, summarizing QA/QC status of the project and any conditions that may interfere with quality.

Checking Your Understanding

1. Identify and describe the elements of the field sampling plan.
2. Identify and describe the elements of the QAPjP plan.

2-2 Types of Samples

Concepts

- Environmental samples are defined by the method chosen for the sampling or by the type of medium collected.
- Biased sampling provides information about particular places of interest at the site, such as areas of known or suspected contamination.
- Unbiased sampling provides an average picture of the entire site.

Approaches to sampling are necessarily contingent upon the type of remedial investigation to be conducted. Is the site an abandoned landfill? An abandoned industrial site? A lagoon or other body of water with unknown contaminants? A contaminated well? The following are some possible sample collections:

—Surface and subsurface soil (used to classify and characterize soil types and analyzed to determine the vertical or horizontal migration of suspected contaminants);

—Groundwater (including water from private wells, monitoring and public wells, industrial and production wells);

—Oily wastes and sludge;

—Surface water (including water from pits, ponds, lagoons, streams, rivers and lakes);

—Sediments (collected from points of suspected deposition in lakes, rivers and streams);

—Gases emanating from soils;

—Drinking water (from residential or public well supplies);

—Air samples.

An evaluation of the intended use of the data must be undertaken in order to ensure that the type of sample obtained provides the necessary information to address the issues of concern. **Environmental samples** are defined by: 1) the method chosen for the sampling (i.e., grab or composite), 2) the type of medium collected (i.e., media or waste) and 3) whether they are **investigative samples** or quality control samples (see section 4).

Biased versus Unbiased Sampling

The sampling plan objectives will specify the extent to which technicians should use **biased sampling** (also called **judgmental sampling**) or **unbiased sampling** (also called **random sampling**) methods during a site investigation. Biased sampling methods are used to collect data about particular places of interest at the site, such as areas of known or suspected contamination. Biased sampling is mostly used during the site investigation process, which is performed to find out whether any contamination is present. This type of sampling provides the determination of critical areas of contamination; for example, downgradient from a contamination source.

Unbiased, or random, sampling methods are used to collect data that provide an average picture of a site, which is best for predicting overall site characteristics. It is performed by collecting samples from statistically pre-selected random targets on a grid staked out at the site by the technician. It is the statistical selection of the sampling points, rather than the subjective one, that defines these samples as unbiased.

Grab versus Composite Samples

Two techniques used for collecting samples are grab sampling and composite sampling. **Grab samples** are single samples collected from a specific location at a given point in time. They offer the most information regarding spatial variability of the contaminant. Because the analytical results are specific to the particular location sampled, grab sampling yields good confidence and is therefore the preferred method for collecting samples in most environmental site investigations. However, since every sample needs to be analyzed separately, the lab costs for grab sampling are higher than for composite sampling.

A **composite sample**, on the other hand, is achieved by mixing samples collected from various locations or from the same location but at different

Figure 2-3: A composite sample is taken from multiple samples and mixed by the technician.

times. Shortcomings of the composite sample include dilution of chemical constituents at one location by the samples from other locations. Also, compositing presents a difficulty in identifying the exact location at which a specific contaminant resides. Therefore, composite data yield less specific contaminant variability and spatial information. On the positive side, compositing lends itself to a qualitative assessment of the average characteristics of the sampled area. Also, since compositing involves combining several samples into one, it represents a less expensive procedure than the grab sampling procedure due to reduced costs in lab analysis. A word of caution: Composite sampling should be avoided if there is any potential risk of dangerous chemical reaction.

The choice between grab or composite samples is determined by the goals of the investigation as well as by the lead regulatory agency. In the following chapters we will describe detailed methods for compositing samples.

Media versus Waste Samples

A **media sample** refers to a specimen taken from the environment: air, water, soil and biota. A **waste sample** refers to a specimen taken from drums, tanks, lagoons, pits, waste piles, fresh spills, impoundments, or any other waste disposal areas. Sampling typically involves investigation of both environmental media and specific waste accumulation areas. Waste samples are therefore different from media samples because the general characteristics of the materials are different, but also because the questions asked of the data are different. In waste characterization we ask:

—What compounds are present?

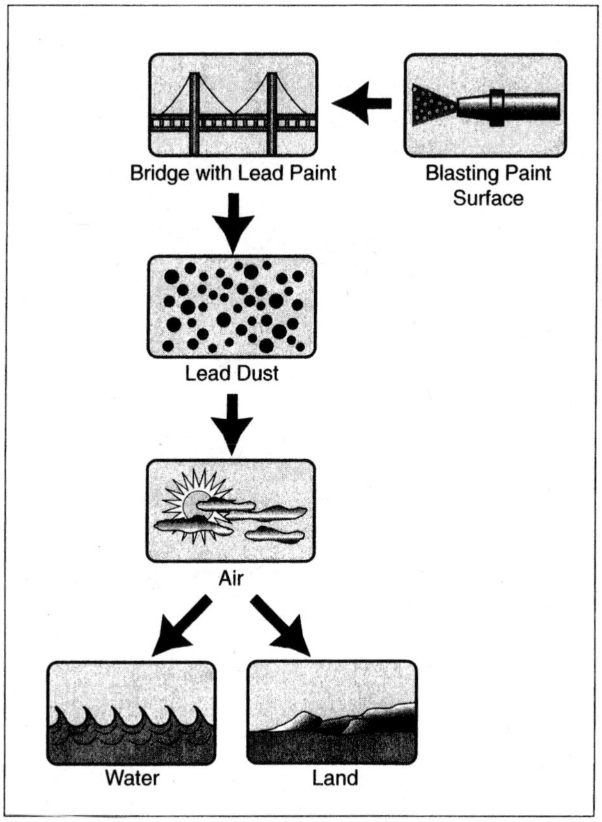

Figure 2-4: A common term within the environmental sampling area is **matrix**. A matrix – a term borrowed from the field of geology – may be thought of as the carrier for a contaminant. In the case of lead contamination due to blasting the paint surface of a bridge, as illustrated here, air is the first matrix, or carrier, for the lead contamination. Soil and surface waters become the subsequent matrices as the lead is deposited upon them. In this sense matrix and medium are interchangeable terms.

—Do these contaminants exceed any criteria or standards?

In the case of environmental media we also ask these questions:

—Which media are contaminated?

—What is the average contamination?

—What is the total contamination (mass, volume)?

—What is the maximum contamination (concentration)?

—What area of the site is contaminated?

—What is the vertical and horizontal extent of contamination?

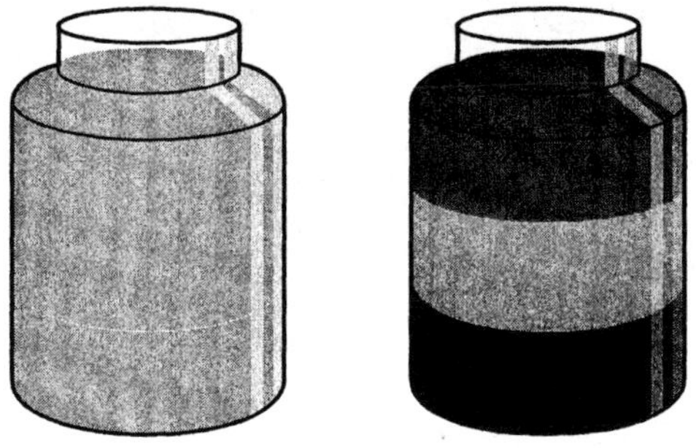

Figure 2-5: The uniform shading indicates homogeneous material, which means it is the same throughout the container. The layers of shading indicate heterogeneous material, which means the composition of the material varies within the container.

Homogeneous versus Heterogeneous Samples

The composition of both media and waste samples may be either homogeneous or heterogeneous. A **homogeneous sample** is apparently uniform throughout; for example, a container holding only sand or lake water. A **heterogeneous sample** varies within its container; for example, a sample drawn from a tank of material that has separated into oily, watery and solid layers, or a soil boring that exhibits signs of sand, clay, rock and even construction debris.

Theoretically, a single sample collected from anywhere within a homogeneous material would yield accurate and precise data on the waste's chemical properties. Such sampling is commonly referred to as **simple random sampling**. A more representative approach is referred to as **systematic random sampling**, in which the first sample is randomly selected and all subsequent samples are taken at fixed space or time intervals; for example, a series of samples collected at one-foot intervals in a tank.

A single sample collected from a heterogeneous material obviously would not yield accurate and precise data on the material's overall characteristics. To obtain representative samples of stratified or layered material, the technician must use a system that collects from different depths and locations within it. Such sampling is commonly referred to as **stratified random sampling**. Later chapters will introduce equipment for sampling various soil strata, surface water and containers.

Sample Collection Order

Samples should be collected from an investigation site in an order that protects against loss of **analyte** to the atmosphere. For example, samples to be analyzed for volatile organic compound (VOC) content must be collected and stored quickly to avoid escape of the vapors. EPA recommends the following collection order:

1. Volatile organic compounds and total organic compounds
2. Dissolved gases and total organic carbon
3. Semivolatile compounds – base and acid extractables, pesticides and PCB
4. Metals and cyanides
5. Major water quality cations and anions
6. **Radionuclides**

It is good management practice to plan collection from the least to the most contaminated area of the site in order to decrease the risk of cross-contamination.

Sample Labeling Designations

Every project or remedial action should have its own project-specific sample labeling system. The project manager should set up the labeling in a way that clearly identifies the particular project and that excludes the possibility of confusion once the samples are all at the lab. A good choice is an alpha-numeric system which combines letters and numbers to identify not only the project but also the matrix. Some common matrix designations include:

—S or SS to indicate surface soil

—MW to indicate a monitoring well

—SW to indicate surface water

—SD to indicate sediment

—BH to indicate sample from a bore hole

Using these designations in the case of a series of samples collected at a landfill known as the Alpha Landfill, the first sample of the series would then be AL-SS-01-01, the second would be AL-SS-01-02, and so on for the rest of that series. For a second series of samples the labeling would be AL-SS-02-01, and so on. For completeness the designation should also contain the date of the sampling event; the first label of the first sample of the first series would then read AL-SS-01-01(10/21/97), if the sample was taken on October 21, 1997.

Checking Your Understanding

1. Differentiate between biased and unbiased sampling techniques.
2. Define and differentiate the following sampling techniques; indicate where and when the techniques should be used.
   a. Simple random sampling.
   b. Stratified random sampling.
   c. Systematic random sampling.
3. Distinguish between homogeneous and heterogeneous samples.
4. Distinguish between grab samples and composite samples; indicate when each one should be used.

2-3 The Statistical Approach to Obtaining Representative Samples

Concepts

- A statistical approach is used to determine the number of sampling points as well as their locations.

This section concerns the statistical methods used to obtain unbiased, or random, samples; it pertains primarily to hazardous waste site Remedial Investigations (RI). The statistical concepts approach emphasized in this chapter is especially appropriate for large, **uncontrolled hazardous waste sites** where a comprehensive approach to characterizing the hazards is necessary and where contractors would be answerable to a federal or state agency.

The selection of random sampling points is made statistically. The particular statistical methods used by a remediation project are selected by the lead regulatory agency. The agency will base its decisions on regulations unless there are compelling reasons for choosing other sampling methods. An example of a situation in which the agency would not choose statistical methods would be the case of prior historical knowledge of the location of the contaminated areas, which would make a random

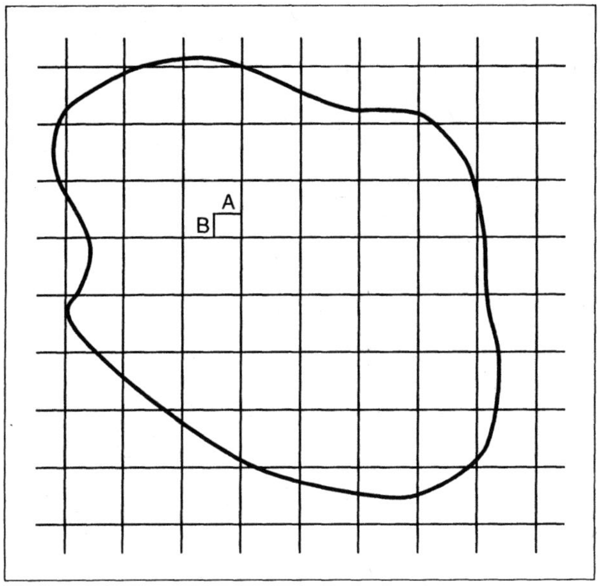

Figure 2-6a: Square grid system.

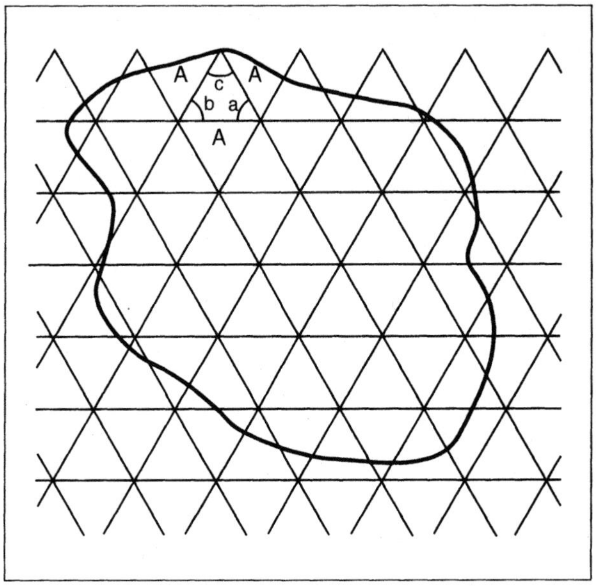

Figure 2-6b: Triangular grid system.

sampling of the site unnecessarily expensive. Another example would be if the site presented barriers to carrying out random sampling, such as a highway or a building; in this case a well-designed biased sampling method may yield the needed information.

This section presents three components of unbiased sampling point selection, two of which involve statistical calculation:

1. A method for gridding a site.
2. A formula for calculating the minimum number of random samples required to accurately characterize a remedial site. This formula is from EPA's guidance document "Test Methods for Evaluating Solid Waste, Physical/Chemical Methods," commonly referred to by its document number, SW-846.
3. A random method of determining where on the grid to collect the number of samples specified by the SW-846 formula.

1. Determining the Grid Spacing

Grid systems are used in developing unbiased sampling plans in which samples are located at consistent distances from one another. A site grid may be rectangular, triangular, or even circular if that is the most effective way to locate where samples are to be collected.

The grid spacing, or distance between the grid lines, is selected according to the site-specific sampling objectives. It may therefore be planned to allow for the low-intensity sampling of a preliminary site investigation or for the high-intensity sampling of a subsequent remedial investigation. For example, suppose the grid lines are spaced at 10-foot intervals. During the preliminary site investigation, samples

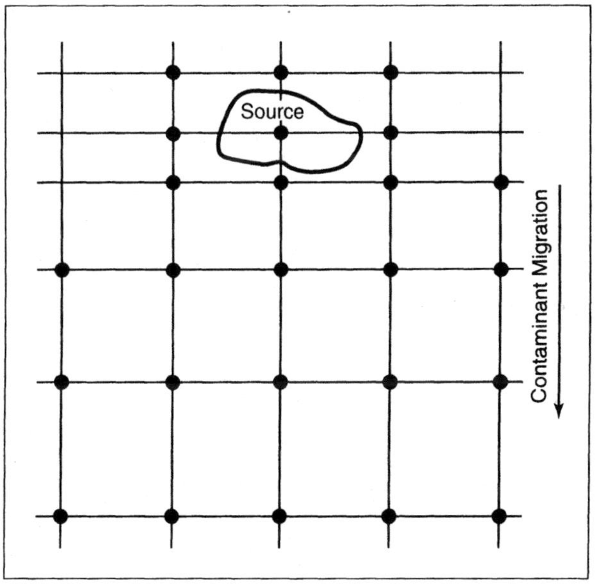

Figure 2-6c: Modified grid system to account for directional correlation.

Technical Corner ■ Model Exercise from SW-846

(reprinted by permission from HMTRI)

The problem:

Calculate the number of samples to be taken from a waste lagoon thought to be contaminated with barium. EPA's regulatory threshold (see TCLP Regulatory Thresholds, Table B, Appendix 4) for barium is 100 ppm, which means that concentrations of 100 or more parts of barium per million parts waste material exceeds EPA's regulatory limit. Preliminary samples were taken from the lagoon, and barium levels were reported as 86, 90, 98, and 104 ppm.

Step One: Calculate the mean (average) concentration of the preliminary samples.

In the formula for finding the mean, sample 1 is s_1, sample 2 is s_2, etc. and the number of samples is n_s.

$$\bar{x} = (s_1 + s_2 + s_3 \ldots) \div n_s$$

$$\bar{x} = \frac{86 + 90 + 98 + 104}{4}$$

$$\mathbf{\bar{x} = 94.50}$$

Step Two: Calculate the variance (s^2) among the samples. This is done through several substeps:

In the formula for finding the variance, sample 1 is s_1, sample 2 is s_2, etc. and the number of samples is n_s. Note to student: To use the formula, compute the functions within parentheses "( )" first. Then compute the functions within braces "{ }". Next, compute the function within brackets "[ ]". Finally, complete the computation.

$$s^2 = [\{s_1^2 + s_2^2 + s_3^2 \ldots\} - \{(s_1 + s_2 + s_3 \ldots)^2 \div n_s\}] \div (n_s - 1)$$

(a) Square each value of the preliminary samples and then find their sum;

$$86^2 + 90^2 + 98^2 + 104^2 = 7396 + 8100 + 9604 + 10{,}816 = 35{,}916$$

(b) Find the sum of the preliminary sample values, square that sum, then divide that result by 4 (the number of preliminary samples);

$$\frac{(86 + 90 + 98 + 104)^2}{4} = \frac{378^2}{4} = \frac{142{,}884}{4} = 35{,}721$$

(c) Subtract the result of (b) from the result of (a), then divide by the number of preliminary samples less one (in this case, 4 - 1);

$$\frac{35{,}916 - 35{,}721}{4 - 1} = \frac{195}{3} = 65$$

$$\mathbf{s^2 = 65}$$

Step Three: Refer to the statistical table commonly referred to as a "t" table (see Table A, Appendix 3).

In the left column, entitled "Degrees of Freedom," find your number of preliminary sample values less one, which is again 4 – 1 in this problem. Look across at the second column. Use this "t" value in your SW-846 formula.

"t" value is 1.638

Step Four: You are now ready to use the SW-846 formula to calculate the number of random samples that should be taken in your next sampling phase (n_1).

To review, the values from the problem and from your calculations are:

n_1	The minimum number of samples to be taken.
t_{20}	1.638
s^2	65
RT	100
$\bar{x}$	94.50

To use the formula:

$$n_1 = \frac{(t_{20}^2\ s^2)}{(RT - \bar{x})^2}$$

(a) Square the "t" value and multiply times s^2.

(b) Subtract the mean concentration ($\bar{x}$) from the regulatory threshold (RT), and then square the result.

(c) Divide (a) by (b).

The calculation (to the nearest tenth):

$$n_1 = \frac{(1.6)^2 \cdot 65}{(100 - 94.5)^2}$$

$$n_1 = \frac{2.6 \cdot 65}{30.3}$$

$$n_1 = \frac{169}{30.3}$$

$n_1 = 5.6$ Round upward to next whole number.

As a result of this calculation, at least six randomly selected samples must be taken, based upon the preliminary samples. Generally, three additional samples are taken to increase sampling precision, in case the preliminary sampling was inaccurate. Therefore, six samples are brought up to a total of nine samples to be taken for analysis.

1	2	3	4	5	6	7	8	9	10
11	12	13	14	15	16	17	18	19	20
21	22	23	24	25	26	27	28	29	30
31	32	33	34	35	36	37	38	39	40
41	42	43	44	45	46	47	48	49	50
51	52	53	54	55	56	57	58	59	60
61	62	63	64	65	66	67	68	69	70
71	72	73	74	75	76	77	78	79	80
81	82	83	84	85	86	87	88	89	90
91	92	93	94	95	96	97	98	99	100

Table 2-1: Example of grid. Each intersection is numbered. Depending upon the size of the site and the number of samples to be taken, the intersections might be 10 to 100 feet apart.

may be taken at the intersection of every tenth grid line, thereby resulting in one sample every 100 feet. If the subsequent review of the preliminary data warrants intensive sampling, it may then be performed at the previously established 10-foot grid intersections.

The grid, plotted on paper or computer by the project planner, is transferred to the actual site by the technician who measures and stakes the area. It is costly, and usually unnecessary, to stake every intersection; in most cases it is sufficient to measure and stake the corners and the sampling points.

2. Determining the Number of Samples to Collect (SW-846 Formula)

The number of samples to be taken from a waste container or site is calculated by applying a formula from EPA's guidance document SW-846. This formula, used in conjunction with the model exercise offered in the Technical Corner, may be applied by people without a background in statistics.

To apply the formula, it is necessary for the project planner to have several preliminary samples analyzed. The analytical results from these few samples provide the data needed to use the **SW-846 formula**.

The formula for calculating the number of samples is:

$$n_1 = \frac{(t_{20}^2\ s^2)}{(RT - \bar{x})^2}$$

Figure 2-7: A technician stakes a grid at a sampling site.

In the formula above the symbols stand for:

n_1 The minimum number of samples to be taken.

t_{20} The "t" value taken from Table A (See Appendix 3).

s^2 The variance of the sample. This value may be calculated by following the procedure included in the model exercise.

RT The **regulatory threshold**, meaning the concentration determined by EPA to be hazardous; this is taken from Table B (See Appendix 4).

$\bar{x}$ The mean (average) concentration of several preliminary samples taken at the site.

Although the SW-846 formula is quick and easy to calculate by the traditional pencil and paper method, personnel who work with these concepts routinely will most likely create a computer spreadsheet for the input of appropriate data and retrieval of results.

3. Determining the Sampling Points (Random Number Method)

Once the project planner has determined the grid size as well as the layout, and he or she has used the SW-846 formula to calculate the number of samples to collect, the next step in selecting random sampling targets is to statistically generate a set of random numbers. These random numbers, which are plotted on the paper copy of the grid, represent the specific locations where the technician will collect the samples; they are called **sampling targets** or **sampling points**. At the sampling site, technicians follow the pattern depicted on the graph paper as they measure and stake out the grid and then mark the sampling points.

The set of random numbers should at least equal the number of samples indicated by the SW-846 formula. It may be a good idea to include several extra numbers in the set in case some grid targets do not work on the actual site because of barriers. Before the technicians go onsite they can compare the grid diagram to what is already known from preliminary site investigation documents, such as sketches, topographic maps, utility easements and underground pipelines. The project planner can anticipate obstacles the technicians may encounter onsite: buildings, trees, underground pipes, or other natural or man-made obstacles.

To generate the needed set of randomly chosen numbers, a project planner may either use a random number table or a computer software program designed to generate statistically the needed components for random sampling.

Using a Random Number Table to Select Sampling Targets – A set of sampling targets may be selected from a random number table (see Table 2-2) by using steps such as the following:

Step 1: Without looking, the project planner drops the pencil point anywhere within the table to establish a starting point. The digit under the pencil

	Column																	
Row	1	2	3	4	5	6	7	8	9	10	11	12	13	14	15	16	17	18
1	9	8	9	6	9	9	0	9	6	3	2	3	3	8	6	8	4	5
2	3	5	6	1	7	4	1	3	2	6	8	6	0	4	7	5	8	4
3	4	0	6	1	6	9	6	1	5	9	5	4	5	4	8	6	8	4
4	6	5	6	3	1	6	8	6	7	2	0	7	2	3	2	1	4	0
5	2	4	9	7	9	1	0	3	9	6	7	4	1	5	4	9	9	7
6	7	6	1	2	7	5	6	9	4	8	4	2	8	5	2	4	2	4
7	8	2	1	3	4	7	4	6	3	0	7	5	0	9	2	9	0	1
8	6	9	5	6	5	6	0	9	0	7	7	1	4	1	8	3	9	7
9	7	2	1	9	9	8	0	1	6	1	6	2	3	6	9	5	9	6
10	2	9	0	7	9	0	8	9	6	3	3	8	5	5	6	5	0	4
11	9	3	5	4	5	7	4	0	3	0	1	0	4	3	3	9	4	3
12	9	7	5	7	9	4	8	6	8	7	6	1	6	8	2	5	1	5
13	4	1	7	8	6	8	1	0	5	8	8	6	1	6	8	2	5	1
14	5	0	8	3	3	4	5	4	4	2	5	3	0	4	9	6	4	3
15	3	5	0	2	9	4	1	0	0	3	9	0	5	8	6	0	4	0
16	0	3	8	2	3	5	1	0	1	0	6	8	5	2	4	8	1	4
17	1	7	2	9	1	2	7	8	4	7	0	3	3	1	5	8	2	5
18	5	0	5	7	9	5	8	7	8	9	3	5	3	4	4	6	8	1
19	7	7	3	3	5	3	6	1	3	2	8	5	4	1	4	8	6	7
20	1	0	9	1	3	8	2	5	3	0	3	8	0	9	3	3	1	3
21	1	3	8	5	1	8	5	9	4	1	9	3	9	3	6	5	0	8
22	8	5	4	7	8	7	5	9	4	1	9	3	9	3	6	5	3	1
23	0	6	9	6	5	1	0	3	2	6	7	7	4	9	6	0	0	5
24	7	6	7	4	7	0	8	3	8	7	3	2	5	1	2	4	2	2

Table 2-2: Random number table.

and any adjacent digit (e.g. one number to the left) become the first random number of the set.

Step 2: Continuing to move the pencil in the direction established in Step 1, the planner takes the next two consecutive digits to form the second random number; the next two digits beyond that form the third random number, and so on, until the edge of the table is reached.

Step 3: If still more numbers are needed, the planner changes the direction of movement (e.g., upward or diagonally) and continues to add more two-digit numbers to the random set. The set is completed when the required number of sampling points is reached according to the SW-846 formula.

The randomly selected numbers are then plotted on a graph paper grid that has consecutively numbered intersections; for example, from 1 to 100 (see Table 2-1). On a large site the grid intersections may reach the three-digit numbers, such as 001 to 200.

The grid is then transferred onto a copy of the site map. The onsite technician uses the site map grid to determine where to stake the corners. Using the staked corners as reference points, the technician may measure to the locations of the sampling points and stake those as well.

Checking Your Understanding

1. Identify three components of random sampling point selection and describe what information each component provides.
2. Identify the three steps for the use of the random number table.
3. Explain the process for transferring sampling points from the grid map to the field.

2-4 Quality Control and Quality Assurance

Concepts

- High quality data in terms of accuracy, precision and completeness can be achieved by applying quality assurance (QA) and quality control (QC) methods.
- Duplicate and blank samples are used to check field collection and laboratory procedures.
- Background samples and critical data point samples are collected for completeness of the sampling event.

In environmental sampling it is imperative to obtain the highest quality data in order to satisfy the goals of the site investigation. High quality data are accurate, precise and complete. Accuracy is the nearness of a measurement to the actual or true value. Precision, on the other hand, is the closeness of repeated measurements. **Completeness**, as it is applied to quality control, means that additional samples, such as background samples and critical data point samples, must be taken as part of a comprehensive sampling event. Accuracy, precision and completeness can be obtained by applying quality assurance (QA) and quality control (QC) methods. QA is the general system of guidelines, standards, and activities put in place to make sure that the sampling event meets the plan's objectives. QC is the actual implementation of the QA standards.

QA/QC procedures involve 1) reviewing the data collected for appropriateness and completeness, and 2) submitting additional samples, called quality control or QC samples, to test the sampling process itself. QA activities include checking field calculations to ensure correct mathematical results; documenting all field and sampling activities (such as testing and calibration of equipment); making sure all signatures are obtained where necessary on field notes and chain-of-custody forms; and using a project-specific sample designation system.

QC activities include collecting the QC samples as a means of checking the quality of both the field collection procedures and the laboratory procedures. These samples may be known by different names. For the evaluation of accuracy and precision the QC samples are duplicate samples (such as split samples and collocated samples); field and trip blanks; and matrix spikes. For completeness, the QC samples are background samples and critical data point samples.

QC samples should be submitted to the laboratory as **blind samples**, which means that their labeling should not reveal to lab personnel that the samples are intended for quality control. This is one of the many actions that makes a sampling event legally defensible.

Duplicate Samples and Blank Samples

The sampling plan will usually specify two types of QC samples to be taken: duplicates and blanks. A duplicate is an additional sample collected along with the investigative sample for control purposes. A blank is not an environmental sample; it is a bottle of deionized water or other medium similar to the investigated medium.

Duplicates – Field samples and field duplicates are equally representative samples but are treated independently for the purpose of evaluating laboratory performance. They may be analyzed at different laboratories, and their respective analyses may be compared. Duplicate samples may be collocated (**collocated sample**), meaning taken virtually from the same place and at the same time; for example, when two air quality analyzers take samples from a common manifold, or when two water samples are collected at essentially the same time and from the same point in a lake. Collocated samples may be used to estimate the overall precision of a data collection activity. Duplicates may also be obtained by splitting one sample (**split sample**) into two or more portions which are then analyzed separately.

Blanks – Several types of QC samples, called blanks, may be devised by the project planner to check some aspect of the sampling activities. Some examples follow.

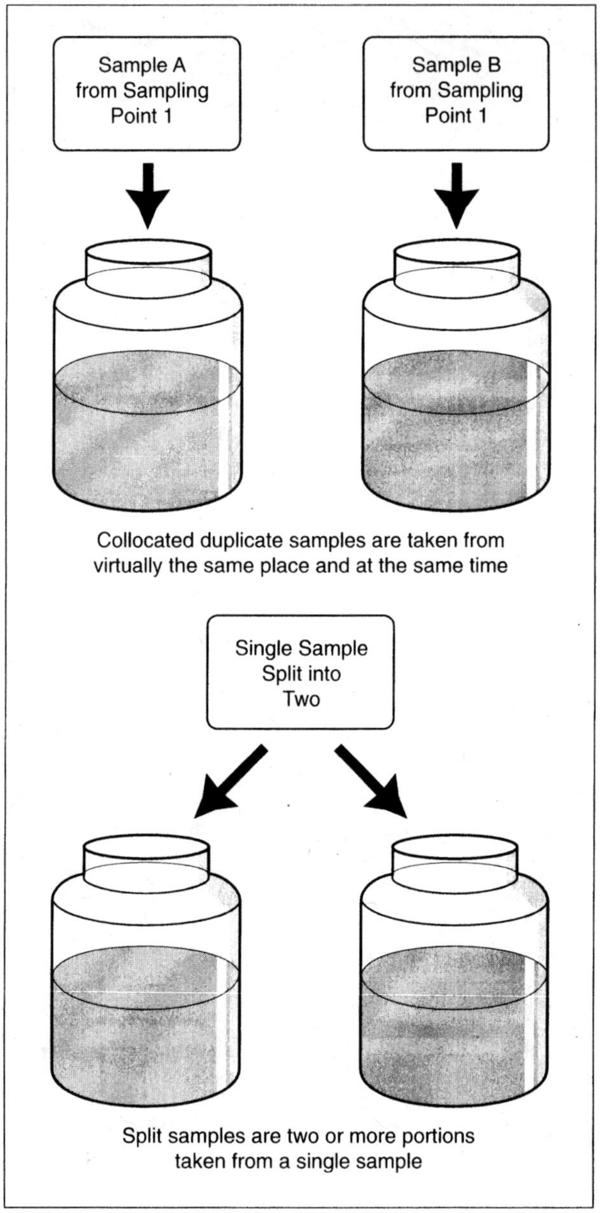

Figure 2-8: Collocated and split samples.

1. *Rinsate or equipment blanks* – **Rinsate blanks,** also called **equipment blanks**, are sample bottles containing deionized water that was used as a final rinse of the equipment. Analysis of the rinse water checks the effectiveness of equipment decontamination. Similar contaminants found both in the rinse water and in the samples suggest a flaw in the collection or decontamination methods.

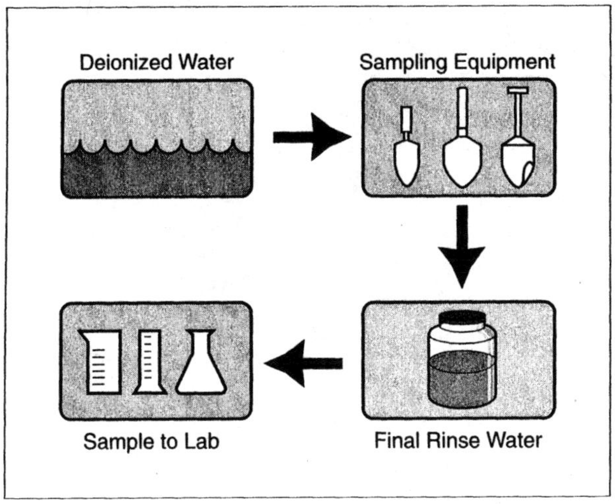

Figure 2-9: Rinsate blanks are collected to check effectiveness of equipment decontamination. Successful decontamination helps avoid cross-contamination.

2. *Field blanks* – These are sampling bottles identical to those used in the investigation and filled with deionized water at the same time the sample is taken. **Field blanks** are sent along with the samples as a check against air-borne contaminants in the handling and decontamination process.
3. *Trip blanks* – These are bottles of analyte-free media that are prepared prior to the sampling event and remain capped throughout the field activities. **Trip blanks** are kept with the investigative samples throughout the event. They are then packaged for shipment along with the investigative samples and sent for analysis. At no time after their preparation are the trip blank containers opened before they reach the laboratory. If analysis shows a contaminant in both the investigative samples and the trip blank, then the bottles might be contaminated, and the sample series should be collected again.

Matrix Spikes

Many samples exhibit matrix effects in which other sample components interfere with the analysis of contaminants of interest. **Matrix spikes** provide the best measurement of this effect. Spiking involves adding a pure form of the contaminant to a QC sample. Comparing the spiked sample with the nonspiked sample enables the analyst to determine interference from other matrix components in the samples. When matrix spikes are prepared in the field, immediately after collection, they also yield a measurement of sampling, handling and preservation error. Field matrix spikes therefore provide the best overall assessment of accuracy for the entire measurement system, as collocated samples do for precision assessment. However, they are generally not recommended because of the high level of technical expertise required for proper use and their sensitivity to environmental variables.

Media	Duplicates: Collocated or Split	Field Blank	Trip Blank	Background Sample	Inter-Lab Split Sample
aqueous (water-based)	1 in 20	1 in 20	1 per day of sampling	minimum of 2 per sampling event/media	when required to meet objectives
soil, sediment	1 in 20	1 in 20		minimum of 2 per sampling event/media	when required to meet objectives
air	1 in 20	not available	1 per day of sampling	minimum of 2 per sampling event/media	when required to meet objectives
source material	1 in 20	not usually required			when required to meet objectives

Note: This table is provided to serve as a guideline only; QA/QC sample requirements must be developed on a site-specific basis. Laboratory blanks and spikes are method-specific and are not included in this table. (Source: EPA - Data Quality Objectives for Remedial Response Activities, Appendix C-13.)

Table 2-3: Guidelines for minimum QA/QC samples for field sampling programs.

Background Samples

Background is defined as the normal ambient environmental concentration of a chemical. **Background samples** are those taken from media characteristic of the site being investigated but outside, and uninfluenced by, the zone of contamination. Monitoring data as well as available literature on natural background concentrations of chemicals in the area should be collected, reviewed and verified to determine background conditions. Background data should be defined as natural or **anthropogenic** chemical contamination resulting from a source other than the site undergoing assessment. The number and locations of required background samples are specified in the sampling plan and documented in the field log.

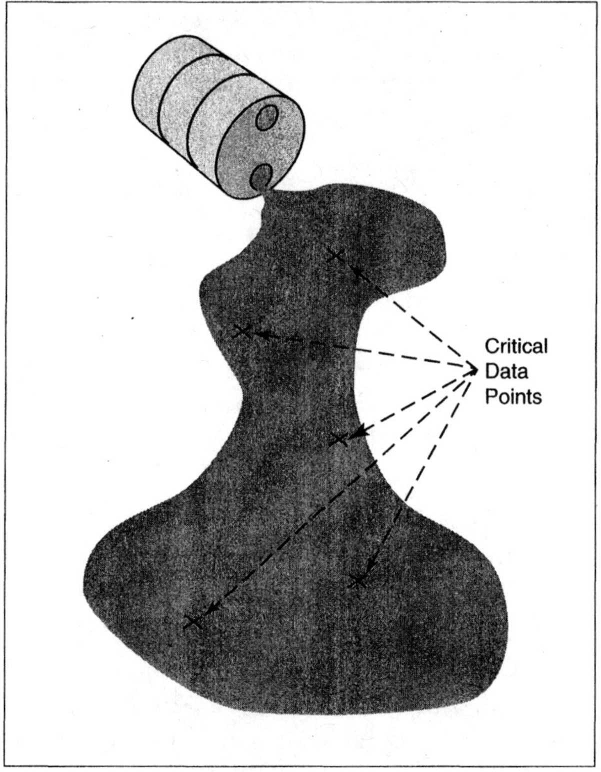

Figure 2-10: Critical data points are sampling points that are likely to reveal contamination; an area downhill from an overturned container, for example.

Critical Data Point Samples

Critical data points are sample locations in the area of the immediate source of contamination for which valid data must be obtained in order for the sampling event to be considered complete. A critical data point sample is therefore collected at any location considered vital for the decision-making process. Critical data points should be carefully considered in the sampling plan design and identified in every completeness statement developed during the DQO process.

Checking Your Understanding

1. Define sampling precision, accuracy and completeness.
2. Identify and describe quality assurance/quality control procedures.
3. Differentiate between duplicates and blanks on one side and background and critical samples on the other.

Summary

The goal of the sampling plan is to produce a document that identifies sampling requirements and needs of the investigation, or the remedial activity, at the site in question. It includes a project management plan, data quality objectives, a field sampling plan, a quality assurance project plan, a health and safety plan and a community relations plan. It is from the guidelines set in the sampling plan that the sampling methodology is defined. The sampling plan document should be written in a clear and readable manner, so as to be easily understood by both the regulatory agency personnel and the onsite technicians.

Depending upon the site situation and the requirements of the lead agency, the planner will specify the collection of biased (judgmental) or unbiased (random) samples or a combination of both types. Two techniques for collecting biased and unbiased samples are grab sampling and composite sampling. The number of unbiased samples to be collected is determined statistically by use of EPA's SW-846 formula. The location of the sampling points on a site may be determined by use of a random number table. A grid system is then used to transfer the selected points from the paper to the actual site.

The planner ensures that all resulting data will be of high quality in terms of precision, accuracy and completeness by specifying the collection of quality control samples in addition to the investigative samples. Duplicates and blanks are used for accuracy and precision, and background samples and critical samples are used for completeness of the sampling event.

Application and Critical Thinking Activities

1. Predict several examples of personnel errors or omissions that could keep a Sampling Plan from being legally defensible.
2. The table of contents in Appendix 5 comes from an actual Sampling Plan. The Sampling Plan is for the RCRA closure of a building in which EPA determined that the owner had mishandled hazardous wastes. Analyze this table of contents to determine if each sampling plan element required by EPA has been included.
3. If 100 investigative samples are to be collected at a site, determine the type and frequency of quality control samples that should be included, according to Table 2-3.
4. Discuss the following statement from the EPA publication "Data Quality Objectives for Remedial Response Activities": "Documentation should be sufficient to allow the persons involved in the site studies to reconstruct the work if necessary."
5. Analysis of preliminary soil samples at a hypothetical site indicates cresol levels of 140, 210, 90, 150, and 225 ppm. Use EPA's SW-846 formula to calculate the number of samples to be collected. Increase the results by three to improve precision.
6. Use the random number table (Table 2-2) to select a random set of two-digit sampling points for the hypothetical situation in Activity 5. Add several sampling points to the set in the event that barriers are found onsite.
7. Copy the grid from Table 2-1. Mark on the grid the sampling points from Activity 6.
8. Select a land site that provides a 100' × 100' space to grid (e.g., on campus). Select a starting point that represents the lower right corner of the grid and place a stake. Measure to the other three corners and place stakes, trying to approximate a square grid.
9. Describe any barriers or difficulties encountered during Activity 8.
10. Arrange with a local environmental laboratory or an environmental services firm to see an example of a sampling plan. Identify the EPA-required elements. Analyze the level of detail of the protocol.

3

Equipment and Sample Preparation

Chapter Objectives

Upon completing this chapter, the student will be able to:

1. **Identify** sampling equipment used to test environmental and waste media.
2. **Describe** problems associated with obtaining samples from the various media.
3. **List** sources of information for recommended procedures.
4. **Describe** procedures for selecting containers and delivering samples to the laboratory.
5. **Define** sample possession (chain of custody) and explain its requirements.
6. **List** and prepare the required documentation for a given sample.
7. **Prepare** field (log) notes.

Chapter Sections

3-1 Preparing to Go Onsite

Concepts

- A detailed equipment list should be compiled prior to going onsite to sample.
- The technician should mentally walk through the sampling plan, visualizing the different tasks and making preparations accordingly.
- Understanding the principles underlying representative sampling is essential to onsite problem-solving.

After describing the elements of the sampling plan, introducing sampling techniques and discussing quality assurance and quality control methods, it is now time to prepare for action.

Prior to going onsite to sample, the technician should compile a detailed equipment list. It is important to work scrupulously and methodically, without relying on memory, in order to produce a comprehensive list. When planning a site visit, the technician should do a mental "walk-through," visualizing each and every task to be performed, so that all contingencies will be anticipated and appropriate preparations made. EPA's SW-846 lists categories of materials to be considered during preparations. A list follows with some examples.

—Personnel equipment – Boots, rain gear, disposable coveralls, face masks and cartridges, gloves.

—Safety equipment – Portable eyewash stations, first aid equipment, safety lines.

—Field test equipment – Air sampling tubes, pH meters, combustible gas indicators.

—Storage equipment – Containers, lids, and liners enough to accommodate a 50 percent increase in sampling requirements or a change in type of media collected.

—Sampling equipment – Clean trier, thief, COLIWASA, weighted bottle, pond sampler, decontamination supplies.

—Shipping and office supplies – Tape, labels, shipping forms, chain-of-custody forms and seals, field notebooks, random-number tables, scissors, pens.

Figure 3-1: An array of field sampling equipment.

—Additional miscellaneous equipment – Tool kit, rope.

Technicians often encounter unexpected barriers to sampling in the field. An example of this is an investigation conducted by sampling personnel who arrived at the site after detailed preliminary work to find that some of their targets were covered by a large pile of unmelted snow. Sometimes other problems arise because of inadequate planning or carelessness; for example, a technician might forget to include container caps in the supplies, neglect to charge meter batteries, or fail to check with utility companies or underground pipeline companies before digging. Planning and onsite problem-solving skills are therefore important assets to technicians. Employers emphasize the importance of the technician's ability to aid in planning by measuring, sketching, and writing about site features and activities that could pose a problem, so that other personnel will be better prepared to act upon them later. Although engineers may be performing the statistical calculations that determine sampling targets, the technician must understand the rationale behind representative sampling. If a selected target is impossible to sample, the technician must be able to decide upon alternatives that will maintain the integrity of the sampling process. Additionally, the technician must be able to communicate the rationale for his or her onsite decisions.

☐ COLIWASA, plastic or glass
☐ Glass Thief
☐ Soil sampler, auger
☐ Grain sampler
☐ Scoop with stainless steel blade
☐ Split spoon sampler
☐ Dip sampler
☐ Trier, single slot
☐ Waste pile sampler
☐ Submersible pump
☐ Other samplers
☐ 1000 mL, 2000 mL HDPE containers
☐ Teflon liners for cap
☐ Coverall, long-sleeved cotton
☐ Neoprene rubber suit, long-sleeved
☐ Neoprene rubber gloves
☐ Self-contained breathing apparatus
☐ Respirator, chemical cartridge type
☐ Cartridges
☐ Goggles
☐ Face shield
☐ Hard hat
☐ Apron, oil and acid proof
☐ Other protective clothing
☐ Eyewash, portable
☐ Fire extinguisher
☐ 5-gallon water container with spigot
☐ Disposable towels or rags
☐ Large polyethylene bags
☐ Polyethylene bags (to store containers)
☐ Technical grade solvent
☐ 5-gallon container for spent solvent
☐ Sample labels
☐ Seals
☐ Sample analysis request sheets
☐ Chain-of-custody forms
☐ Field log book
☐ Waterproof pens
Sources: Appendix C, *Samplers and Sampling Procedures for Hazardous Waste Streams* and *How to Sample Ground Water and Soils*, by Peter Garrett.

Table 3-1: Checklist of items required in field sampling.

The technician must be able to use various sampling strategies depending on the location and the types of hazardous material, environmental media or waste media being sampled. The materials may be in a container, barrel, tank, waste pile, lagoon, pond, lake or even traveling through a pipe. The environmental media or wastes may be difficult to reach.

According to EPA's "A Compendium of Superfund Field Operations Methods," several general procedures are applicable to the collection of all surface water, sludge, soil, or sediment samples, regardless of the methods or equipment selected:

—Before commencing collection of samples, thoroughly evaluate the site. Observe the number and location of sample points, landmarks, references and routes of access or escape.

—Record pertinent observations. Include a sketch. Identify sample locations where appropriate.

—Prepare all sampling equipment and sample containers prior to entering the site. Provide protective wrapping to minimize contamination.

—Place sample containers on flat, stable surfaces for receiving samples. Use **sorbent** materials to control spills.

—Start sample collection from the least contaminated areas and end with the most contaminated areas in order to minimize the risk of cross-contamination.

Figure 3-2: Environmental technology class practices taking a soil sample with a hand auger. Turning the auger can be a two-person task!

—Collect samples and close containers as quickly possible. Make field observations (pH, temperature, conductivity) at the source rather than from containers.

—Follow the sampling plan in every detail. Document all steps in the sampling procedures.

—For potentially hazardous samples, dispose of sampling gear as specified in the sampling plan, or carry it back to the contamination reduction corridor in a plastic bag for decontamination and cleaning; deliver the sample containers and equipment to the decontamination station for cleaning before further handling.

—Always be attentive to the potential hazards posed by the sampling procedures and the material sampled.

Checking Your Understanding

1. List potential barriers to carrying out representative sampling points at a site.
2. Recognize procedures that are applicable to the collection of all types of samples.
3. State the abilities that employers value in sampling technicians.

3-2 Selecting Sampling Equipment

Concepts

- Sampling occurs in various media: air, soil and sediment, groundwater, surface water, containers (drums, tanks, bags), and other media including hard surfaces and biota.
- Each medium requires particular sampling equipment especially designed to enable collection of representative samples.
- EPA provides Standard Operating Procedures for sampling each medium.

A technician should be able to sample a variety of media, which occur in the natural environment – air, water or land – or may be human-made, such as waste piles or storage containers. EPA provides standard operating procedures for sampling each medium including:

Air	Atmospheric and vapor/gas emissions
Land	Sediment, soil and soil gas
Water	Groundwater and surface water
Containers	Drums, tanks and bags
Other	Waste piles; surfaces (chip, wipe and sweep sampling); biological materials; physical objects (artifacts).

The technician needs to be able to select appropriate equipment for a sampling situation. This may be compared to selecting kitchen equipment: a cook would not try to lift peas with a knife or take a cherry from a small jar with a soup ladle; a hostess serving a three-layer cake would use a spatula that could successfully lift a slice representing all three layers. Likewise, the technician must know the advantages and purposes of each sampling device available to him or her and must select the device that is appropriate for the situation.

The following pages offer an overview of sampling for each of the above media, while future chapters will present sampling principles and procedures for each of these media in greater detail.

Preview of Air Sampling

Air sampling is performed for two reasons: 1) to identify hazardous exposures to humans, and 2) to determine sources of environmental pollution. Relative to those two purposes, the technician might:

—survey an entire waste site or check a particular waste site area for potential hazards to workers;

—monitor an entire workplace area or an individual worker's breathing zone for potentially hazardous exposures to contaminants;

—monitor particular sources of emissions that may contribute to air pollution or monitor the ambient atmosphere for air pollution.

Figure 3-3: Student with air monitoring equipment.

Employees must wear personal air samplers at workplaces that indicate exposure to hazardous air contaminants. The technician must study each individual's motions during tasks, so that the personal samplers may be appropriately positioned on the worker's person to obtain samples from the breathing zone.

A protective air surveillance program for hazardous waste site workers involves monitoring the atmosphere with portable direct reading instruments (combustible gas indicators, organic vapor analyzers, oxygen meters, etc.). The advantage of using direct reading instruments is that they yield almost immediate analysis of conditions, usually within minutes. They also provide a picture of changing conditions. Organic vapor analyzers may be used in conjunction with a **gas chromatograph** for real-time identification of specific gases and vapors. However, according to EPA's "A Compendium of Superfund Field Operations Methods," the most accurate method for evaluating any air contaminant is to collect samples and analyze them at a reliable laboratory. Therefore, measurements obtained with direct reading instruments are supplemented by collecting and analyzing air samples.

Air sampling systems are commonly composed of a sampling pump that draws air into a sorbent tube containing a solid or liquid medium. The sorbent must be selected according to the contaminant to be sampled whenever it is possible to know in advance; for example, activated carbon is a good sorbent for collecting most organic vapors. **Aerosols**, which are defined as solid or liquid particulates, may be collected on a particulate filter or with an **impinger**, which is a collection bottle connected to a pumping system.

Resources Available to Sampling Technicians

A list of key organizations that guide or regulate air sampling follows:

—Occupational Safety and Health Administration (OSHA)'s regulations in 29 CFR 1910.1000, Air Contaminants

—EPA's Office of Air Quality

—American Society for Testing and Materials (ASTM)

—Air and Waste Management Association (AWMA)

—American Conference of Governmental Industrial Hygienists (ACGIH)

Preview of Soil and Sediment Sampling

Sediment is the geologic material that settles below flowing or standing surface waters. Both soil and sediment may include coarsely grained material like gravel or sand, medium grained material like silt, finely grained material like clay, or decaying organic matter like **humus**.

Sampling Equipment for Sediment and Soil

The spade, trowel and scoop are commonly used for sampling the top few inches of soil or for sediments at shallow depths. The auger and the split spoon sampler allow sampling at greater depths.

Soil Sampling – Sampling soil at depths of less than two feet can be done with a hand-held **auger**, which is a tool for boring into soil (see Figure 3-4). This tool has a crank-type handle attached at right angle to the auger shaft; at the bottom of the shaft is a hollow cylinder for collecting the sample. The sharp, pointed lower edge of the cylinder acts as a bit as it bores into the soil.

Figure 3-4: A hand auger with a sample.

Figure 3-5: Truck-mounted hollow-core flight auger.

For greater depths the technician may use a larger, truck-mounted hollow-core **flight auger** (see Figure 3-5). The bottom of this power-driven auger is a five-foot long hollow steel tube. The outer surface of the tube is a single blade (the flight) that twists around and up the tube. The soil is loosened as the blade turns and moves down into the earth. Additional five-foot sections may be added to the auger if deeper samples are desired. Sampling tubes can be placed inside the hollow auger to collect a core.

The **split spoon sampler** is a device that allows the technician to collect an undisturbed core of soil. This device is a two-foot tube that is split lengthwise into two halves with closing mechanisms that keep the tube halves together during boring. The split spoon sampler is driven into nonaugered, or undisturbed, soil and then withdrawn. The closing mechanisms are released to open the halves of the tube so the soil core is accessible for transfer to a sample container. This device is available in both hand-held and truck-mounted versions.

Sediment Sampling – In shallow water, sediment may be scooped with a shovel. In deeper water, sediment is sampled with a **dredge**, which scoops up or grabs the sediment. Like the bailer in groundwater sampling, the dredge is lowered by cable. A trigger component allows the jaws of the dredge to be

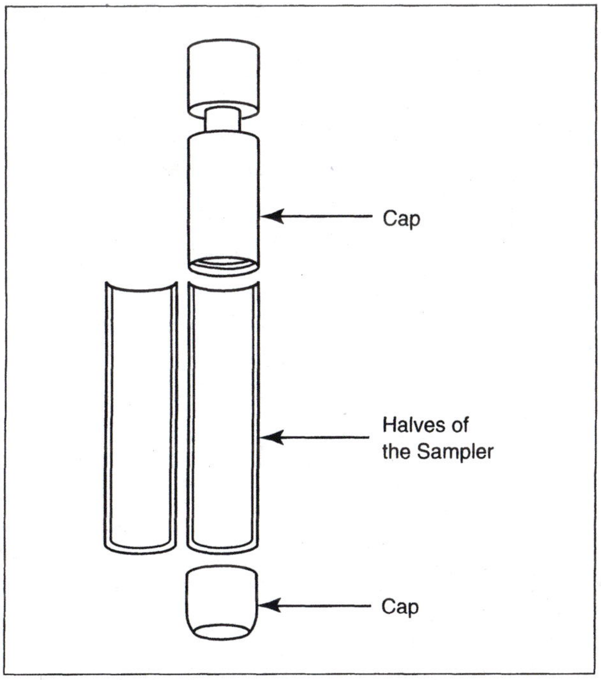

Figure 3-6: The two halves of the split-spoon sampler with screw-on end caps.

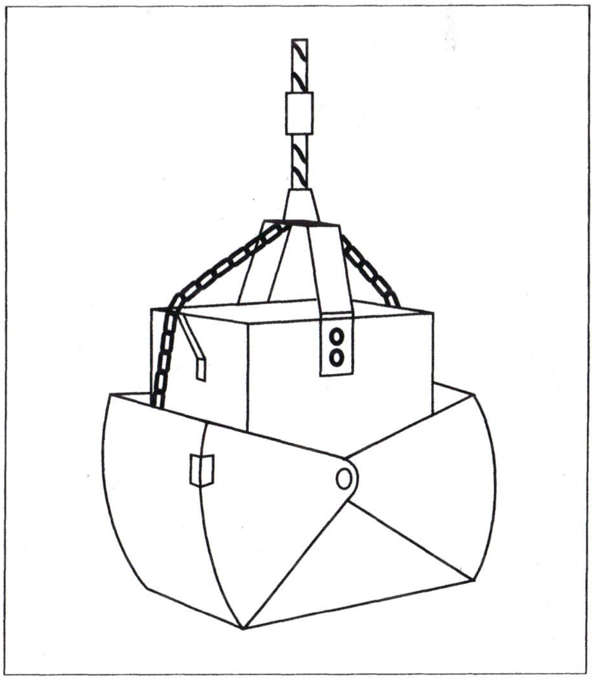

Figure 3-7: Ponar dredge sediment sampler.

opened and then closed around the sediment before the sampler is raised. The dredge cable may be mounted on a winch in a boat.

A **gravity corer** is a metal tube sampler that is dropped forcefully by cable into soft sediment. The core is brought up to the surface in the tube. Power-driven split spoon samplers may also be used for sediment sampling and may be mounted on barges or other vessels.

Soil Gas Sampling – Contaminated soil may cause the accumulation of soil gases (air is the naturally occurring soil gas). To sample for soil gas, a tool such as a flighted auger or a manual jackhammer is used to create a hole that is four to five feet deep and 3/8-inch in diameter. A 1/4-inch stainless steel tube is inserted into the hole as a small sampling well, and the area around the well is sealed with modeling clay. The soil gas is extracted through a series of connections to the well, which is referred to as a **sampling train**. It consists of a length of 1/4-inch Teflon® tubing, which in turn is connected either to a gas monitoring instrument or to a special gas sampling bag, such as a **Tedlar® bag**, for collection of the soil gas sample.

Resources Available to Sampling Technicians

A list of the key organizations that guide or regulate soil and sediment sampling are:

—EPA

—American Society for Testing and Materials (ASTM)

—U.S. Geological Survey (USGS)

—State-level environmental agencies (many states require that groundwater sampling personnel be state-certified or registered)

Preview of Groundwater Sampling

Groundwater, which is the water that lies beneath the earth's land surface and provides about 25 percent of our nation's residential, commercial, agricultural and industrial water, is subject to contamination by pollutants. In our recent history, groundwater has become the target for greatly increased sampling and monitoring activity as a result of various regulatory requirements.

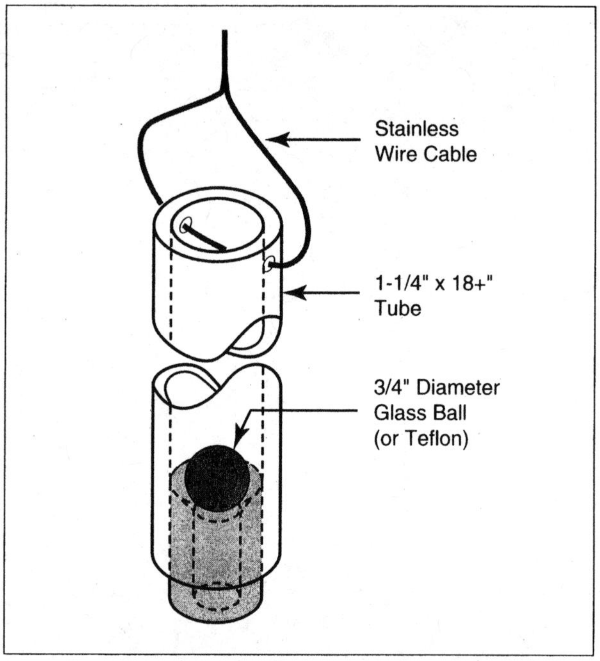

Figure 3-8: Bailer with a glass ball used as a stopper.

Sampling Equipment for Groundwater

Groundwater is typically sampled at selectively placed wells drilled at specific points on a site. These points are selected by an engineer or scientist to determine the movement and extent of groundwater contamination. Before a well is sampled it must be purged; **purging** means that stale water is removed prior to sampling in order to ensure that the sample is representative of the groundwater. Periodic sampling can be done at these wells by lowering a groundwater sampling device into the well. During sampling sessions the technician also takes measurements such as water level, pH, conductivity and water temperature, according to the requirements of the sampling plan.

The groundwater sampling device may be as simple as a **bailer**, which is a tube constructed of an inert material such as Teflon®, stainless steel, or PVC (polyvinylchloride). A bailer usually has a one-inch interior diameter and a length of 18 to 36 inches. The bailer shown in Figure 3-8 has a hole at the bottom, which allows water to enter as the tube is lowered by cable into the well. As the tube is lifted, a glass ball slightly larger in diameter than the hole drops onto the opening, sealing the contents within the bailer. The bailer is widely used by environmental consultants because it is portable, inexpensive, easy to clean, and because it does not require an external power source. More complex groundwater sampling systems might utilize a pump to retrieve water from the well.

A type of groundwater sampler that does not involve the expense and intrusion of drilling sampling wells is a probe that is driven into the ground. After a sample is drawn, the probe may be removed and used again at other sampling points. This is possible when the groundwater is relatively close to the soil surface.

Preview of Sampling Surface Waters

Surface waters lie on the earth's surface and may occur either naturally as streams, rivers, lakes and ponds, or as human-made lagoons or other surface impoundments. Such waters may be standing, as in lakes or lagoons, or flowing, as in rivers. Methods for sampling surface waters are used not only for aqueous, or water-based, environments but also for nonaqueous environments, such as lagoons and surface impoundments intended for the storage of chemical waste. Whether the technician uses surface waters sampling methods for aqueous or nonaqueous material, the safety and chemical compatibility of all equipment used – from samplers to PPE – must be considered.

Samples may be collected either from the water's depth or from its surface. (Note that in this section the term "surface waters" is used in two contexts: to describe the waters on the surface of the earth, as opposed to groundwater, and referring to the waters as the topmost layer of a body of water or of other substances.) Since surface waters have varying characteristics – shallow or deep, still or flowing, broad or narrow – several types of devices are available to

Figure 3-9: Students using a pond sampler with telescoping pole are careful not to disturb the sediment as they take a surface water sample.

suit different sampling situations. Sampling might be performed from the edge or shore of surface water, from a boat, from a catwalk or other construction, or wading in the water itself.

Sampling Equipment for Surface Waters

The bailer used for groundwater sampling may also be used for sampling surface waters. Other devices that work for sampling surface waters are pond samplers and specially constructed bottles. In shallow flowing water the technician may simply stand downstream of the flow and dip the sample container itself into the water. If the container is quickly closed, any VOC content is likely to be preserved. This method eliminates the transfer of sample from collection device to container, which also helps retain any VOC content. In addition, dipping with the container helps retain oil and grease components. The portion clinging to the wall of the device would be lost if the contents were transferred, and the sample would therefore produce incorrect analytical results.

A **pond sampler**, also called **dipper**, is simply a sample container that is attached by clamp to a telescoping pole. The pole enables the technician to extend his or her reach when sampling from the edge or shore of a surface water. This method is especially useful if the technician is trying to reach from the shore to the surface water below the end of a wastewater discharge (outfall) pipe. One drawback of this method is that the dipper's open top can allow the escape of volatile components during sampling.

Special surface waters sampling bottles are also available. Unlike the bailer, which remains open at the top, these bottles have closing (stopper) mechanisms that allow a sample to be taken at a specific

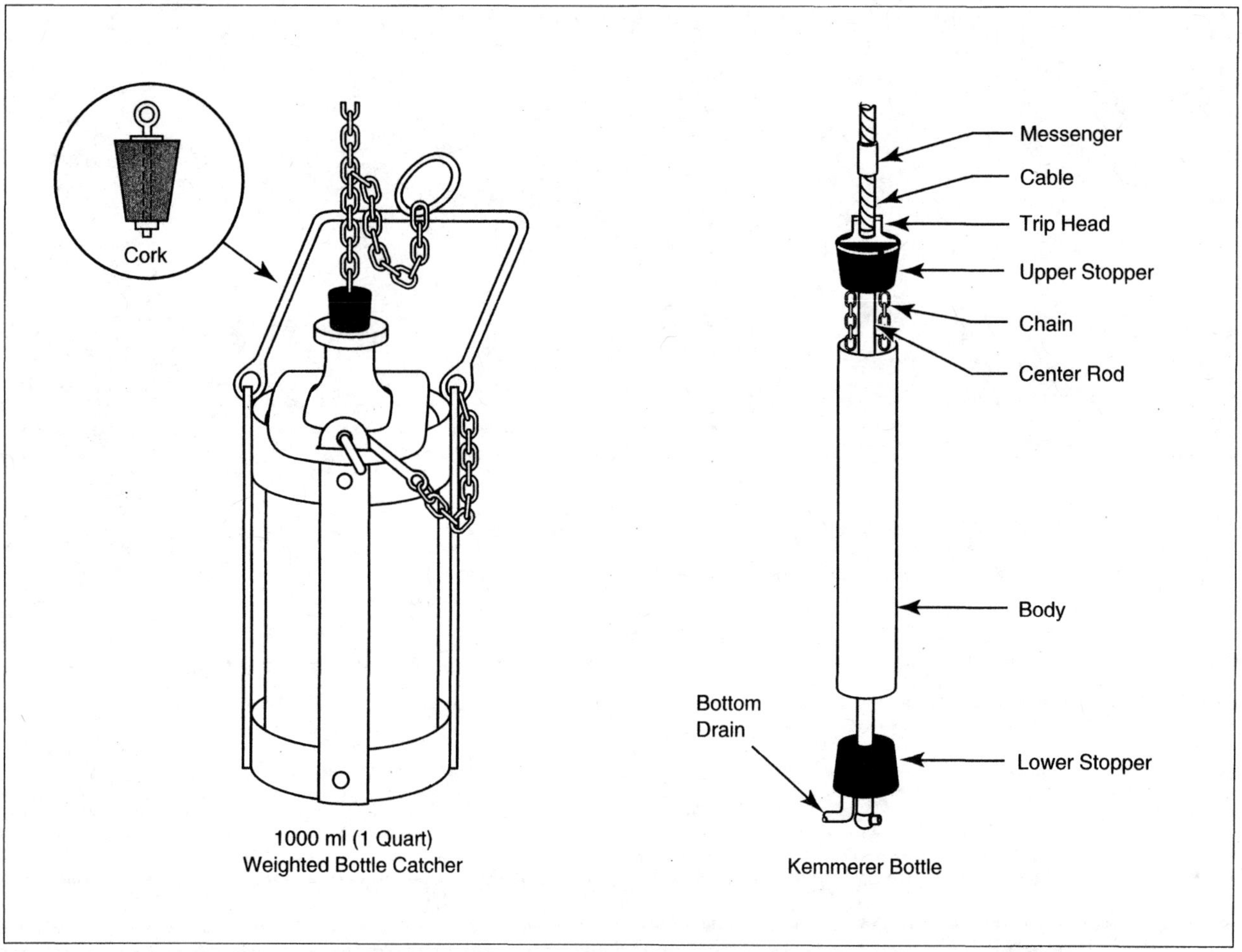

Figure 3-10: The weighted bottle and the Kemmerer bottle samplers can be lowered to a specific depth in surface water. Their stopper mechanisms can be activated after the sample is taken at the desired depth.

depth. The ability to close the bottle after the sample is taken at the desired depth prevents contamination from other depths as the bottle is drawn back up to the surface. A closed bottle also makes it less likely that any volatile components will be lost from the sample during retrieval.

Resources Available to Sampling Technicians

A list of key organizations that guide or regulate groundwater and surface water sampling follows:

—EPA

—American Society for Testing and Materials (ASTM)

—National Groundwater Association (NGWA)

—U.S. Geological Survey (USGS)

—State-level environmental agency (Many states require that groundwater sampling personnel be state-certified or registered)

Preview of Container Sampling: Drums, Tanks and Bags

Containers can range from 5-55 gallon drums to bulk storage tanks of thousands of gallons. Nonpressurized containers may hold liquid, solid or semi-solid materials, which may also yield gases and vapors. Metal or plastic drums must be used for liquids, while

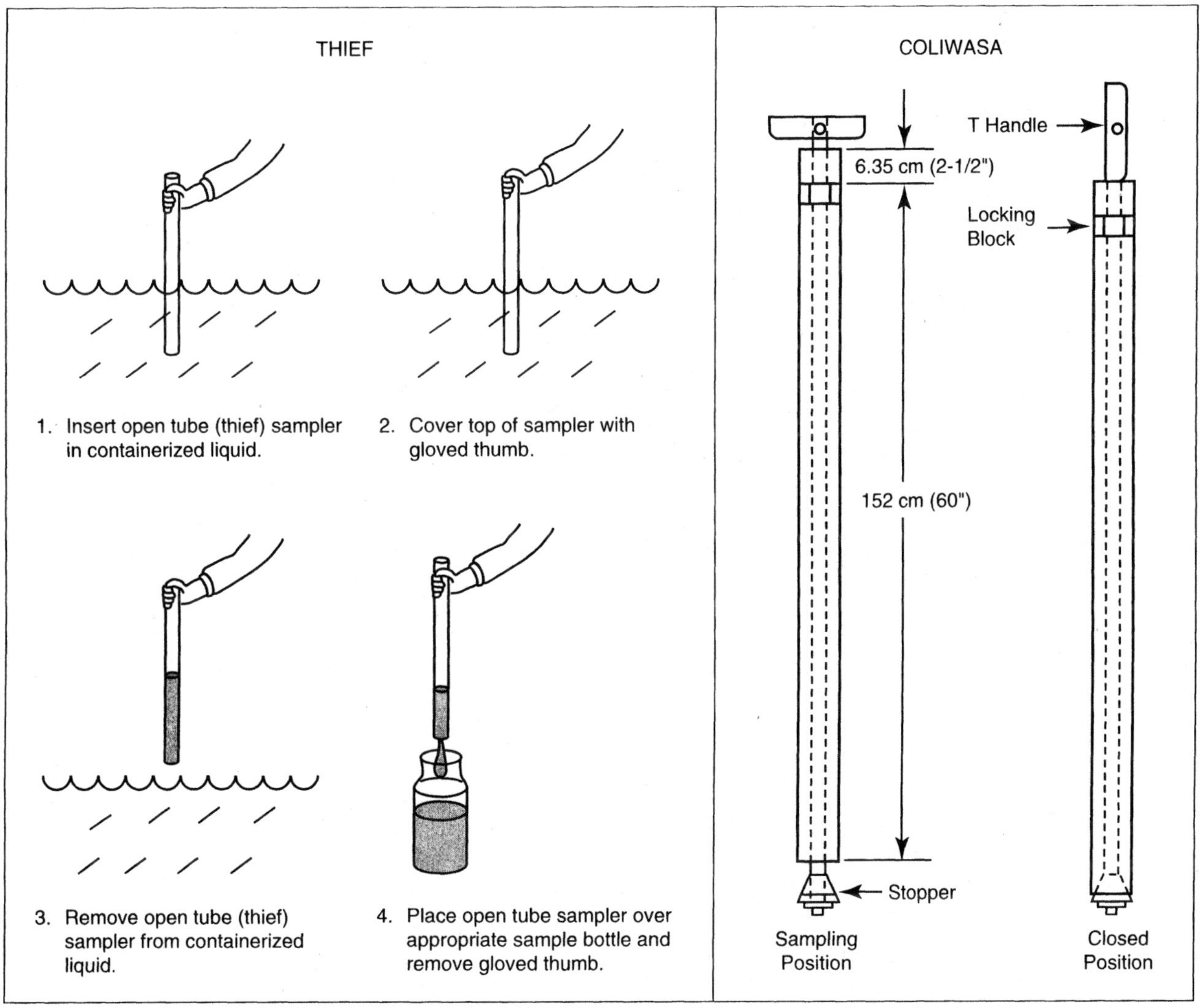

Figure 3-11: Drum and tank sampling equipment.

fiber drums are used to hold dry contents. The contents of a container may be homogeneous or heterogeneous. However, when sampling a container of unknown material, the technician should always assume the contents are heterogenous and select the sampling equipment accordingly in order to collect a representative sample.

Drum Sampling – Drum samplers are glass or Teflon® tubes long enough to reach down through layers of material to retrieve a composite sample. The tubes, commonly called thieves, are usually 48 inches long and open at both ends. To retrieve a sample the technician inserts the **thief** almost to the bottom of the drum, or until a solid layer is encountered. After the waste fills the tube, the technician stoppers the top end of the tube and withdraws it. Placing the open end of the thief over the sample collection container, the technician unstoppers the top end to release the waste into the container. The process is analogous to putting a drinking straw into a glass of liquid, stoppering the top of the straw with the thumb, and then lifting the straw out of the glass. Removing the thumb allows the beverage to be released from the straw.

A device designed especially for collecting a representative sample from a drum or tank of liquid is the **COLIWASA** (pronounced kol-ah-wah-sah), which stands for Composite Liquid Waste Sampler. The COLIWASA is a tube made of inert material, such as Teflon®, that is long enough to sample the depth of a 55-gallon drum. Because it has a stopper component, the COLIWASA may also be used to sample a larger storage container at depths specified in the sampling plan. The ability to open and close the COLIWASA at the specified depth prevents contamination from other depths during retrieval. The closing device also makes it less likely that any volatile components will be lost from the sample during retrieval.

The technician may choose from a variety of manual and remote drum opening devices. The preferred devices are of nonsparking construction. According to 29 CFR 1910.120, manual methods of opening are not to be used if the drum contents are 1) unknown, 2) flammable, 3) reactive, explosive or shock sensitive. Manual equipment includes the **bung** wrench, drum deheader, manual spike and bolt cutter. Remote equipment includes pneumatic bung removers and drills, hydraulic drum piercers, backhoe spikes and robot units.

Tank Sampling – Although a storage tank may be sampled by a bailer, a glass thief, a COLIWASA, or a bottle, other devices are available that allow the collection of samples from a chosen depth: the subsurface grab sampler or the bomb sampler for collecting liquids and the sludge judge for collecting solids that have settled in the tank.

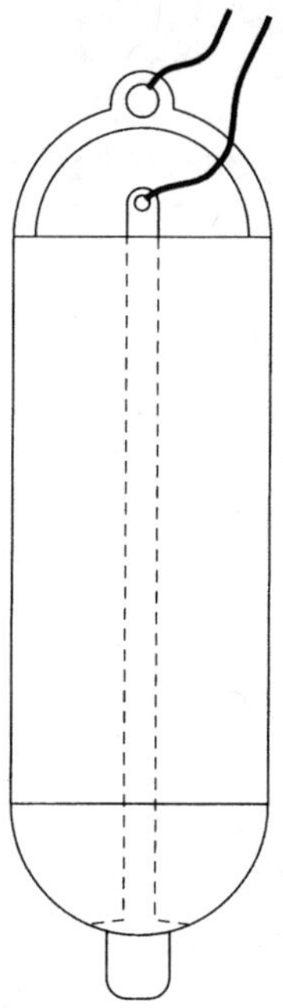

Figure 3-12: Tank sampling equipment.

The subsurface **grab sampler** is usually constructed of aluminum or stainless steel tubing with a head assembly that attaches to a one-liter sample container. The head assembly contains a spring-loaded plunger for opening and closing the sample container when the sampler is lowered to the desired depth.

The **bomb sampler** is weighted to enable it to drop through the contents of a container. It is attached to two cords – one to suspend the sampler and the other to activate an internal tapered plunger that opens and closes the sampler at the desired sampling depth. It is a more versatile type of equipment in that it can be used for collection at greater depths than the subsurface grab sampler due to its heavy weight.

The **sludge judge** consists of 3/4-inch pipe in five-foot sections that are marked at one-foot intervals. The markings enable the technician to determine the amount of sludge in the container. The bottom of the sludge judge has a check valve that seals it to stop the substance from escaping while it is being lifted.

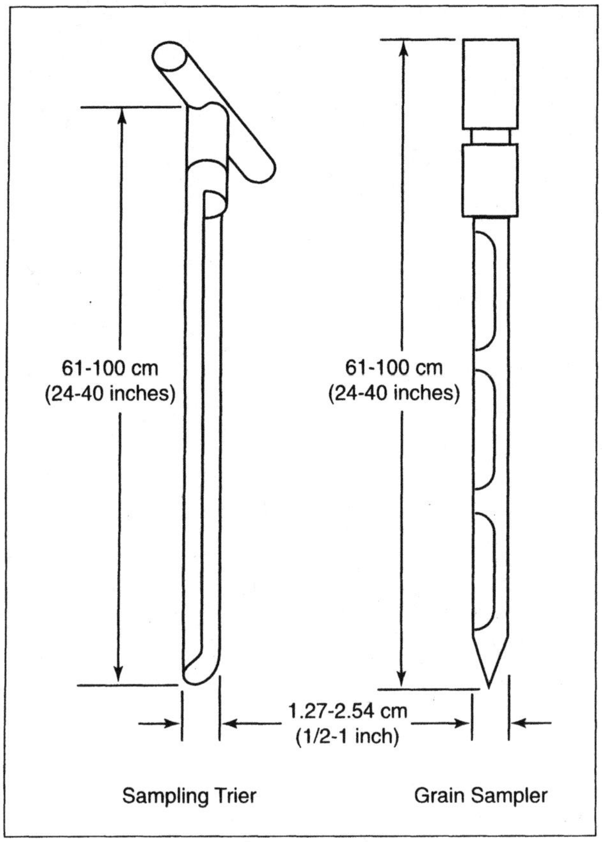

Figure 3-13: Sampling trier and grain sampler.

Sampling Bags, Sacks, Fiber Drums and Uncontained Piles – Shallow samples of granular or powdered materials within bags, fiber drums, sacks, or within uncontained piles may be collected with a shovel, scoop, trowel or spoon. Samples at depth may be collected with a **grain sampler** or a **trier**. The trier looks like a long thin scoop and is used as such. Materials sampled with the trier may easily fall out. The grain sampler's construction, on the other hand, keeps the material from falling out; it consists of an interior and an exterior stainless steel or brass tube with lengthwise slots in each. To take a sample at a specified depth within a pile or a container, the technician inserts the grain sampler at an angle (as specified by the sampling plan), rotates the inner tube so that the slots line up to allow material to settle into the tube, and then rotates the tube again to close off the slots. The grain sampler may then be withdrawn from the pile or container without contamination of the sample from surrounding material.

Resources Available to Sampling Technicians

A list of key organizations that guide or regulate container sampling methods and drum handling methods follows:

—EPA

—Occupational Safety and Health Administration (OSHA)'s regulations in 29 CFR 1910.120, known as HAZWOPER.

—EPA Environmental Response Team Standard Operating Procedures and other related guidance documents

—American Society for Testing and Materials (ASTM)

—National Fire Protection Association standards (NFPA)

—Air and Waste Management Association (AWMA)

—American Conference of Governmental Industrial Hygienists (ACGIH)

Sampling Other Media

Sampling Hard Surfaces (Chip, Wipe and Sweep Sampling) – Chip, wipe and sweep sampling are used to sample contamination of various surfaces, especially human-built surfaces. Wipe sampling is also used to check equipment decontamination. All three types of sampling are carried out with laboratory-cleaned or **dedicated equipment** on a measured and marked off area, typically one square foot in size.

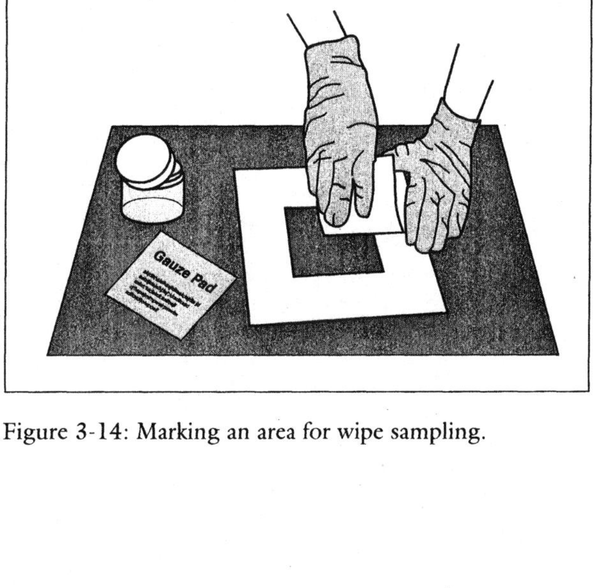

Figure 3-14: Marking an area for wipe sampling.

In chip sampling of porous surfaces, the technician loosens material with a hammer and chisel or with an electric hammer. The chips are then placed in a sample container. In wipe sampling the technician samples contamination of smooth surfaces by wiping a solvent-soaked sterile gauze pad horizontally and then vertically across a marked off area. The pad is then placed in a sample container. In sweep sampling the technician uses a dedicated broom and dust pan to sweep dust, dirt or residue from a marked area. The sweepings are then placed in a sample container.

Taking Biological Samples (Biota) – Biological test materials collected from sites can include vegetation, aquatic invertebrates, tissues from terrestrial animals, and whole fish or tissues from fish. According to "A Compendium of Superfund Field Operations Methods," approximately 30 grams of vegetative material and 100 grams of animal tissue are needed to run most tissue analyses. Vegetation and animal droppings or tissue may be transferred to a sampling jar. Live animals, fish and invertebrates may be collected by using live trapping, lethal trapping, or hunting, as specified by the sampling plan. Appropriate fishing, trapping and hunting permits must be obtained as necessary.

Taking Physical Object Samples (Artifacts) – An **artifact** is a physical object made by a human, which may be intact objects or fragments of objects that are of interest in a site investigation if they are suspected of contributing to contamination. An example of an artifact that may be sent to a lab for analysis is material suspected to contain asbestos, such as ceiling or floor tiles. The presence of artifacts should be described in the field log and perhaps photographed, whether or not they are to be collected as samples. To avoid contaminating a collected artifact, it must be handled with clean gloves and clean or dedicated retrieval equipment; it must then be placed in appropriate sampling containers for transfer to the laboratory.

Problems Associated with Obtaining Representative Samples

A number of problems may occur during sampling events. Some are common to most media, for example:

—Equipment failure, such as cables disconnecting, pumps breaking down, batteries or devices failing to respond, and so on.

—Adverse weather conditions, such as heavy winds making it difficult to maintain grid sampling position on surface water.

—Accessibility, such as difficulty collecting samples from a boat or a catwalk, or from the water's edge, or also while maneuvering a truck-mounted device on uneven terrain.

—Unforeseen barriers, such as utility lines or other structures, that prevent collection at selected sampling points.

In the case of air sampling at an uncontrolled hazardous waste site the following problems may occur:

—Technicians may face unidentified hazards while performing air monitoring;

—Some real-time air monitoring devices, such as combustible gas indicators, do not yield instantaneous readings. The lag in timing may place the technician in danger before he or she knows it;

—Unless used in conjunction with a gas chromatograph, a toxicity monitor identifies neither the specific chemicals to which technicians may be exposed nor the individual concentrations of multiple air contaminants;

—The technician must select respirators and other personal protective equipment (PPE) that will provide adequate protection from potential exposure to air contaminants; furthermore, working in PPE can be uncomfortable and restrictive as the technician goes about the work of obtaining samples.

In the case of container sampling, technicians may need to inventory, stage (arrange drums in groupings according to their contents), and open containers (typically drums) prior to the sampling event. Where contents are unknown, drum handling presents potential hazards:

—An accumulation of explosive vapors in the headspace of the container could ignite on contact with a spark. Nonsparking drum openers and tools are necessary when handling such containers. Opening devices that can be used from a remote location are preferred. It is advisable to attach grounding and bonding wires to drums to prevent the conduct of static electricity during the sampling event.

— The vapors may also create a buildup of pressure within the container, which can cause violent rupture. Bulging, swollen drum heads as well as groaning or whistling sounds indicate a critical situation that must be handled with emergency response procedures or perhaps evacuation.

— Chemical incompatibility of the container and its contents may have resulted in the container's deterioration. One effect could be the failure of the container during sampling, which would directly expose personnel to the contents.

— Some chemical contents may be reactive with components of the atmosphere (oxygen or moisture) if the container is opened.

— Some chemical contents may be shock-sensitive.

Unmarked containers of unidentified material can present a hazardous situation for sampling personnel. Personnel must be properly trained and certified in drum handling and emergency response procedures according to federal regulations found in 29 CFR 1910.120, OSHA and EPA's HAZWOPER standard. Personnel must also comply with all OSHA construction safety regulations in 29 CFR 1926. The site safety plan and site safety officer must be consulted in all potentially hazardous situations.

Checking Your Understanding

1. Identify sampling equipment used to test environmental and waste media:
   a. air
   b. soil and sediment
   c. groundwater and surface water
   d. containers (drums, tanks, bags)
   e. chips, wipes, sweepings; biota; artifacts
2. Describe problems associated with obtaining samples from the various media.
3. List sources of information for recommended procedures.

3-3 Preparing Samples for the Laboratory

Concepts

- Early contact with the laboratory assures proper procedures, containers and additional supplies, as well as timing for a sampling event.
- EPA's SW-846 publication contains standard field and laboratory methods for environmental sampling.
- Documentation is an essential element of sampling activities, not only because it ensures reliable results, but also because it supports the technician's actions in the case of legal or regulatory problems.

After a sample has been collected it must be transferred to an appropriate sample container, properly documented and protectively packaged for shipping. It is customary for sampling personnel to communicate with the chosen environmental laboratory before the actual sampling event and for the laboratory to provide suitable containers, preservatives, packaging materials and instructions for the sample collection. The laboratory's instructions will specify items such as the volume of sample necessary for analysis, the correct preservative or storage method, and the maximum **holding time** before the sample must be analyzed. The information provided by an environmental laboratory is drawn from EPA's SW-846 publication. In principle, a sampling technician could study the thick volumes of SW-846 to determine how to prepare each type of sample for the lab. Common practice is, however, to plan the sampling requirements in communication with laboratory personnel, whose specific responsibility it is to know and apply SW-846 procedures. This ensures the laboratory's analytical needs will be met.

The technician may also refer to a **sampling chart** to determine EPA's requirements and recommendations. A sampling chart conveniently summarizes **preservation**, storage and holding time information. Table 3-2 is a brief excerpt from a much longer chart (for complete chart see Appendix 6).

Selection of Containers

According to SW-846 guidelines the most important factors to consider when choosing containers for hazardous waste samples are:

Methods *EPA600 **SW846	Parameter	Recommended Container***	Preservation	Maximum Holding Time	Minimum Volume Required for Analysis
*413.1 **9070	Oil and grease	G only	Cool, 4°C H_2SO_4 to pH<2	28 days	100 ml
*625, 1625 **8250, 8270	Semivolatile nonvolatile organics	T, G	Cool, 4°C	14 days	60 ml
*624, 1624 **8240	Volatiles	G, T-lined	Cool, 4°C	14 days	60 ml
*909.A, 909.C, STD **9131, 9132	Coliform bacteria	PP,	Cool, 4°C G (sterilized)	6 hours	200 ml
**9320	Radium	P, G	Field acidified to pH<2 with HNO_3	6 months	1 gallon (3.7851l)

* Methods for Chemical Analysis of Water and Wastes (EPA-600)
** Test Methods for Evaluating Solid Waste (SW-846)
*** Container Types: P = Plastic, G = Glass, T = Fluorocarbon resins (PTEF, Teflon®, FEP, PFA, etc.) PP = Polyproylene

Table 3-2: Sampling chart.

1. Compatibility with the waste
2. Cost
3. Resistance to breakage
4. Volume

The laboratory provides the technician with sample containers certified to be manufactured and cleaned according to EPA procedural standards. The sampling plan might specify that containers must receive full quality assurance treatment, which includes precleaning by EPA protocols as well as lab analysis to verify the cleaning. Or, the sampling plan might specify precleaning without the lab analysis, which would be less costly. The lab may also supply clean containers that have not been precleaned by EPA protocol, if the situation warrants it in terms of future liability and levels of hazard.

The material of the selected containers must be chemically compatible with the components of the waste samples. Therefore, the technician should have some idea of the characteristics of the substances to be sampled. The containers must have enough wall thickness to withstand handling during sample collection and transport to the laboratory. Containers with wide mouths make it easier to transfer samples from sampling equipment. Also, the containers must be large enough to hold the amount of sample needed by the lab for analysis.

Containers for collecting and storing hazardous waste samples are primarily made of HDPE (high-density polyethylene), Teflon®, FEP (fluorinated ethylene propylene), clear glass, amber (brownish-yellow) glass, or Pyrex® (borosilicate glass). The advantages of Teflon®, FEP, and Pyrex® are that they are sturdy and inert to most chemical reactions. The disadvantage is that they are costly. The advantage of glass is that it is relatively inert to most chemicals except strong alkalies and hydrofluoric acid. The disadvantage is its potential for breakage, although some glass bottles are available in a safety-coated model. The advantage of HDPE is its lightness for handling and shipping. The disadvantage is its incompatibility with a number of chemicals.

English, fl oz	Metric, mL
1.28 oz	40 mL
4 oz	125 mL
8 oz	250 mL
16 oz	500 mL
32 oz	1000 mL

Note: The figures in the left and right column correspond to sizes commonly used in each system, but are not directly equivalent to each other.

Table 3-3: Typical container sizes in fluid ounces and milliliters.

All containers must have tight, screw-on lids. Plastic bottles usually have screw caps made of the same material as the bottles. Glass containers usually have rigid plastic screw caps and Teflon® cap liners. Other cap liners that may be specified or supplied by the laboratory are neoprene, polyethylene and polypropylene plastics.

The technician should verify with the laboratory any instructions for filling containers. Generally, liquids containing VOCs should be overfilled, leaving no air space when the container is capped. VOC container caps are a two-part design, which includes a membrane through which the analyst extracts an **aliquot** without having to open the cap and release VOCs. Unlike VOCs, corrosive liquids should not be overfilled, because of the dangers associated with spillage. Drinking water samples should have some air space left above the liquid so the analyst can shake the container before testing.

Sample Holding Time, Preservation and Storage

According to EPA's document "Samplers and Sampling Procedures for Hazardous Waste Streams," Section 8, the ideal time to submit hazardous waste samples for analysis is immediately after collection, thus enabling maximum reliability of analytical results. The technician should deliver samples to the laboratory the same day they are taken, if that is possible.

If transport by mail or other carrier is necessary, the technician must ensure that specified holding times are not exceeded. The laboratory staff should be informed of the approximate time of arrival so they can carry out analysis within the specified holding times. Some materials, such as fecal coliform, must be analyzed within hours; most must be analyzed within days, and a few may be held for months.

EPA's "Samplers and Sampling Procedures for Hazardous Waste Streams" explains that hazardous wastes are such complex mixtures it is difficult to predict exactly the physical, biological and chemical changes that might occur in the samples over time. As an example, the following things could happen after collection of samples:

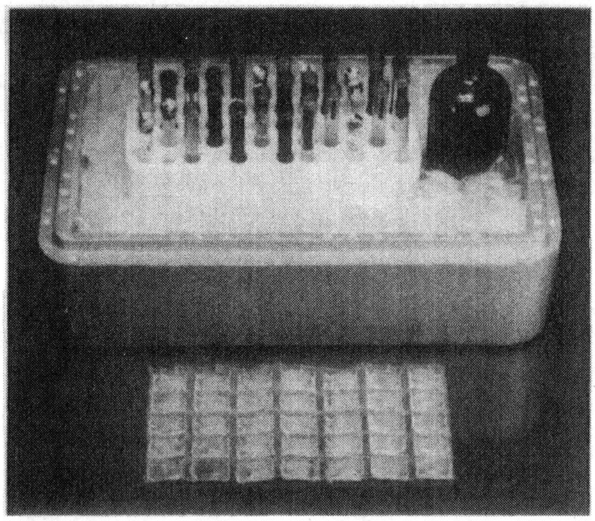

Figure 3-15: Refrigerant sealed between layers of plastic film provides greater thermal efficiency than wet ice without mess or damage to samples; it also protects the samples against breakage.

—pH may change significantly in a matter of minutes.

—Growth of microorganisms may cause changes to certain constituents of the sample.

—Volatile compounds may be rapidly lost.

—The air may oxidize one or more of the sample's components.

To decrease the likelihood of sample deterioration during storage and transport to the laboratory, the technician may be directed to preserve them by refrigeration and/or chemical preservation. Samples are refrigerated at 4° to 6°C (39° to 42°F); the 4°C temperature should be certified and documented when the samples are placed in the cooling chest and ice packs, and must be maintained throughout transit. A maximum-minimum thermometer is placed into the chest to document the temperature. A possible disadvantage of refrigeration is due to the fact that refrigeration might cause some dissolved substances to precipitate at the lowered temperature. Upon warming the sample to room temperature for analysis, the substances may not redissolve; this would result in a change in the actual concentrations of dissolved sample constituents.

Chemical sample preservation involves the addition of a chemical to the sample container, such as sulfuric acid, nitric acid or sodium hydroxide. The environmental laboratory typically specifies the chemical preservative that will stabilize one or two constituents of interest in a sample. The laboratory may also supply sample containers that already have the appropriate preservative in them. Chemical preservation methods are carefully implemented to avoid the conversion of some compounds to other forms due to the added chemical preservative, which would obviously reflect in the subsequent analysis.

Shipping Requirements

Shipment should be planned to ensure that samples are neither lost nor damaged en route to the laboratory. The samples must be packaged in a proper shipping container (i.e., a clean, marked, one-gallon paint can) to avoid leakage and/or breakage. An outer cardboard box must provide at least 10 cm (four inches) of tight packing around the sample container, such as polystyrene chips, crumpled newspapers, or vermiculite. A 4-mil polyethylene bag around the sample container should provide extra protection in the case of leakage.

For samples needing refrigeration, the technician may select from equipment that ranges from the basic picnic-type cooler filled with dry or wet ice to custom-designed mailer systems in which sample containers are surrounded by freezer gel-packs fitted into a specially molded polystyrene foam box, which in turn is enclosed in a corrugated cardboard outer carton. Hazardous materials shipping mailers may also be designed with an inner containment system to handle potential spills within the packaging.

All transported sample packages must be accompanied by proper documentation, which includes a sample analysis request sheet and the chain-of-custody record. The complete address of the sender and the receiving laboratory must be written legibly on each package, which is sealed with shipping tape.

Packages handled by the U.S. Postal Service are sent by registered mail with return receipt requested. If the package is to be sent by common carrier, like a trucking company for example, the technician retains a copy of the **bill of lading**, or shipping paper. Normally, hazardous waste that is to be shipped must be accompanied by an EPA hazardous waste manifest, which is a special shipping paper. Hazardous waste samples, however, are exempt from this requirement (see 40 CFR part 261).

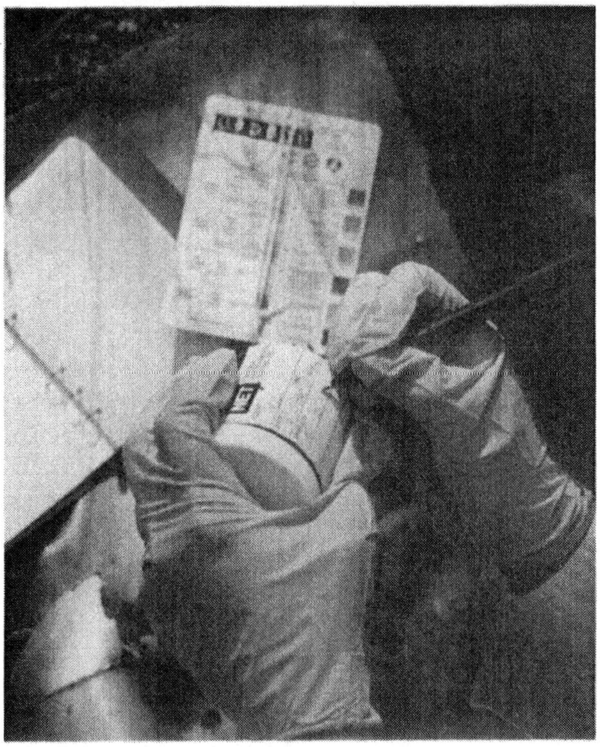

Figure 3-16: Filling out a label on a rounded surface may be challenging!

Documentation

Considerable paperwork is involved in a sampling effort. Documentation is necessary not only to ensure organized and reliable results of the sampling efforts, but also to support the technician's actions in the case of legal or regulatory problems. Documentation of sampling activities consists of several components:

—Identification of Sample

—Sample Analysis Request Sheet

—Chain-of-Custody Record

—Field Log Book

—Other records such as post office receipts and bill of lading copies

Identification of Sample – The technician attaches a seal and a tag, or adhesive label, to each sample immediately after its collection. On the seal he or she writes identifying information, such as his/her name, the date and the sample number. This waterproof and adhesive seal is then fastened over the lid and bottle in such a way that it would reveal tampering if the container were opened before analysis.

Tags are preferred to labels since the tag can be easily kept as part of the permanent documentation. If the tags or labels are not large enough to contain all the necessary information, the technician should record the remaining data in the log book. The American Society of Testing and Materials specifies that the following data be entered on the tag or label with waterproof pen:

—Project identification code

—Place of collection

—Name and signature of the person collecting the sample

—Date and time

—Sample number

—Observations and remarks

—Safety considerations

—Company or agency name (if not already printed on the tag or label)

—Type of sample (composite or grab)

—Preservative added (if specified in method)

—Storage temperature (if applicable)

Sample Analysis Request Sheet – The sample analysis request sheet accompanies the sample to the laboratory. The field portion of the form must be completed by the person collecting the sample and should include most of the pertinent information noted in the log book. The laboratory portion, which is completed by the analyst, includes:

—Name of person receiving the sample

—Laboratory sample number

—Date of sample receipt

—Sample allocation

—Analyses to be performed

Chain-of-Custody Record – To establish the documentation necessary to trace sample possession from the time of collection through the time of analysis, a completed **Chain-of-Custody (COC)** record must accompany samples. Each time the samples are transferred to a different person, the technician who surrenders them and the person who

California Department of Health
Hazardous Materials Laboratory

Collector's Sample No. ______

CHAIN OF CUSTODY RECORD
Hazardous Materials

Location of Sampling ____ Producer ____ Hauler ____ Disposal Site
____ Other

Company's Name ______________ Telephone ( ) ______

Address ______________
number street city state zip

Collector's Name ______________ Telephone ( ) ______

Date Sampled ______ Time Sampled ______ hours

Type of Process Producing Waste ______________

Waste Type Code ______ Other ______________

Field Information ______________

Sample Allocation:

1. ______________
name of organization
2. ______________
name of organization
3. ______________
name of organization

Chain of Possession

1. ______________
signature title inclusive dates
2. ______________
signature title inclusive dates
3. ______________
signature title inclusive dates

Figure 3-17: Example of chain-of-custody record.

receives them sign the COC record. The COC record becomes especially important when the sample is to be introduced as evidence in court.

Laboratory Custody – The sample is delivered to the **sample custodian**, a lab employee authorized to receive samples. Upon arrival of the sample, the sample custodian inspects it to see that the seal is not broken and the container is not leaking. The custodian also ensures that the chain-of-custody form and the analysis request documents are in order. A laboratory number is assigned to the sample. The number, the receiving date, a description of the sample, information from the analysis request form and the collector's identification number are entered into a log book. Subsequent analytical data and calculations are entered into a log.

A sample is considered to be in a person's custody if it is in his or her physical possession, in his or her line of sight, secured so no one can tamper with the sample, or placed in a restricted area off limits to unauthorized personnel. The analyst who receives the sample should be prepared to testify in court that the sample was in his or her possession or secured in the laboratory at all times from the moment it was received from the custodian through performance of the analysis. The same chain-of-custody procedures must be used for all portions of the sample if it is split between laboratories.

Field Log Book – All information pertinent to a field survey and/or sampling must be recorded in a log book, which is protected and kept in a safe place. The log must be bound, preferably with consecutively numbered pages, 8-1/2 by 11 inches in size.

As sampling situations vary widely, no general rule can be given about the extent of information to be entered in the log book. A good rule, however, is to record enough information for someone to reconstruct the entire sampling situation from it. Entries in the log book must include at least the following:

—Purpose of sampling

—Location, date and time of collection

—Name and address of field contact

—Number and volume of samples taken

—Description of sampling points (targets)

—Collector's sample identification numbers

—Sample distribution (e.g., laboratory, hauler)

—References such as maps or photographs of the sampling site

—Field measurements and observations

If documenting waste sampling they should also include:

—Name and address of producer of waste

—Type of process (if known) producing waste

—Type of waste (e.g., sludge or wastewater)

—Declared waste components and concentrations

Checking Your Understanding

1. Describe procedures for preparing samples for the laboratory:
   a. selection of containers

b. preservation, storage and holding times
c. packaging
d. shipping requirements

2. Define sample possession (chain of custody) and explain its requirements.

Summary

Prior to going onsite to sample, the technician should mentally "walk through" each procedure in order to compile a detailed equipment list. He or she should also call utility companies to locate underground lines. Even after careful planning, barriers to sampling may occur at the site, requiring the technician to be skilled in problem-solving.

Several general procedures are applicable to the collection of all air, surface water, sludge, soil or sediment samples, regardless of the sampling methods or equipment selected; during sample collection the technician should always:

—Document procedures

—Avoid cross-contamination

—Quickly close filled sample containers

—Follow the sampling plan in every detail

—Decontaminate all equipment

—Beware of potential hazards

The technician selects from a variety of sampling equipment depending upon whether he or she shall be collecting samples from soils and sediments, surface or ground water, air, or containers. Each type of sampling can present its own set of problems, which the technician must be able to address safely. Guidance for performing sampling procedures is available from a number of resources, such as professional associations and EPA.

After collection the technician prepares the samples for transport to the laboratory by using appropriate containers, by properly preserving and storing the samples, by operating within the specified holding times, by packaging the samples for safe transport, and by fully documenting the sampling activity through log entries, shipping papers, sample request forms and chain-of-custody procedures.

Application and Critical Thinking Activities

1. Contact an official of the local gas and electric utility company and discuss the implications of digging and collecting soil samples without communicating with them.

2. From at least one environmental equipment catalog, study models of equipment available for sampling a particular medium. Report the results of your study, including sketches or clippings. Try to show the diversity of available sizes, prices, designs and sophistication. Compare the models by assessing their limitations and advantages.

3. Suppose you are using COLIWASAs to collect waste samples from drums known to have a VOC content. The samples will be sent by carrier to a laboratory.
   —Select an appropriate container and explain considerations in filling the container.
   —Determine the preservation, storage and holding time of the samples.
   —Prepare the documentation.

4. Given a site by your instructor, prepare field (log) notes.
   —Sketch and measure natural and human-constructed site features.
   —Note all site conditions and observations.
   —Date and initial the entry.
   —Compare your results with those of other class members and the instructor. Identify your strengths and areas for improvement in both the sketched and the verbal communications.

4

Sampling Air Quality

Chapter Objectives

Upon completing this chapter, the student will be able to:

1. **Distinguish** between air sampling and air monitoring.
2. **Describe** the equipment and calibration techniques used in sampling air.
3. **Discuss** sampling techniques used to obtain air samples.
4. **Indicate** the selection process for determining containers, preservatives and holding times for air samples.
5. **Summarize** procedures for making field observations and taking measurements to accompany air samples.
6. **Specify** the care and cleaning, including decontamination techniques, necessary to maintain the equipment used in sampling air.
7. **Describe** safety precautions recommended in sampling air.

Chapter Sections

4-1 Introduction

Concepts

- Technicians support the engineer or site manager by monitoring the atmosphere with direct reading instruments and sampling with pump and collection media.
- OSHA and EPA together require monitoring and sampling programs to evaluate air quality for hazardous waste site workers. EPA is also concerned with air quality affecting the public.
- Air is typically sampled for aerosols and particulates, for organic and inorganic vapors and gases as well as for combinations of these.

The previous chapter introduced two reasons for performing air sampling: 1) to identify hazardous exposures to humans, and 2) to determine sources of environmental pollution. Recall that the technician might

—monitor an entire waste site or check a particular waste site area for potential hazards to workers and the public;

—sample an entire workplace area or sample an individual worker's breathing zone for potentially hazardous exposures to contaminants;

—sample particular sources of emissions that might contribute to air pollution or sample the **ambient atmosphere** for air pollution.

To assess air quality, a technician should be able to support the engineer, site manager, or other qualified environmental professional in two site characterization strategies:

—Monitoring the atmosphere with **direct reading instruments (DRI)** that provide real-time data for almost immediate use, and

—Sampling the atmosphere with a pump and collection media for analysis in a laboratory.

This chapter will introduce the equipment, procedures, and safety precautions associated with monitoring and sampling air quality to meet OSHA's requirements for protecting workers and EPA's requirements for protecting the environment.

Figure 4-1: Technicians in Level A equipment monitor a drum.

Regulatory Authority for Air Sampling and Monitoring

Because EPA and OSHA (Occupational Safety and Health Administration) are both concerned with the safety of workers whose responsibilities include field monitoring and sampling, the two agencies cooperated in developing the Hazardous Waste Operations and Emergency Response (HAZWOPER) standard, which is found in OSHA's regulations at 29 CFR 1910.120 and in EPA's regulations at 40 CFR Part 311. EPA is also concerned with the migration of contaminated air, which could expose the public downwind of a site.

The Occupational Safety and Health Administration

Two paragraphs in OSHA's 29 CFR 1910.120 specifically address sampling and monitoring. A requirement of 29 CFR 1910.120(c), Site Characterization and Analysis, is that workers monitor visually and with appropriate direct reading instruments "...during initial site entry when a preliminary site evaluation shows the potential for ionizing radiation or IDLH conditions (conditions that are Immediately Dangerous to Life and Health), or when site information is not sufficient to eliminate those possible conditions." Furthermore, 29 CFR 1910.120(h) requires that a monitoring program using direct reading instruments be ongoing during site activities.

OSHA also sets forth requirements in various sections of 29 CFR 1910 Subpart Z, Toxic and Hazardous Substances, for any employer whose workers could potentially be exposed to hazardous materials during work duties. For example, Appendix A of 29 CFR 1910.1001 is the mandatory Asbestos Sampling and Analytical Procedure.

Environmental Protection Agency

The Clean Air Act (CAA) Amendments of 1990, which the EPA has responsibility for enforcing, requires the assessment of air quality so that the applicability of CAA standards may be determined. Title I, Attainment and Maintenance of National Ambient Air Quality Standards (NAAQs), sets standards for the quality of the ambient, or surrounding, atmosphere. If air quality testing shows that a geographic area exceeds specified limits of certain pollutants, then that region is designated as a **nonattainment area**, and EPA restrictions and requirements take effect.

While Title I applies to general geographic areas, Title III, Hazardous Air Pollutants (HAP), applies instead to the air emissions from particular sources such as stacks, pipes or industrial processes. In 1990 EPA designated about 190 toxic substances as HAPs for the first time; this increased the air monitoring responsibilities of owners/managers who generate emissions of these toxic substances.

EPA's air quality regulations are found in 40 CFR Parts 50 to 85. However, additional EPA

Figure 4-2: The problem of pollution is not limited to the big cities. Here we have a photograph taken at the Great Smoky Mountains National Park on a clear day (left) and one of the same location on a day when the air was quite polluted (right). The visibility impairment is primarily due to sulfates, which come from combustion of fossil fuels.

regulations that specifically address air quality may result in air sampling and monitoring programs as well. For example, the Superfund Amendments and Reauthorization Act (SARA) Title III requires toxic emissions reporting by large companies as part of EPA's Community Right-to-Know regulations.

Following the Sampling Plan

Before site monitoring or sampling begins, the technician is responsible for studying and following the sampling plan, including the health and safety plan, which is developed by the project planner. The plan normally requires preliminary field observations and measurements, such as meteorological measurements, which may affect monitoring and sampling activities.

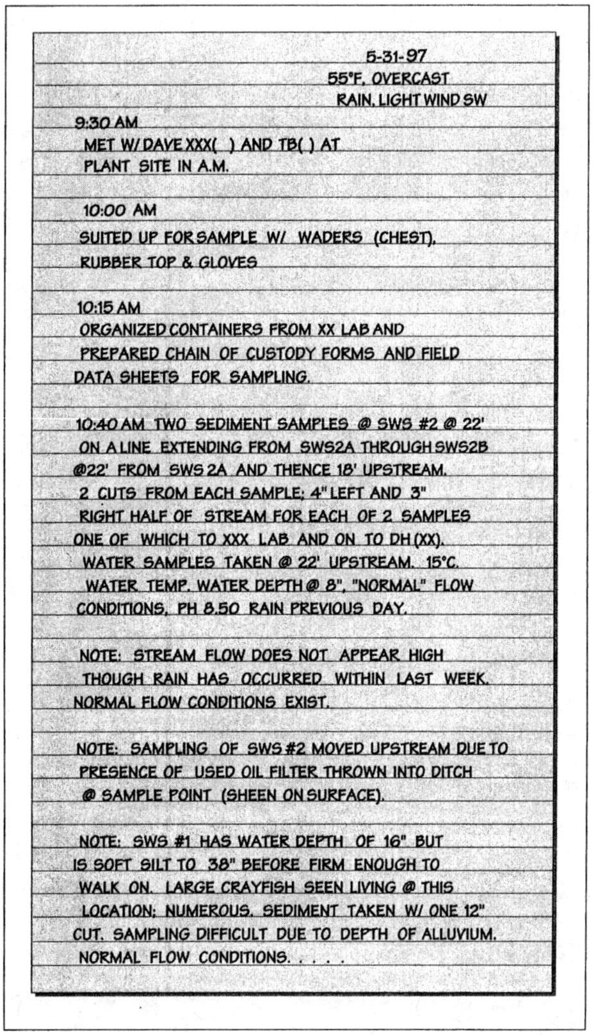
5-31-97
55°F. OVERCAST
RAIN. LIGHT WIND SW
9:30 AM
MET W/ DAVE XXX( ) AND TB( ) AT
PLANT SITE IN A.M.

10:00 AM
SUITED UP FOR SAMPLE W/ WADERS (CHEST),
RUBBER TOP & GLOVES

10:15 AM
ORGANIZED CONTAINERS FROM XX LAB AND
PREPARED CHAIN OF CUSTODY FORMS AND FIELD
DATA SHEETS FOR SAMPLING.

10:40 AM TWO SEDIMENT SAMPLES @ SWS #2 @ 22'
ON A LINE EXTENDING FROM SWS2A THROUGH SWS2B
@22' FROM SWS 2A AND THENCE 18' UPSTREAM.
2 CUTS FROM EACH SAMPLE; 4" LEFT AND 3"
RIGHT HALF OF STREAM FOR EACH OF 2 SAMPLES
ONE OF WHICH TO XXX LAB AND ON TO DH (XX).
WATER SAMPLES TAKEN @ 22' UPSTREAM. 15°C.
WATER TEMP. WATER DEPTH @ 8", "NORMAL" FLOW
CONDITIONS, PH 8.50 RAIN PREVIOUS DAY.

NOTE: STREAM FLOW DOES NOT APPEAR HIGH
THOUGH RAIN HAS OCCURRED WITHIN LAST WEEK.
NORMAL FLOW CONDITIONS EXIST.

NOTE: SAMPLING OF SWS #2 MOVED UPSTREAM DUE TO
PRESENCE OF USED OIL FILTER THROWN INTO DITCH
@ SAMPLE POINT (SHEEN ON SURFACE).

NOTE: SWS #1 HAS WATER DEPTH OF 16" BUT
IS SOFT SILT TO 38" BEFORE FIRM ENOUGH TO
WALK ON. LARGE CRAYFISH SEEN LIVING @ THIS
LOCATION; NUMEROUS. SEDIMENT TAKEN W/ ONE 12"
CUT. SAMPLING DIFFICULT DUE TO DEPTH OF ALLUVIUM.
NORMAL FLOW CONDITIONS. . . .

Figure 4-3: Log notes include meteorological data, site observations and activities.

As technicians approach a site, they should note in their logs any out-of-the-ordinary details such as dead fish or animals; dust, spray, or vapors in the air; pools of liquid, and oil or foam on those pools; deteriorated containers, pipes or structures. Technicians should also log meteorological data. Meteorological factors such as the following may affect monitoring instruments: temperature, atmospheric pressure, humidity, wind direction and speed.

Temperature – Manufacturers recommend operating temperature ranges for their instruments; for example, the recommended range for an oxygen monitor is 32 to 120 degrees Fahrenheit (0 to 49 degrees Celsius).

Atmospheric Pressure – Atmospheric pressure is defined as pressure caused by the weight of the atmosphere (nitrogen, oxygen and other trace gases) and is measured by a **barometer**. Atmospheric pressure is at maximum at sea level and becomes lower with increasing altitude. Since a change in altitude changes atmospheric pressure, it stands to reason that an extreme change in pressure could produce an inaccurate reading by an oxygen monitor. Therefore, an oxygen monitor may need to be calibrated again (i.e., checked and adjusted) during a sampling event if the technician makes a significant altitude change or if the weather (barometric pressure) changes dramatically.

Humidity – Humidity, which is the moisture content in the atmosphere, can affect not only the operation of monitoring equipment but also the dispersion of air contaminants. Humidity is measured with the **hygrometer**, **humidistat** or **sling psychrometer**.

Wind Direction and Speed – Like humidity, the wind's direction and speed affect the dispersion of air contaminants. The technician uses wind data to determine placement of samplers. The technician might also enter wind data into computer software to model the movement of air contaminant **plumes**, or clouds. The **anemometer** measures wind force and velocity.

The collection of meteorological data may be informal or formal, depending upon the requirements of the sampling plan. For informal data collection, the technician might simply note in the log "8:00 a.m., sunny and dry; about 15°C; light breeze from the southwest" and further note any weather changes during the sampling event. For formal data collection, the sampling plan may specify that the technician take readings from thermometers as well as other meteorological instruments.

Substances Sampled and Monitored

Air is typically sampled for aerosols and particulates, for organic and inorganic vapors and gases, and for combinations of these.

Aerosols/Particulates – **Aerosols** include liquids (mists and fogs) and solids or particulates (dusts, fumes and smoke) dispersed in the air. Solids, or particulates, are among the environmental contaminants that EPA regulates in Title I of the Clean Air Act Amendments of 1990. OSHA also regulates workplace exposure to certain particulates such as asbestos, cotton fibers and lead. Since aerosol monitoring provides a combined reading of all contaminants, sampling becomes necessary when the components and the particulate size of the contaminants need to be identified.

Organic Gases and Vapors – Organic gases and vapors are the gaseous phase of compounds that contain the carbon atom. Carbon compounds may exist in the gaseous state, such as methane; in the liquid state, such as octane; or in the solid state, such as tar. Organic vapors are produced when a carbon compound in a solid or liquid state is evaporated. Concentrations of organic vapors and gases are almost always explosive when exposed to an ignition source.

Inorganic Gases and Vapors – Inorganic gases and vapors are those gaseous substances that typically do not contain carbon atoms. They may be elements that naturally occur as a gas, such as oxygen, nitrogen and hydrogen; or they may be compounds in a gaseous or liquid state, such as ammonia, hydrogen chloride and sulfur dioxide. Mercury is a liquid metal at room temperature, but produces toxic vapors. Many inorganic and organic gases and vapors in air have toxic or corrosive effects.

Combination (Semivolatile) Materials – Air samples can contain any combination of organic and inorganic vapors as well as gases and particulates. Sampling equipment and media, such as particulate filters and activated charcoal, can be combined in a series to collect the combination of materials.

Checking Your Understanding

1. Distinguish between air sampling and air monitoring.

2. Define and identify examples of the following:
   —aerosols and particulates
   —organic gases and vapors
   —inorganic gases and vapors

4-2 Air Monitoring

Concepts

- Characterizing unknown hazards in the air is accomplished using various monitoring techniques and equipment for oxygen, combustible gas and vapor, toxic gas and vapor, radiation and aerosols.
- Combustible gases and vapors are monitored using a Combustible Gas Indicator (CGI) which reports the concentration of a substance as a percentage of its Lower Explosive Limit (LEL).
- Concentrations of specific substances can be measured by specific toxic gas and vapor monitoring equipment: 1) Specific Chemical Instruments, 2) Colorimetric Detector Tubes, and 3) Total Vapor Survey Instruments.
- Gas Chromatography is a technique for separating components of a sample and then identifying and measuring those components with a detection instrument.
- Hazardous atmospheres are defined by Class, Group and Division, as established by the National Electrical Code (NEC).

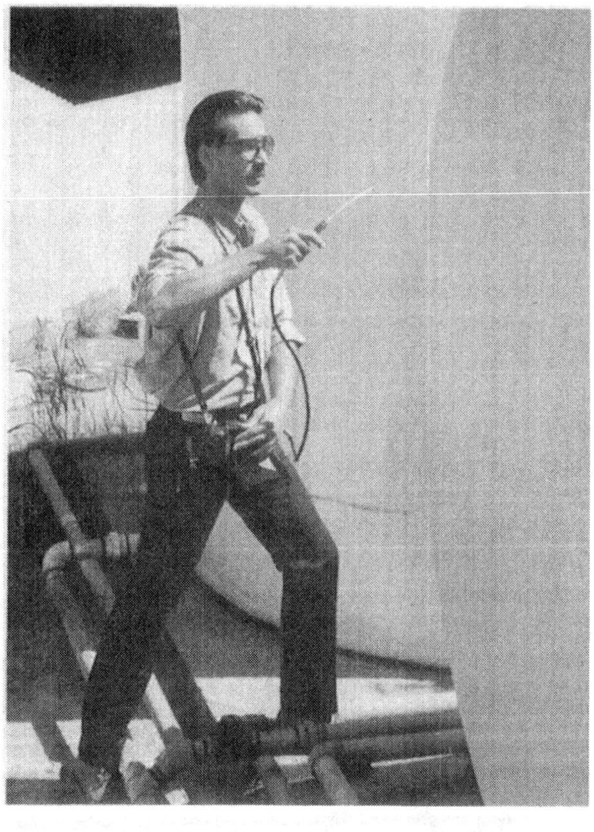

Figure 4-4: This technician routinely monitors storage tanks at the paint manufacturing company where he is employed.

If a situation requiring monitoring involves a known hazardous material, the technician selects a particular type of monitoring based on knowledge, experience or research. Examples of this situation would be monitoring for combustible gas levels within and around a tank during a gasoline tank removal, or monitoring for lethal hydrogen sulfide from decomposing organic matter during a trenching operation.

Uncontrolled waste sites may require several types of air monitoring as technicians begin to characterize the site. It may be necessary to perform 1) oxygen monitoring, 2) combustible gas monitoring, 3) toxic gas/vapor monitoring, 4) radiation monitoring, and/or 5) aerosol monitoring.

The technician must select personal protective equipment (PPE) appropriate for the monitoring situation. If the technician will be monitoring an uncontrolled hazardous waste site that presents unidentified hazards, then "...the ensemble of clothing and equipment referred to as Level B protection is generally the minimum level recommended for an initial entry until the site hazards have been further identified and the most appropriate protective clothing and equipment chosen." (NIOSH/OSHA/USCG/EPA, Occupational Safety and Health Guidance Manual for Hazardous Waste Site Activities, 1985. See Table 4-1 for a description of PPE levels.)

Oxygen Monitoring

Oxygen availability is critical for survival and is therefore a monitoring priority. Air at sea level contains approximately 20.8 percent oxygen. In an emergency response the technician should be concerned if monitoring indicates an oxygen level lower or higher than normal. If the oxygen monitor is calibrated and working properly, a lower reading should indicate to the technician that something is displacing the oxygen. That something may be a hazardous concentration of a toxic or flammable air

Level of Protection	Equipment	Protection Provided	Should be used when:	Limiting Criteria
A	RECOMMENDED: • Pressure-demand, full-facepiece SCBA or pressure-demand supplied-air respirator with escape SCBA. • Fully-encapsulated, chemical-resistant suit. • Inner chemical-restistant gloves. • Chemical-resistant safety boots/shoes. • Two-way radio communications. OPTIONAL: • Cooling unit. • Coveralls. • Long cotton underwear. • Hard hat. • Disposable gloves and boot covers.	The highest available level of respiratory, skin, and eye protection.	• The chemical substance has been identified and requires the highest level of protection for skin, eyes, and the respiratory system based on either: — measured (or potential for) high concentration of atmospheric vapors, gases, or particulates. or — site operations and work functions involving a high potential for splash, immersion, or exposure to unexpected vapors, gases, or particulates of material that are harmful to skin or capable of being absorbed through the intact skin. • Substances with a high degree of hazard to the skin are known or suspected to be present, and skin contact is possible. • Operations must be conducted in confined, poorly ventilated areas until the absence of conditions requiring Level A protection is determined.	• Fully-encapsulating suit material must be compatible with the substances involved.
B	RECOMMENDED: • Pressure-demand, full-facepiece SCBA or pressure-demand supplied-air respirator with escape SCBA. • Chemical-resistant clothing (overalls and long-sleeved jacket; hooded, one- or two-piece chemical splash suit; disposable chemical-resistant one-piece suit. • Inner and outer chemical-resistant gloves. • Chemical-resistant safety boots/shoes. • Hard hat. • Two-way radio communications. OPTIONAL: • Coveralls. • Disposable boot covers. • Face shield. • Long cotton underwear.	The same level of respiratory protection but less skin protection than Level A. It is the minimum level recommended for initial site entries until the hazards have been further identified.	• The type and atmospheric concentration of substances have been identified and require a high level of respiratory protection, but less skin protection. This involves atmospheres: — with IDLH concentrations of specific substances that do not represent a severe skin hazard. or — that do not meet the criteria for use of air-purifying respirators. • Atmosphere contains less than 19.5 percent oxygen. • Presence of incompletely identified vapors or gases is indicated by direct-reading organic vapor detection instruments, but vapors and gases are not suspected of containing high levels of chemicals harmful to skin or capable of being absorbed through the intact skin.	• Use only when the vapor or gases present are not suspected of containing high concentrations of chemicals that are harmful to skin or capable of being absorbed through the intact skin. • Use only when it is highly unlikely that the work being done will generate either high concentrations of vapors, gases, particulates or splashes of material that will affect exposed skin.

Table 4-1: Sample protective ensemble.

Level of Protection	Equipment	Protection Provided	Should be used when:	Limiting Criteria
C	RECOMMENDED: • Full-faceplate, air-purifying, canister-equipped respirator. • Chemical-resistant clothing (overalls and long-sleeved jacket; hooded, one- or two-piece chemical splash suit; disposable chemical-resistant one-piece suit). • Inner and outer chemical-resistant gloves. • Chemical-resistant safety boots/shoes. • Hard hat. • Two-way radio communications. OPTIONAL: • Coveralls. • Disposable boot covers. • Face shield. • Escape mask. • Long cotton underwear.	The same level of skin protection as Level B, but a lower level of respiratory protection.	• The atmospheric contaminants, liquid splashes, or other direct contact will not adversely affect any exposed skin. • The types of air contaminants have been identified, concentrations measured, and a canister is available that can remove the contaminant. • All criteria for use of air-purifying respirators are met.	• Atmospheric concentration of chemicals must not exceed IDLH levels. • The atmosphere must contain at least 19.5 percent oxygen.
D	RECOMMENDED: • Coveralls. • Safety boots/shoes. • Safety glasses or chemical splash goggles. • Hard hat. OPTIONAL: • Gloves. • Escape mask. • Face shield.	No respiratory protection. Minimal skin protection.	• The atmosphere contains no known hazard. • Work functions preclude splashes, immersion, or the potential for unexpected inhalation of or contact with hazardous levels of any chemicals.	• This level should not be worn in the Exclusion Zone. • The atmosphere must contain at least 19.5 percent oxygen.

Source: NIOSH/OSHA/USCG/EPA "Occupational Safety and Health Guidance Manual for Hazardous Waste Site Activities", U.S. Government Printing Office, Washington DC, 1985.

Table 4-1: Sample protective ensemble (continued).

contaminant. A higher reading indicates the presence of a source of added oxygen (e.g., the leak of an oxidizer), which would heighten the flammability of objects within the vicinity.

The correct level of oxygen in the atmosphere is necessary not only for human safety but also for the proper functioning of the combustible gas indicator monitor (CGI). Too little (below 10-14 percent) or too much (above 25 percent) oxygen interferes with the operation of the CGI, which could result in dangerously incorrect combustible atmosphere readings. Oxygen monitoring should therefore occur before or at the same time as combustible gas monitoring. Some monitoring devices have multiple capabilities indicating, for example, both the percentage of oxygen and the concentration of a combustible gas.

Air is made up of about one part oxygen to four parts nitrogen (with only traces of other components, such as carbon dioxide).

Suppose a monitoring instrument indicates a one percent drop in oxygen concentration at a site. This would suggest that something is displacing the oxygen. Since this something is probably displacing all the components in the air, most notably nitrogen, one could assume an overall five percent displacement.

Such a drop may not seem too serious. However, if we convert the five percent to a parts-per-million equivalent ($0.05 \times 1,000,000$), we have a 50,000 ppm displacement of air.

Consider the consequences if that displacing substance happened to be colorless, odorless natural gas!

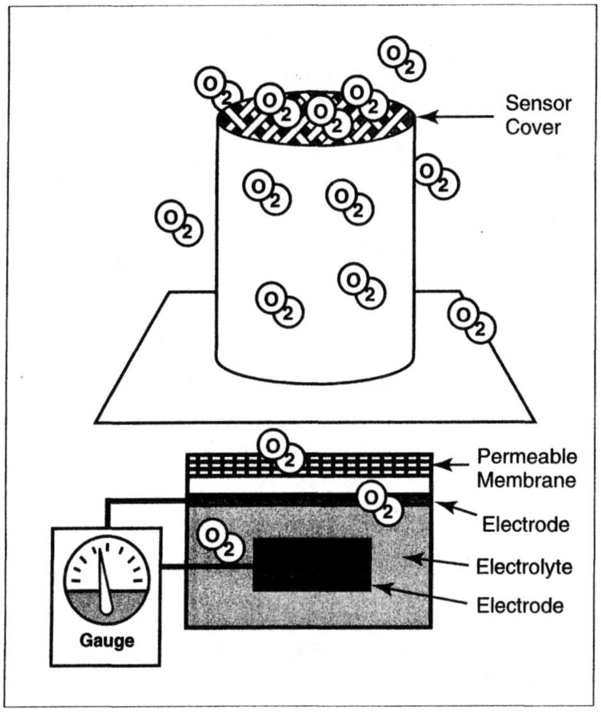

Figure 4-5: Diffusing-type monitor: on their own, oxygen molecules enter the sensor and move through the permeable membrane into the electrolytic solution.

Oxygen Monitor Principle of Operation

The principal components of the oxygen monitor are the sensor and the readout. Air is either drawn into an internal sensor with a bulb or battery-operated pump or it simply diffuses into the sensor. The workings of the electrochemical sensor typically consist of two electrodes, a housing containing a basic electrolytic solution, and a semipermeable Teflon® membrane. Oxygen molecules (O_2) diffuse through the membrane into the solution. Reactions between the oxygen, the solution and the electrodes produce a small electric current proportional to the oxygen content. The current passes through the electronic circuit and results in a signal, which is shown as a needle deflection on a meter or as a digital reading. Optional components of the oxygen monitor include a sampling line or tube mounted over the sensor. The line has a probe at the end, through which the sample is pulled by pumping. Lines are usually around ten feet long, enabling the technician to sample from a distance or through the opening of a container or confined space. It is preferable that the sampling line have a water trap to prevent moisture from contacting the sensor.

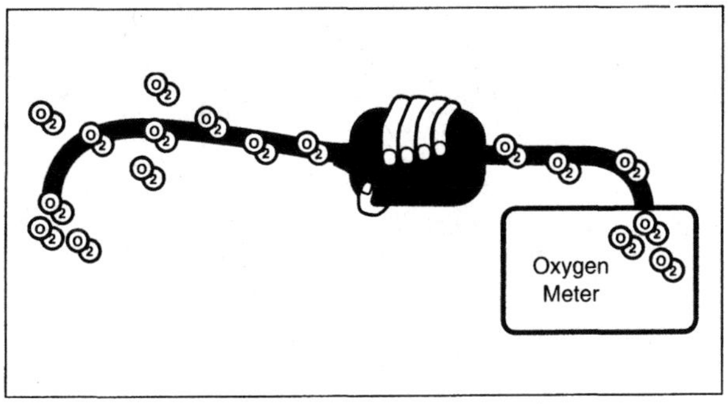

Figure 4-6: Pump-type monitor: oxygen molecules are drawn into the sensor by a hand-operated or electrically operated pump.

Oxygen Monitor Limitations and Considerations

First, the operation of oxygen meters varies according to the atmospheric pressure at the site. Therefore, it is necessary to calibrate the instrument for the altitude at which it is being used. Second, high concentrations of carbon dioxide may shorten the useful life of the oxygen sensor. As a general rule, sensors should be replaced frequently if used in atmospheres of greater than 0.5 percent carbon dioxide. Strong oxidizing chemicals, such as ozone or chlorine, can distort oxygen monitor readings, and indicate high or normal O_2 content when the actual content is normal or even low.

Combustible Gas/Vapor Monitoring

The **Combustible Gas Indicator (CGI)** is a direct reading instrument that measures the concentration of flammable gases or vapors in the air. The result of that measurement is expressed on the monitor's readout as a percentage of the **Lower Explosive Limit (LEL)**.

LEL, UEL and Flammable Range – The LEL (alternately known as the **LFL** or **Lower Flammable Limit**) of a combustible gas or vapor is the minimum concentration of the material in the air that will support a flame upon contact with an ignition source. The corollary measurement is the **Upper Explosive Limit (UEL)**, or **Upper Flammable Limit (UFL)**, which is the maximum concentration in air that will support a flame upon contact with an

Box 4-1 ■ Calibration of Detection Instruments

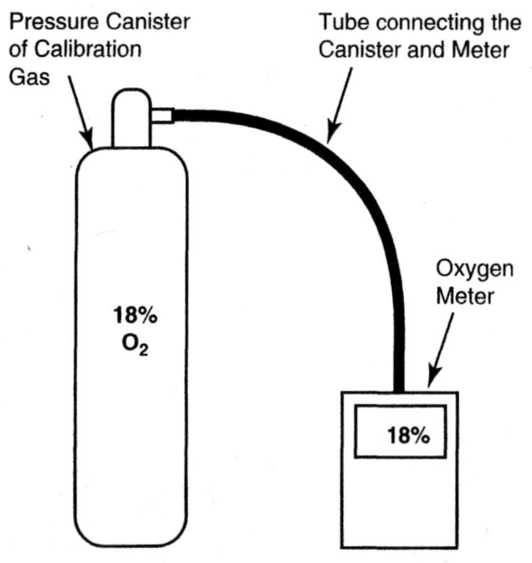

The reading of a calibrated monitor should match the concentration of the calibration gas, plus or minus the error of measurement factor specified by the manufacturer.

A detector's accuracy must be periodically checked with a calibration kit. Common calibration kits consist of a pressure canister containing a known concentration of the calibration gas. For example, a canister may contain 18 percent oxygen. To check the calibration of the oxygen meter, its sensor is exposed to a release of the gas from the canister. The meter readout should match the concentration of the canister's contents, in this case 18 percent. If the reading does not match, the technician should follow the manufacturer's instructions to adjust. If the instructions for adjustment still do not work, a new sensor is needed.

The manufacturer's specifications for the detector will provide a percentage of accuracy; for example, one manufacturer specifies an accuracy of ±0.5 percent for oxygen measurement by a certain detector. In the calibration example a reading within the range of 17.5 percent to 18.5 percent may be considered acceptable when checked against the calibration gas.

ignition source. A reading below the LEL indicates insufficient fuel (i.e., too little flammable gas or vapor) to support combustion, a concentration described as too lean. A reading above the UEL, on the other hand, indicates too much fuel (in other words, too little oxygen) to support combustion, a concentration described as too rich. Concentrations between the LEL and UEL are considered within the **flammable range**.

Propane's flammable range, for example, is 2.2 percent to 9.5 percent propane in the air. A CGI readout of 100 percent LEL thus means that the air contains a concentration of at least 2.2 percent propane. The technician does not want to be in this extremely hazardous atmosphere!

A CGI's audible and visible alarms are typically factory-set to go off at 10 percent LEL, which is the action level specified in OSHA's Permit-Required Confined Spaces standard (29 CFR 1910.146). It is certainly not necessary for the technician to pause and do the math – that is to multiply propane's LEL by the 10 percent to determine the concentration of propane in the atmosphere (2.2 × 10 percent = 0.22 percent). Rather, it is necessary for the technician to regard a 10 percent LEL alarm as a warning to take action. The site health and safety plan will specify the action level and procedures for technicians who detect a concentration of flammable gas in the air. Some want their personnel to withdraw at readings as low as five percent LEL. A reading, even a low one, may mean a leak is causing gas to accumulate.

Further Interpretation of a Percent of LEL Reading – An instrument's LEL percentage reading should not be considered an absolute or perfect measurement. The manufacturer may specify in the instrument's written documentation an error of measurement factor such as ±25 percent. Suppose, for instance, the ±25 percent factor is applied to a CGI reading of 10 percent LEL: 25 percent × 10 percent = 2.5 percent. A reading of 10 percent LEL would therefore be more precisely interpreted as a propane concentration ranging from 7.5 percent to 12.5 percent.

The technician should also be mindful that the CGI is calibrated to a particular gas, such as pentane or methane. Therefore, a reading for a different compound, such as propane, should not be considered an absolute measurement. The CGI

Point of View ■ Sally Gaines

Sally Gaines teaches in a Hazardous Materials Technology Program. She authored the soil and water sampling chapters of this textbook. Sally received her B.S. in geology from Indiana University and her M.S. in geology from University of Kentucky. Four years of working in the field has provided Sally with experiences to share with her students.

Although some of her students may find themselves working in higher-risk situations, such as Superfund or RCRA cleanups or emergency response, many will be associated with sites that represent the more routine, lower-risk world of environmental monitoring and sampling such as the assessment of gas station underground storage tank locales, real estate properties and industrial plants. As employees in the field they will typically wear the protective equipment that EPA calls Level D. They will consistently use safety glasses, steel-toed footwear and work gloves. To be ready for some common hazards such as splashes, falling objects or respiratory irritants, their gear will also include a hard hat, Tyvek coveralls, and half-face respirators with canisters to protect against dusts or vapors.

Although their low-risk sites will seldom require higher level protective equipment such as supplied air respirators or expensive Level A or B suits, these workers must operate under a site safety plan that includes monitoring before entry and during site activities.

Sally says, "Just because you're not suited up doesn't mean you shouldn't be careful." Even when a site has undergone a historical study, surprises still may occur, and technicians can run into things that have not been anticipated. She describes an instance where groundwater monitoring wells were installed at a small Superfund cleanup site, which comprised a "hot" zone of two buildings holding a total of about nine hundred drums. Two team members in Level B suits characterized the site surrounding the hot zone by systematically monitoring with an oxygen meter, a CGI and a toxic monitor (e.g., PID or FID), beginning upwind from the buildings and working gradually in toward the buildings. It was determined after this initial monitoring that there were areas of the site where Levels C and D were adequate protection for the technicians' activities. Throughout the activities, however, the technicians continued to use supplied oxygen as well as their CGI and toxic vapor meters to ensure that they were not exposed to hazards.

Sally recommends that students learn not only the technical procedures for sampling and monitoring equipment but also the appropriate personal behaviors for site work, such as avoiding kicking up dirt while walking and resisting the urge to pick up and smell contaminated soil.

manufacturer usually provides in its written documentation a table of **relative responses**, which are usually expressed as conversion factors. For example, the manufacturer of a certain CGI calibrated with pentane states that the relative response conversion factor for propane is 0.8. Thus, any reading that occurs while monitoring for propane indicates a slightly lower concentration when multiplied by the 0.8 factor.

For example, multiplying the 0.8 conversion factor by propane's LEL reading of 2.2 percent yields an adjusted 1.76 LEL reading for propane, when pentane is the **calibration** gas. Suppose, on the other hand, a technician monitors for toluene, which has a conversion factor of 1.1. Relative to the pentane calibration gas, the toluene concentration would be 1.1 times the meter reading. Whether the technician performs the quick calculation or rapidly refers to a conversion table, it is essential to remember that a reading without the conversion factor is an incomplete reading. Furthermore, most instruments are equipped with alarms that will be set off when toxic levels are reached; no calculation is necessary at that point, and leaving the area immediately is the thing to do.

CGI Principle of Operation

As on the oxygen monitor, the CGI's sensor may be either external, which allows the diffusion of gas into the monitor, or internal, requiring a hand-operated or battery-operated pump. Once inside the monitor, the flammable gas or vapor to be measured contacts a hot filament (part of a circuit called a Wheatstone Bridge). The resulting rise in resistance in the circuit leads to the percentage of LEL reading.

Readouts in currently manufactured CGIs are typically displayed digitally by LCD (Liquid Crystal Display); however, readouts in older CGIs are analog (i.e., meters and needle indicators). A technician using one of the older analog CGIs needs to be aware of a readout problem that occurs with a gas or vapor concentration above the UEL. The meter needle may rise above the 100 percent level and then return to zero or may even lock in at 100 percent, depending on the model. Since the actual concentration cannot then be determined, the technician must interpret the situation and probably withdraw from the area to reset the CGI.

A "Supersensitive" CGI includes circuitry that amplifies the signal so the instrument can provide a parts per million (ppm) measurement. A ppm measurement capability enables the Supersensitive CGI to monitor for toxic air contaminants at levels that a regular CGI cannot detect.

CGI Limitations and Considerations

—Organic lead vapors (e.g. tetraethyl lead used in leaded gasoline), sulfur compounds and silicone compounds will foul the CGI's filament. Acid gases such as hydrogen chloride and hydrogen fluoride can corrode the filament. Most units have an optional filter that protects the sensor from leaded vapors.

—The CGI's response is temperature-dependent, which means that the monitor should be calibrated at the sampling site. Hotter temperatures raise the temperature of the filament and produce a higher-than-actual reading. Cooler temperatures can reduce the reading.

—The CGI is also sensitive to moisture. An optional moisture trap protects the workings.

—The CGI is intended for use only in normal oxygen atmospheres.

—The CGI does not differentiate between petroleum vapors, such as the gaseous phase of gasoline, and combustible gases, such as methane. The CGI readout will thus indicate the percentage of LEL for combined vapors and gases in an atmosphere. For example, if an atmosphere contains both gasoline vapors and methane, the CGI will not differentiate between the two substances but will give a reading for their combined presence.

—The technician must calculate relative response if the CGI is used to monitor a gas or vapor other than the calibration gas.

Toxic Gas and Vapor Monitoring

Concentrations of toxic air contaminants are expressed as ratios: for example, so many parts of air contaminant per million (ppm) parts of air, or so many milligrams of air contaminant per cubic meter (mg/m^3) of air. During site characterization it is important to identify the presence of toxic chemicals so their concentrations can be compared to exposure guidelines published by OSHA, EPA, the National Institute for Occupational Safety and Health (NIOSH), or the American Conference of Government Industrial Hygienists (ACGIH). For example, a glance through OSHA's 29 CFR 1910.1000 exposure limit tables (of toxic and hazardous air contaminants) shows hundreds of substances with exposure

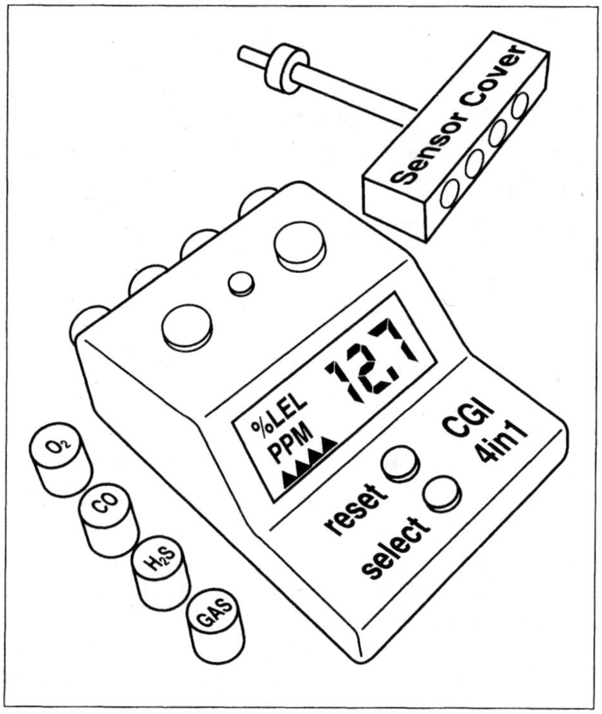

Figure 4-7: Combination monitor with sensors for oxygen, carbon monoxide, hydrogen sulfide and flammable gas/vapor.

levels ranging from less than one ppm to more than 1,000 ppm. The percentage readings of the oxygen monitor or the CGI are not sensitive enough to detect toxic contaminants; after all, it takes an O_2 or CGI reading of one tenth of one percent (0.001) to be equivalent to 1,000 ppm, so imagine trying to measure 1 or 2 ppm with a CGI!

Concentrations of specific substances can be measured by specific toxic gas/vapor air monitoring equipment:

—Specific Chemical Instruments

—Colorimetric Indicators

—Total Vapor Survey Instruments with a Gas Chromatograph Option

Specific Chemical Instruments

Specific chemical instruments measure concentrations of individual toxic gases; they are designed for and used by personnel who know the exact substance they wish to monitor. One type of toxic gas monitor is based on the electrochemical cell (as is the oxygen sensor). The particular chemical solution and electrodes composing an electrochemical cell determine which toxic gas contaminant may be monitored. Electrochemical sensors are available, for example, for ammonia, carbon monoxide, carbon dioxide, chlorine, hydrogen chloride, hydrogen cyanide and hydrogen sulfide. Electrochemical cells will gradually deteriorate; therefore, the equipment maintenance program must include periodic factory rejuvenation or replacement according to the manufacturer's specifications.

Another type of toxic gas monitor is based upon the **metal oxide semi-conductor (MOS)**, also called a solid-state sensor. This monitor's detector is composed of a metal oxide film that coats a ceramic substrate, which in turn is fused around a platinum wire coil. The particular metal oxide and the temperature of the heater determine which toxic gas contaminant may be monitored. MOS detectors are available for ammonia, carbon monoxide, hydrogen chloride, hydrogen cyanide, hydrogen sulfide, methyl chloride, nitrogen oxides and sulfur dioxides among others. Because the principle of operation involves oxygen in the air combining with the detector's metal oxide film, the sensor needs at least 14 percent oxygen to function properly. On the other hand, a high concentration of oxygen can saturate the sensor, resulting in a slow recovery of the sensor during monitoring.

MOS monitors may be used with a nonselective sensor, meaning that a mixture of toxic vapors/gases may be detected. With the nonselective sensor, the MOS monitor's readout gives a total concentration of the substances it detects. Instruments combining several functions are also available; a typical combination monitor might include, for example, sensors for oxygen, carbon monoxide, hydrogen sulfide and flammable gas/vapor.

Colorimetric Indicators

A **colorimetric indicator** is a device that contains a chemical reagent. Upon exposure of the reagent to a contaminant, a chemical reaction occurs that causes the reagent to undergo a color change. The technician uses the device's printed directions to interpret that color change. The type of color change indicates the level of contamination.

One form of the colorimetric indicator consists of a glass tube several inches long containing the reagent. The particular reagent determines which toxic gas contaminant may be monitored. The technician breaks the ends off the glass tube, inserts the tube into a hand operated pump, and then squeezes the pump a specific number of times, according to the manufacturer's directions, in the contaminated air to be monitored. When pulled into the tube, the contaminant reacts chemically with the reagent, causing a color change within the tube. The length and/or intensity of the color change is proportional to the contaminant's concentration. Concentration ranges may be expressed in percentage or as ppm.

Another form of the colorimetric indicator is a badge saturated with reagent. The technician visually inspects the color change on a badge and compares it to the manufacturer's color change chart. The technician may also send the badge to a laboratory for an instrumental analysis.

Total Vapor Survey Instruments

While the colorimetric indicator tube or a specific chemical instrument might be selected to detect the level of a known or suspected toxic contaminant, a **Total Vapor Survey Instrument** might instead be chosen when it is a question of detecting the presence of unknown air contamination, be it a single material or a mixture of materials. As the name implies, this instrument's readout will provide a total measure of the vapors present whatever they are.

Technical Corner ■ Detector Tubes

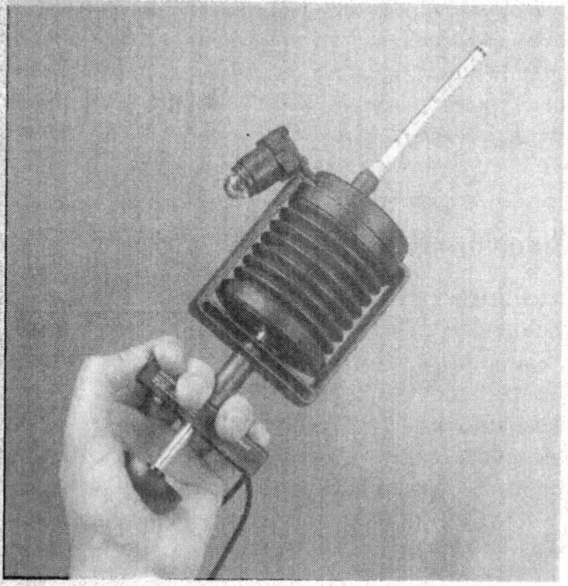

MSA detector tube inserted in Deluxe Kwik-Draw™ pump.

Several factors determine the effective use of **detector tubes**. These factors are found in the instructions issued with each box of tubes.

—Chemical Group: Some tubes are for a specific chemical, some for a group of chemicals.

—Lot #: The instructions for the tubes may change with different model numbers of different lots. Thus, the instructions should be matched with the proper tubes.

—Expiration Date: The chemicals used in the tubes deteriorate over time. The tubes are therefore assigned a shelf life and the expiration date is printed on the box.

—Pump Strokes/Volume/Time: The total volume of air to be drawn through the tube varies with the type of tube. The volume needed is given as the number of pump strokes; i.e., the number of times the piston or bellows is manipulated. Also, the air does not instantaneously go through the tube. It may take one to two minutes for each volume (stroke) to be completely drawn. Therefore, sampling times can vary from 1 to 30 minutes per tube. This can make the use of detector tubes time-consuming.

—Color Change: The instructions will give the appropriate color change for indicating the chemical of concern. Other color changes may be noted for potential interferences. Information about interferences may be used to check for presence of other chemicals.

—Interferences: As mentioned previously, not every tube is chemical-specific. For example, an acetone tube will also respond to other ketones. Thus, methyl ethyl ketone would be considered an interference if one were checking for acetone. The instructions will give known interferences or color changes that are not for the chemical of interest.

—Temperature/Humidity/Pressure: The length of color change (stain) can be affected by temperature, humidity and barometric pressure. If this is a problem, the instructions will note it and may give correction factors. Cold weather slows the chemical reaction in the tube and reduces the reading. Hot temperatures increase the reaction and can cause a problem by discoloring the indicator even when a contaminant is not present. This can happen even in unopened tubes. Therefore, the tubes should be stored at a moderate temperature or even be refrigerated during storage.

—Reusability: Most tubes can only be used once, even if the result is negative. Some tubes, however, can be reused during a single day until a positive result is obtained.

—Accuracy: The accuracy of detector tubes varies. Some tubes are certified by the manufacturer to be ± 25 percent accurate at readings from one to five times the OSHA Permissible Exposure Limit (PEL) and ± 35 percent at concentrations 1/2 the PEL. Some studies have reported error factors of 50 percent and higher for some uncertified tubes. One factor that affects accuracy is the interpretation of the end of the color change. Some color changes are diffused and the endpoint is not definite; others may have an uneven endpoint. When in doubt the technician should use the highest value that would be obtained from reading the different aspects of the tube.

While many limitations and considerations exist for using detector tubes, the tubes allow the versatility of being able to measure a wide range of chemicals with a single pump. Also, detector tubes are the only direct-reading indicators for some chemicals.

Source: EPA, Air Surveillance for Hazardous Materials, Environmental Response Team, 1990.

Suppose, for example, the total reading is 10 ppm. That may indicate 10 ppm of one substance, or it may indicate 3 ppm of propane plus 4 ppm of methane plus 3 ppm of gasoline (which itself has numerous constituents). There is no way of knowing what chemicals comprise the total reading unless further instrumentation is used (see Gas Chromatography).

Here we will discuss two types of Total Vapor Survey Instrument: the **Photoionization Detector (PID)** and the **Flame-Ionization Detector (FID)**. Although the PID and the FID each detect toxic gases/vapors, they differ in their principles of operation. In the operation of a PID the contaminants are exposed to an ultraviolet light source, which removes an electron and results in positively charged particles or ions; as the ions collect on a negatively charged plate, the electrons strike the positively charged plate producing a current that yields the measurement. In the operation of a FID the contaminants are exposed to a hydrogen flame, which results in positively charged ions; as the ions collect on a negatively charged plate, they likewise produce a current that yields a measurement. The measurement on each instrument's readout is a ratio of parts of total contaminant per million parts air (ppm).

PID Principle of Operation – The operation of the PID is based on a concept called **Ionization Potential (IP)**. Ionization potential is defined as the energy required to remove the outermost electron from a molecule. That energy, or IP, is a specific value for a given compound and is measured in **electron volts (eV)**. Table 4-2 lists the IPs of a number of compounds.

Chemical	Ionization Potential (eV)
Carbon monoxide	14.0
Hydrogen cyanide	13.9
Methane	13.0
Hydrogen chloride	12.5
Water	12.6
Oxygen	12.1
Chlorine	11.5
Propane	11.1
Hydrogen sulfide	10.5
Ammonia	10.1
Acetone	9.7

Table 4-2: Ionization potentials for several compounds.

Ultraviolet radiation is able to provide the necessary energy for ionization and is, in fact, called **ionizing radiation**. Within a Photoionization Detector is an ultraviolet (UV) lamp. When the lamp's UV radiation strikes a chemical compound drawn into the detector, the molecule becomes ionized if the UV lamp's energy is at least equal to the compound's ionization potential (that is, its eV value). The negatively charged electrons removed during ionization are collected on a positively charged plate where they produce a measurable current, which then yields an instrument reading.

The energy levels of available PID lamps are in eV (electron volt) increments of 8.4, 9.5, 10.0, 10.2, 10.6 and 11.7. If the technician is trying to determine the presence of a known compound, he or she selects an appropriate level of lamp by referring to a chart of chemical IPs. The lamp's eV must be higher than the IP of the compound the technician intends to detect. Suppose, for example, the technician wishes to detect the presence of hydrogen sulfide, which has an IP of 10.5 eV. Of the available PID lamps, the 10.6 eV lamp and the 11.7 eV lamp can both detect hydrogen sulfide's lower IP of 10.5.

PID Considerations and Limitations – Dust drawn into the instrument may block the transmission of UV light which would cause a misreading. The technician should be able to identify this type of problem during calibration and should maintain the PID's dust filter and lamps to prevent interference.

Humidity may cause two problems. A cold instrument taken into a warm moist atmosphere may cause moisture to condense on the lamp. As with dust, this reduces the transmission of UV light. Moisture in the air also reduces the ionization of chemicals and causes a reduction in readings.

If a compound's IP is only slightly below the lamp's eV level, the PID's ability to detect the compound is optimum if the lamp is clean and in good shape. For example, the 11.5 IP of chlorine is very close to the 11.7 eV of the highest level lamp. Using a new lamp would improve the PID's ability to detect chlorine.

It is possible for the technician to differentiate between two known chemicals by first installing the level of lamp that would detect the higher IP chemical and then installing the lamp that would detect the lower IP chemical. However, using the PID with a gas chromatography option is preferred when it is necessary to differentiate chemicals, especially if unknown contaminants may be present.

The highest level lamp, (at 11.7 eV), is below the IPs of many compounds – for example, methane or hydrogen cyanide, which have IPs of 13.0 and 13.9 respectively. Therefore, the PID is not an appropriate instrument for detecting those compounds. The technician should consult an instrument's operating instructions for other considerations and limitations of a specific model.

FID Principle of Operation – When organic vapor contamination is drawn into the FID, a hydrogen flame ionizes the vapors, which produces positively charged carbon-containing ions. As the ions are collected by a negatively charged electrode in the detector chamber, a measurable current is produced, which the instrument then amplifies to yield a reading. The flame responds to virtually all organic compounds but does not detect inorganic compounds like chlorine, hydrogen cyanide or ammonia.

FID Considerations and Limitations – Since hydrogen gas fuels the FID's operation, the technician must be able to work safely with a compressed gas supply. Inadequate oxygen as well as high concentrations of organic chemicals may cause the FID's flame to go out. Cold weather may also cause the flame to go out, or it may interfere with ignition of the instrument. Flame-ionization is a destructive form of monitoring: that is, an organic vapor sample contacts the flame and causes the material to decompose into combustion products such as carbon monoxide and water. If the material contains a substituted hydrocarbon (e.g., a chlorinated compound), then decomposition may produce toxic or corrosive byproducts such as hydrogen chloride. The technician should also consult the instrument's operating instructions for other considerations and limitations of a specific model.

Radiation Monitoring

A radiation monitor, also known as a Geiger-Muller Counter, GM Counter, or simply as a **Geiger Counter**, determines the presence of ionizing radiation, which includes alpha particles, beta particles, gamma rays, X-rays, neutrons, high-speed electrons, high speed protons and other subatomic particles. The technician must ensure compliance with employee radiation protection regulations in OSHA's 29 CFR 1910.96 and with transportation requirements in DOT's 49 CFR Parts 173-177. If monitoring detects radiation levels that exceed OSHA's and DOT's thresholds, the employer must secure the area and notify the proper authorities as specified in the site health and safety plan.

A radiation survey meter is usually sent back to the manufacturer for recalibration because of the high radiation of the calibration source. However, low radiation sources might be used to check meter response. Technicians must follow the manufacturer's guidelines as well as the Standard Operating Procedures (SOPs) of their own work situations.

Personal Radiation Badges and Dosimeters – Each employee who may be exposed to radiation during work duties or emergency response must wear film badges and pocket dosimeters. A film badge measures accumulating radiation dosage and is sent to a laboratory for periodic evaluation. A dosimeter enables the employee to read a real-time measurement of radiation dosage. Film badges and **self-reading dosimeters** should be worn on the upper front part of the body, usually on a shirt pocket, with as little shielding material as possible between the devices and the source of radiation. A

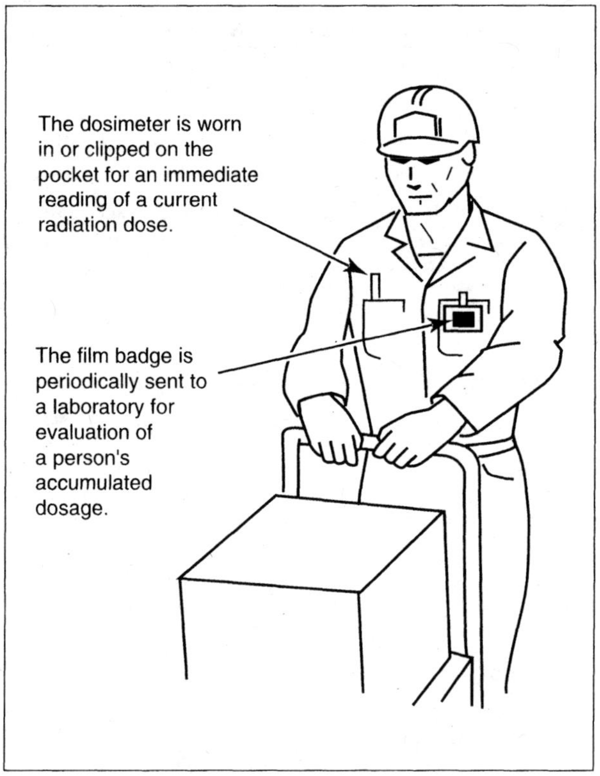

Figure 4-8: The radiation dosimeter gives an immediate reading whereas the film badge is analyzed by a lab.

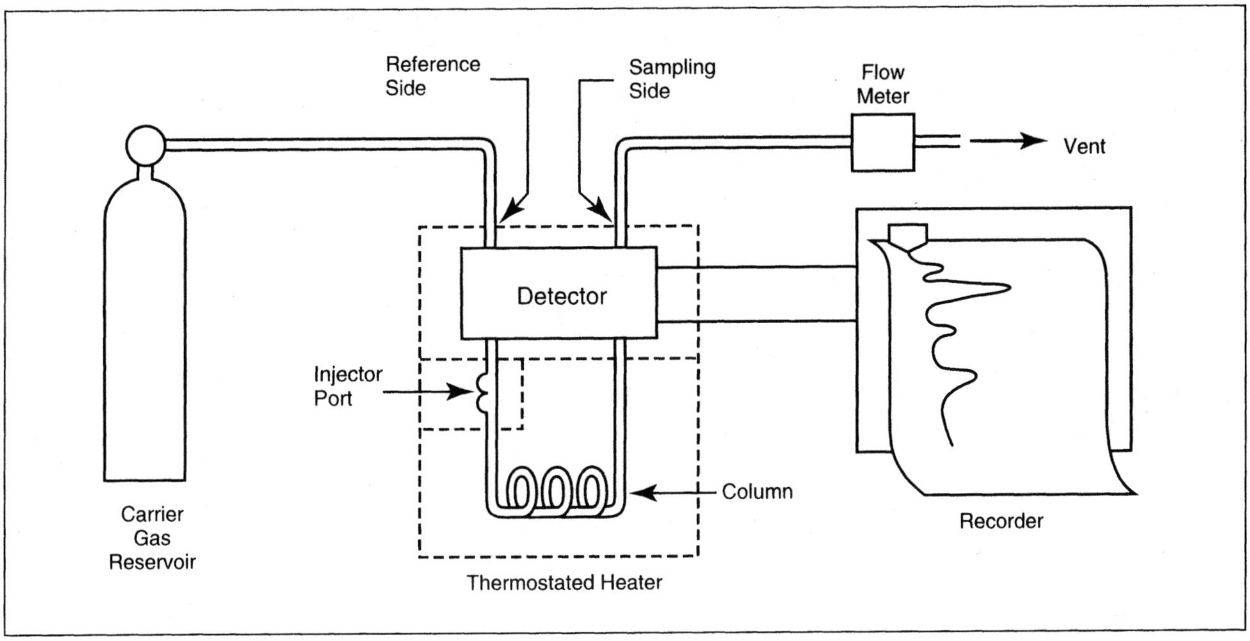

Figure 4-9: Schematic of a gas chromatograph.

site health and safety plan might require that two real-time dosimeters be worn at the same time as an extra precaution. Since a dosimeter is insensitive to beta and neutron radiations, a complete radiation monitoring program must include the monitoring of an area by survey instrument as well as by periodic laboratory analysis of film badges.

Gas Chromatography (GC)

A gas chromatograph (GC) is an analytical instrument that can separate and identify components of a sample. The GC can then be linked with one or more detectors, such as the PID or FID, to measure those components. The GC's principle of operation is based on the injection of an air sample into a column packed with an inert solid material. Each of the sample's separate components is retained in the inert material for a different period of time. As a component exits the column, it flows into the detector. The output of the detector is connected to a strip chart recorder that graphs each component as a separate peak. This graphed readout is called a gas chromatogram.

The high cost of the gas chromatograph makes it less likely to be generally available as a field instrument, although it might be found at large hazardous waste cleanup sites that have mobile lab facilities. Instead, GC analysis typically is applied to samples that have been collected in the field, stored and preserved in containers, and then transported to a laboratory.

Aerosol Monitoring

Not all toxic materials dispersed in the air are in the form of a gas or vapor. Solids and liquids can become suspended in air by combustion, by splashing liquids, or if soil is disturbed.

Aerosol monitors are direct-reading instruments. Most of them use a light source and a light sensor, which measures the amount of light scattered by the airborne substance. Readouts are in milligrams per cubic meter (mg/m^3).

Characteristics of Air Monitoring Instruments

To be useful, air monitoring instruments must be 1) portable and rugged; 2) easy to operate; 3) able to generate reliable and useful results; and 4) inherently safe.

Portability

Since the technician must be able to carry instruments in the field, their design should include shoulder straps, backpacks or belt fasteners. The instrument must be constructed to withstand field conditions and activities as it is carried about; thus, it should have a reinforced shell, shock-mounted housing for its electronics as well as an anodized or coated finish. A portable instrument needs an internal power supply, such as replaceable or rechargeable batteries. A remote sensor option improves the accessibility of some hard-to-sample targets.

Ease of Operation

The technician must be able to operate the instrument while he or she wears Personal Protective Equipment. Gloves can present a special challenge. Many instruments are now actually small computer workstations. They display menus that cue the technician about each step of operation. Preparation time for using the instrument, such as the calibration procedure, the warm-up time, the attachment of accessories and the instrument checks, should be brief and easy. Also desirable is a rapid response time – that is, the number of seconds or minutes it takes the instrument to sense a contaminant and then yield data.

Reliable and Useful Results

The sensitivity, accuracy and precision of an instrument contribute to the reliability and usefulness of its data.

Sensitivity – **Sensitivity** is the ability of the instrument to detect even small changes in concentration. Relative to its sensitivity, an instrument has a particular operating range. The lowest concentration of a substance that can be measured by the instrument is called the **lower detection limit**. The highest concentration that can be measured by the instrument is the **upper detection limit**, which is also referred to as its **saturation concentration**. One manufacturer, for example, states that a particular instrument detects toxic concentrations from 0 ppm (lower detection limit) to 2,000 ppm (upper detection limit).

Also relative to an instrument's sensitivity is its **selectivity** – that is, its ability to detect and measure a specific chemical (or a group of similar chemicals). Selectivity becomes an issue for the technician who is trying to monitor for a particular compound, but gets interference from another compound with a similar response.

Accuracy – An instrument's **accuracy** is a measure of how close readings are to true values. The technician may determine the relative accuracy of an instrument by comparing its readings to prepared samples for which the true values are known. An instrument's accuracy is expressed as accuracy or bias percentage. For example, if detector tubes for a particular chemical are tested numerous times against a standard, and the average results are 15 percent higher than the true concentration of that standard, then that type of tube is said to have a bias of +15 percent.

Precision – An instrument's **precision** is its ability to repeatedly yield the same results for a chemical. An indication of the instrument's precision would be its ability to consistently yield the same measurement when exposed to a standard concentration of a calibration chemical.

Inherent Safety

The term **inherent safety** refers to the instrument's level of protective construction. Monitoring instruments to be used in potentially hazardous atmospheres must be constructed so that their components (switches, batteries, etc.) do not provide a spark that could ignite explosive vapors or gases. Instruments are certified by standards organizations for particular types of hazardous atmospheres.

The National Electrical Code (NEC), which is established by the National Fire Protection Association (NFPA), classifies hazardous atmospheres and lists types of materials that generate these atmospheres. The hazardous atmospheres are divided into Classes, which are further subdivided into Groups and Divisions:

— Class I consists of flammable vapors and gases, such as gasoline and hydrogen. Class I is further divided into Groups A, B, C and D on the basis of similar flammability characteristics. One example for each Group is acetylene (A), hydrogen (B), hydrazine (C) and gasoline (D).

— Class II consists of combustible dusts. Class II is further divided into Groups: E, conductive dusts; F, semivolatile dusts; and G, nonconductive dusts.

Selected Class I Chemicals by Groups

GROUP A Atmosphere	GROUP B Atmospheres (not sealed in electrical conduit 1/2 inch or larger)	GROUP C Atmospheres (selected chemicals)	GROUP D Atmospheres (selected chemicals)
acetylene	allyl glycidyl ether n-butyl glycidyl ether 1,3-butadiene ethylene oxide formaldehyde (gas) hydrogen manufactured gas (containing greater than 30 percent H2 by volume) propylene oxide propyl nitrate	acetaldehyde carbon monoxide chloroaldehyde diethyl ether ether acetate ethyl mercaptan ethylene hydrogen cyanide hydrogen sulfide nitropropane tetraethyl lead tetrahydrofuran vinyl chloride xylenes ethylene glycol ethanol fuel oils acetic acid ethane hexane gasoline liquified petroleum gas	acetone methane acetonitrile methanol acrylonitrile methyl ethyl ketone ammonia naphtha benzene propane butane styrene

Selected Class II Atmospheres by Groups

Group E Conductive Dusts – Atmospheres containing metal dusts, including aluminum, magnesium and their commercial alloys, as well as other metals of similarly hazardous characteristics.

Group F Semi-Volatile Dusts – Atmospheres containing carbon black, coal or coke dust with more than 8 percent volatile material.

Group G Non-Conductive Dusts – Atmospheres containing flour, starch, grain, carbonaceous, chemical thermoplastic, thermosetting and molding compounds.

Class III Atmosphere

Ignitable fibers such as cotton dust.

Source: "Classification of Gases, Vapors and Dusts for Electrical Equipment in Hazardous (classified) Locations" Nat'l. Fire Prot. Assoc., 1986. ANSI/NFPA 497M.

Table 4-3: NEC hazardous atmospheres.

One example for each Group is aluminum (E), coal dust (F) and flour (G).

— Class III consists of ignitable fibers, such as cotton dust.

Whereas the Class and Group indicate the classification of the chemicals themselves, the Division indicates a type of location and dispersion of the hazardous material.

— Division 1 is a location where a release is continuous, intermittent or periodic into an open, unconfined area under normal conditions. Instruments certified for Division 1 locations are referred to as intrinsically safe.

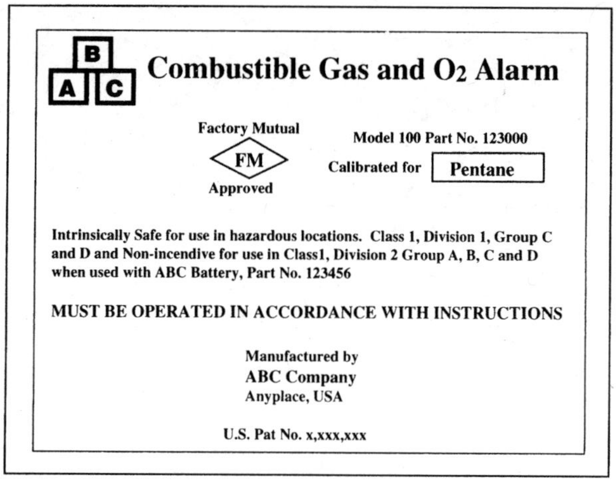

Figure 4-10: Certification label from a combustible gas and oxygen indicator.

—Division 2 is a location where a release would occur only because of a rupture, leak, or other failure of a closed system or container.

Using the NEC Classes, Groups and Divisions, the technician can define a hazardous atmosphere. For example:

—An abandoned waste site containing intact closed drums of xylene would be considered a Class I Division 2 Group D environment. Class I indicates flammable vapors or gases. Division 2 indicates a closed system; Group D is the list containing the xylenes.

—An industrial facility that periodically releases xylene vapors during an open-system cleaning operation would be considered a Class I Division 1 Group D environment. The Division 1 classification indicates a location where a release is continuous, intermittent or periodic into an open area under normal conditions.

—A textile operation that continuously releases cotton dust into the plant is Class III for ignitable fibers and Division 1 for the open location.

—Dust from an aluminum processing operation is exhausted into a baghouse. Conductive combustible dust is Class II Group E and the closed system is Division 2.

The technician is responsible for selecting the appropriate instrument for the characteristics of the atmosphere that will be monitored. When investigating a site that has unknown chemical contaminants, the technician should select an instrument certified for the most hazardous Class I atmosphere; in other words, a Class I Group A Division 1 instrument. The technician is furthermore responsible for using, maintaining and servicing the device according to the manufacturer's instructions so that it will not contribute to ignition.

Checking Your Understanding

1. Identify the different types of air monitoring techniques for an uncontrolled waste site.
2. Explain why an instrument's LEL reading should not be considered absolute.
3. Identify and describe the specific monitoring devices for measurement of toxic gas and vapor.
4. Explain the difference between a PID and a FID.

4-3 Air Sampling

Concept

- A laboratory may measure contaminant levels in samples to the parts per billion or parts per trillion range, while field instruments generally do not measure beyond the parts per million range.
- Air sampling usually consists of actively drawing the air contaminants by pump into collection media or into containers.
- The choice of collection media and/or containers depends on the purpose of the sampling and the nature of the substances to be sampled.
- Sampling pumps must be approved for gas/vapor and dust atmospheres and calibrated to the desired flow rate.

Whereas air monitoring instruments are used to obtain real-time data about air contaminants, sampling equipment and storage containers are used to collect air samples for later analysis at a laboratory. The technician should be able to sample 1) a workplace, including an individual worker's breathing zone, 2) particular sources of emissions, and 3) the ambient atmosphere.

There are three advantages of laboratory analysis for air samples:

1. Depending upon the sensitivity of a laboratory's instrumentation, contaminant levels in samples can be measured to the parts per billion or parts per trillion range. The sensitivity of field instruments generally allows detection to the parts per million range, although some field instruments can detect to the parts per billion range.
2. Laboratory instruments are usually able to identify and quantify the specific components of a sample.
3. Stationary air sampling equipment makes sample collection possible over a span of time (e.g., days) without human intervention.

However, compared to real-time field monitoring, laboratory analysis is costlier and slower in delivering results. Also, accuracy may decline if volatile samples experience some loss of analyte during transport to the laboratory.

Purposes for Sampling: Personal and Area Assessment

Air sampling programs are conducted to assess 1) potential worker exposures to hazardous materials and 2) the general quality of the atmosphere in an area.

Personal Assessment – The technician periodically conducts **personal assessment**, which is personal sampling of individual workers who are exposed to potentially hazardous workplace materials. The laboratory should be informed before the sampling event so a timely analysis can be scheduled. The lab can supply the technician with appropriate filters, absorbent media and related equipment along with instructions for their use.

The sampling device is attached to the worker's clothing so the battery-operated pump can draw in air from the breathing zone, which is defined as a twelve-inch (30 cm) hemisphere around the worker's nose. The personal sample is usually collected for an eight-hour period so the results may be compared to OSHA's Time-Weighted Averages in 29 CFR 1910.1000. Upon completion of a sampling session, the technician seals the sample according to the manufacturer's or laboratory's instructions and sends it for analysis.

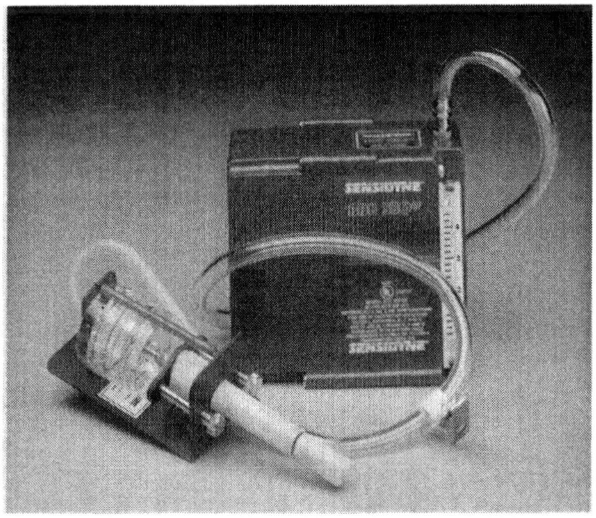

Figure 4-11: The personal sampler pump may be worn on a belt, while the sorbent portion of the unit is clipped onto the shirt within the wearer's breathing zone.

Area Assessment – At a workplace where potentially hazardous materials exposures exist, the technician may conduct **area assessment**, in addition to personal sampling to obtain more data on the site's air quality. Area sampler pumps, which are AC-powered, take a cumulative air sample over a time period ranging from 24 hours to several days or more.

Area sampling is used to assess not only a work site but also environmental air quality. For example, state and local air quality offices may install samplers in selected locations to collect ambient air samples over an extended period. Analytical results are compared to EPA and state air quality standards. Samplers may furthermore be installed next to or downwind of a specific source of air emissions, such as a smokestack. The owner of the emissions source (or a state air quality inspector) can use the results to check compliance with regulations such as the Hazardous Air Pollutants list in the Clean Air Act Amendments.

Principle of Operation

Air sampling usually consists of actively drawing the air contaminants by pump either into collection media – such as filters, sorbents, or liquid – or into containers such as gas bags or canisters. (An exception to pulling contaminants through by pump includes the diffusion method, called passive sampling, see Figure 4-5.)

The choice of collection media and/or containers depends on both the purpose of the sampling (personal vs. area sampling) and the nature of the substances to be sampled (aerosols and particulates, inorganic gases, organic vapors, or a combination of contaminants).

The site sampling plan specifies for the technician what type of handling and storage is necessary for each type of sample. The developer of the sampling plan selects collection procedures that follow the methods described by agencies and professional organizations: EPA, OSHA, NIOSH and ASTM.

Collection Media and Containers

Air sampling equipment must collect the contaminant for analysis either by trapping it within a special **filter medium** or sorbent (absorbing) medium or by containing it within a sample-collection bag or a sampling canister.

Filter Medium

Two types of filters are used: fiber and membrane. Fiber filters are irregular glass or cellulose meshes that form openings of 20 micrometers (mm) diameter or less. (A micrometer is a millionth of a meter.) If a filter is constructed, for example, to collect particulate matter of at least 10 mm, smaller particles will pass through as the larger particles are pulled into and trapped by the fibers. At the conclusion of sampling, the technician seals the filter and sends it to the laboratory. Analysis involves microscopically inspecting the particles on the filter, dividing the filter into a grid for a particle count and possibly weighing the particles. Cellulose filters have a tendency to retain moisture, which becomes problematic if weighing is part of the analysis.

A membrane filter is a plastic film with pores that are 0.01 to 10 mm diameter. This filter acts as a sieve; for example, if the filter is meant to collect particulate matter of at least 10 mm, then smaller particles will pass through the pores while the larger particles remain on the filter surface. Like the fiber filter, the membrane filter is microscopically analyzed by a laboratory.

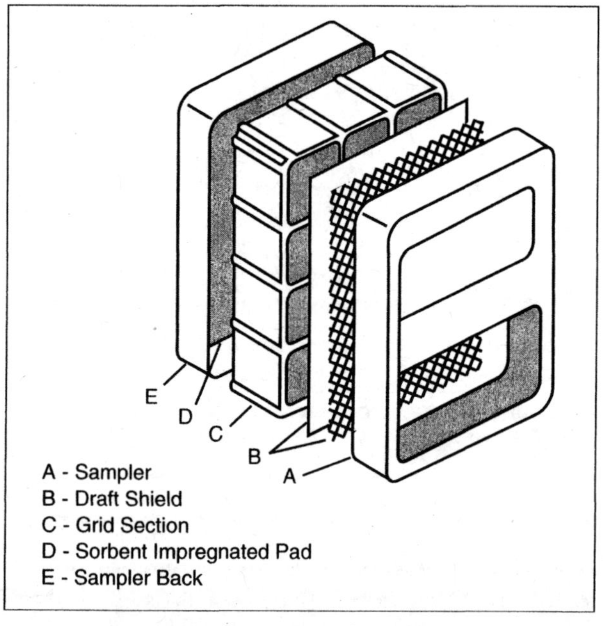

Figure 4-12: Parts of a sorbent sampler.

Sorbent Medium

Sorbents may be in solid or liquid form. A solid sorbent can hold (adsorb) gas or vapors on its surface. Examples of solid sorbents are activated charcoal and silica gel. When a gas or vapor is drawn by pump into the sampler, it is held to the sorbent. Upon conclusion of sampling, the technician packages and seals the sorbent and sends it to the laboratory for analysis.

Since a solid sorbent is limited in the amount of chemical it can hold, the technician must be aware of the potential for **breakthrough**, which occurs when the sorbent has exceeded its capacity to hold the sampled gas or vapor. Because the sorbent cannot adsorb more sample material, the excess material passes through; this will result in an error in determining the concentration of the sampled gas or vapor. To detect breakthrough, the technician may use a back-up sorbent and have it analyzed separately. To avoid breakthrough, the technician might increase the amount of sorbent or limit the size of the sampling pump to reduce the volume of air pulled through the sorbent.

The technician follows procedures provided by the sampling plan and the analytical laboratory; for example, the lab will specify proper handling as well as maximum holding time, within which the samples must be returned for analysis. This is important because a sorbent holding a gas or vapor sample cannot be stored indefinitely.

Liquid sorbents are suitable for collecting gases and vapors that are soluble or reactive. The liquid medium is held in a sampler that consists of a glass vial, a stopper containing a vapor inlet valve, and tubing that connects to a pump. The pump pulls the air sample into an absorbent liquid such as sodium hydroxide or hydrochloric acid. Since these absorbent liquids have hazardous characteristics and their samplers are glass, breakage presents a possible disadvantage. Other disadvantages of using liquids as a sorbent medium involve their limited shelf life, their need for careful handling and storage, and possible evaporative loss during sampling.

Any sorbent should remain sealed until its moment of use. If the sorbent is prematurely exposed to the atmosphere, substances other than the analyte of interest may be absorbed and could interfere with the analysis.

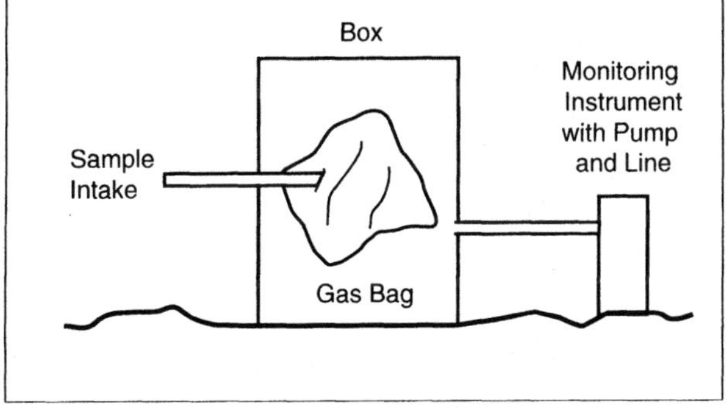

Figure 4-13: Gas sampling bag. To avoid cross-contamination, the bag is placed in a box that is connected to a sampling pump. Pumping air from the box creates a partial vacuum. Positive pressure in the bag results in its inflation as the sample is drawn from the atmosphere.

Sample Collection Bag

Atmospheric samples may be collected in sample collection bags, which are specially designed bags constructed of polyethylene, Saran™ , Mylar™ , Tedlar™ or Teflon®. An on/off valve fitting allows the bag to be connected to monitoring instruments or to sampling equipment such as a photoionization detector; a gas chromatograph; a Geoprobe™ (a steel probe hammered into the ground for soil gas sampling); or a calibration kit for a monitoring instrument (this use does not require a separate sampling pump, since the calibration canister valve is simply opened to discharge gas into the bag). Some models of sampling bags feature a **septum**, which is a partition that allows a syringe to be inserted to draw out a sample for analysis.

A sampling pump used for drawing an air sample into the bag could contaminate the sample. To avoid direct contact of the air sample with the pump, the technician may install the bag in a sealed box that has fittings for tubes. One tube connects the bag to the sample source, such as a Geoprobe™ ; the other tube connects the sampling pump to the interior of the box. Operation of the pump creates a partial vacuum in the box; positive pressure in the sample bag results in the bag's inflation as the sample is drawn in.

Because of the potential for chemicals in the sample bag to permeate the bag's material, samples should be analyzed as quickly as possible. Although bags might be advertised as cleanable and reusable, some professionals in the field prefer to dedicate a bag to a particular use rather than try to clean them in the field.

Sampling Bottles and Canisters

Atmospheric samples may be collected in glass bottles or in special metal canisters, which may be referred to as **vacuum cans**. A vacuum is created within the **gas sampling bottle** or **canister** so that when an inlet valve is opened, an air sample is drawn in for a one-time grab sample. A metal canister may be connected to an air flow device for long-term sampling.

The disadvantage of glass bottles is the potential for breakage. A feature of the metal canister is its ability to be pressurized so that a sample volume greater than the canister volume may be obtained. The canister is sent to the laboratory for tests to identify the air sample components.

Sampling Pumps

Earlier in this chapter we introduced toxic vapor detector tubes that are inserted into a hand-operated pump for collecting a real-time grab sample. **Sampling pumps** for collecting cumulative samples over hours or days are electrically powered. Personal sampler pumps operate for six to ten hours on batteries, and area sampler pumps operate as long as needed on AC current. The pump must pull a certain amount of air during sampling; therefore, some type of air flow rate regulator is necessary:

— Low flow pumps, which are suitable for personal samplers, have a flow rate range from 10 cubic centimeters of air per minute (cc/min) to about 750 cc/min.

— Medium flow pumps, which are suitable for personal or area samplers, have a flow rate of about one to six liters of air per minute (Lpm).

— High volume (Hi-Vol) pumps, which are used for area sampling, are AC powered and can achieve up to forty cubic feet per minute (cfm), which is equivalent to 1,130 Lpm. Hi-Vol pumps are not suitable for use with sorbent media, because the pumps might pull more contaminant than the sorbent can handle, causing inaccurate results from breakthrough of the sampled substance.

The technician should be sure that all sampling pumps used are approved for gas/vapor and dust atmospheres.

Calibration

Pumps, as well as the entire sampling system, should be calibrated to the desired flow rate when newly purchased, before and after sampling, after any repair, and after any suspected abuse. The primary calibration standard is based upon a method called the **soap bubble meter**, which is available electronically as well as in the form of basic chemistry equipment (i.e. a beaker and burette configuration that is connected to the pump being calibrated).

The principle of calibration by soap bubble meter is based on the time it takes a soap bubble to rise through an inverted **burette** (a long glass tube with measurement marks), which is connected by tube to the sampling pump as the technician adjusts the pump's flow rate:

$$\text{Flow rate} = \frac{\text{volumetric distance traveled by bubble (mL)}}{\text{travel time of bubble (sec)}}$$

If the technician rents personal air sampling pumps from a laboratory or industrial hygienist, the calibration is left to the laboratory. If the sampling pumps are owned and operated in-house, then the technician calibrates the equipment.

In an organization that owns and maintains its own sampling pumps, the technician gets to know the "personality" of each pump. If a pump is slightly off in calibration, the technician will know whether it is at least consistent during each calibration. The technician may learn to troubleshoot repairs and to do annual performance tests, such as checking a pump's operation during a pressure drop. When using and maintaining battery-operated pumps, the technician must follow the manufacturer's instructions for proper recharging of batteries.

Sampling and Monitoring Equipment Maintenance

Sampling and monitoring equipment must be subject to a program of preventive maintenance as well as procedures to prevent or minimize contamination during use.

The technician must follow the Standard Operating Procedures (SOP) established in the sampling plan for reducing contact and contamination. The SOPs might include the following suggestions:

— Keep sensor probes out of water and dirt.

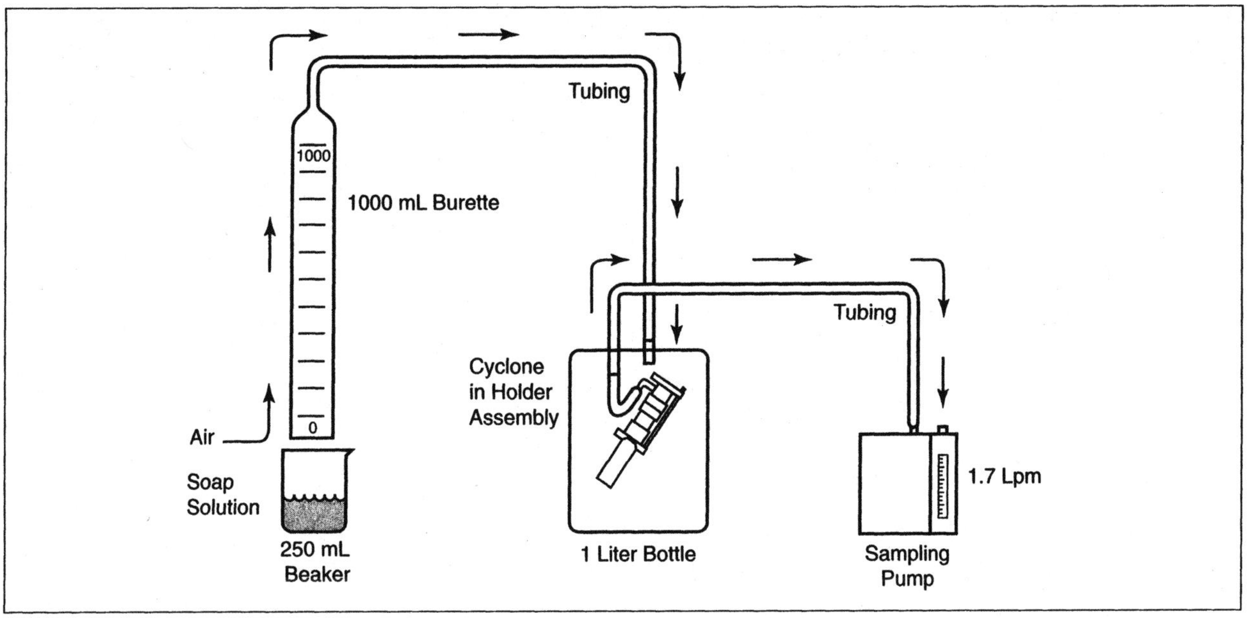

Figure 4-14: Calibration of cyclone respirable dust sampler using a bubble meter. The cyclone, which is the sample collection filter assembly, is placed into the liter bottle to enable an air flow to be maintained through the sampling train during the calibration procedure. After the pump is turned on to create air flow up into the burette, the soap solution beaker is lifted so that the surface of the solution contacts the burette. The bubble that is formed is timed as it travels a specific distance up the burette. The formula in the text is then applied to determine the flow rate. Changes in the flow rate are achieved by adjusting the regulator on the pump.

—Wrap monitoring equipment in plastic bags and seal with duct tape, leaving an opening for the sensor. Dispose of the wrappings at the decontamination area.

—Encase the source of contamination with plastic sheeting or overpacks.

—Use disposable sampling equipment where appropriate.

—Monitor from a distance with a sampling line.

The technician should decontaminate all equipment that is moved from a sampling site or from an **exclusion zone** (hot zone) of a hazardous materials incident. Contamination may be removed by physical or chemical methods or a combination of methods. Physical methods include scraping, brushing or wiping equipment and sample containers. Chemical methods include rinsing several times with a solution. Biological samples may be disinfected. Dusts and vapors that cling may be rinsed with water. If the equipment cannot be completely decontaminated in the field, then it should be placed in plastic bags for later decontamination procedures. The technician should place all decontamination fluids, wipes and debris into containers for proper hazardous waste disposal under applicable local and federal regulations.

Safety Procedures

The protective requirements of OSHA and EPA's HAZWOPER standard (29 CFR 1910.120) and OSHA's Permit-Required Confined Spaces standard (29 CFR 1910.146) are relevant to air sampling and monitoring activities. The technician and his or her employer are responsible for knowing and complying with the requirements of these and all other pertinent OSHA general industry standards. The technician is also responsible for complying with the procedures written in the Site Safety Plan developed for a particular sampling operation.

Among the safety procedures the technician should follow are:

—Be certain of the proper functioning of monitoring and sampling devices through maintenance and calibration before and after each monitoring event.

—Select the appropriate level of PPE.

—Be aware of the onset of an explosive or IDLH atmosphere and be ready to leave the area if necessary.

—Take precautions if the site presents accessibility problems, such as uneven terrain or water hazards.

—Use the buddy system as required for Confined Space and Exclusion Zone work.

—Do not eat, drink or smoke on the site or before going through decontamination.

—Do not use matches or lighters onsite.

—Follow SOPs for bringing in equipment that may cause a spark (back hoes, trucks, etc.).

—Use intrinsically safe, explosion-proof electrical devices with audible and visible alarms.

—Use nonsparking tools.

—Be cautious around compressed gas cylinders. Look for signs of stress such as a change in appearance of part of its metal; a bulging appearance; a visible, unplanned release of material; or a whistling or groaning sound.

—If opening a tank or vault in which volatile substances are stored, first vent excess pressure.

—Do not perform **hot work** in hazardous atmospheres. (Hot work produces heat or a static electricity ignition source; for example, burning, welding, riveting, drilling, abrasive blasting and space heating.)

—Obey lockout/tagout procedures.

Checking Your Understanding

1. Describe the equipment and calibration techniques used in sampling air.
2. Discuss sampling techniques used to obtain air samples.
3. Indicate the selection process for determining containers, preservatives and holding times for air samples.
4. Specify the care and cleaning necessary to maintain the equipment used in sampling air, including decontamination techniques.
5. Describe safety precautions recommended in sampling air.

Summary

OSHA and EPA require monitoring programs for protection of workers and the environment. Air monitoring may be achieved in real-time with electronic instruments taken to,, or installed at, a site to be assessed. Air sampling usually consists of actively drawing the air contaminants by pump into collection media – such as filters, sorbents or liquid – or into containers such as gas bags or canisters. The air samples are then analyzed by a laboratory. Air is typically sampled for aerosols and particulates, for organic and inorganic vapors and gases, and for combinations of these.

Compared to real-time field monitoring, laboratory analysis is more accurate, but takes longer to yield data about samples, and it is expensive. The results of laboratory analysis are meaningful only when onsite personnel have taken representative samples, avoided breakthrough, and have properly followed quality assurance procedures.

Before site monitoring or sampling begins, the technician is responsible for studying and subsequently following the Sampling Plan and the Health and Safety Plan. The site Sampling Plan specifies what type of handling and storage is necessary for each type of sample. The project planner selects collection procedures that follow the methods described by agencies and professional organizations: EPA, OSHA, NIOSH and ASTM.

Uncontrolled waste sites may require that several types of air monitoring be performed prior to technicians beginning site characterization. The types of monitoring that may be required include:

—Oxygen Monitoring

—Combustible Gas Monitoring

—Toxic Gas/Vapor Monitoring

—Radiation Monitoring

—Aerosol Monitoring

To be useful, air monitoring instruments must be portable and rugged; easy to operate; able to generate reliable and useful results; and inherently safe. Sampling and monitoring equipment must be subject to a program of preventive maintenance as well as procedures to prevent or minimize contamination during use.

Application and Critical Thinking Activities

1. In the Code of Federal Regulations, locate and study the OSHA citations discussed in this chapter.
2. Study a weather measurement instrument introduced in this section, and prepare a training session or written guide for the other students. Include the instrument's operation and the interpretation of its measurements. Include English or metric conversions of the measurements as appropriate.
3. Browse through the operating manual for a direct-reading air monitoring instrument to find meteorologically related instructions. Report findings to the class.
4. Interview emergency response practitioners about their experiences with meteorological effects upon air contaminant plumes.
5. Given a selection of monitoring equipment catalogs and other promotional literature from safety equipment vendors, compare the costs and features of various direct reading instruments. Arrange with a vendor the demonstration of direct reading instruments.
6. Given chemical references, find the flammable range (LEL and UEL) for substances of interest to you. Suppose one of your selected substances was accidentally released from a bulk storage container. Suppose you were monitoring for one of the substances. How would you interpret a reading of five percent LEL? of 10 percent LEL? of 50 percent LEL? of 100 percent LEL?
7. Using available direct reading instruments, try some simple monitoring situations to test the instruments' responses:
   - —Breathe into the sensor of an oxygen meter. (Note: High concentrations of carbon dioxide may shorten the useful life of an oxygen sensor.)
   - —Monitor the atmosphere around a butane lighter with a CGI.
   - —With a total toxics detector that has a carbon monoxide sensor, monitor the atmosphere around an idling car. (Caution: Avoid direct human exposure to the exhaust.)
   - —With a colorimetric indicator tube for ammonia, determine the contamination level indicated when the tube is exposed to a container of ammonia. (Caution: Avoid direct human exposure to the vapor.)
8. Create a concrete example of the concepts of ppm or mg/ m^3. For example, 1 ppm equals $0.01 in $10,000. Check chemical references to determine the exposure limits of several substances of interest to you. Use your concrete example to visualize and to emphasize the seemingly small concentration of each substance that is potentially hazardous.
9. Arrange to watch a technician analyze a substance by gas chromatography. Observe the GC's operation and study the resulting graphical data as the technician interprets it for you.
10. Arrange for demonstration of sampling pump calibration by soap bubble meter, by either the burette and beaker apparatus or by an electronic device.
11. Arrange to inspect the assemblage of a filter cassette type of sampler. Properly place the pump and filter on your clothing.
12. Given a selection of safety catalogs determine the cost of an order for the following air sampling and monitoring equipment:
    - —One package of gas sample bags (capable of holding about one liter of sample).
    - —Ten direct reading personal dosimeters for detecting gamma and x-ray radiation.
    - —Ten film badges for detecting gamma and x-ray radiation.
    - —Ten badges for detecting organic vapors, preferably with a backup for breakthrough.
    - —Ten direct reading personal dosimeter detector tubes for carbon monoxide.
    - —One personal air sampling pump and filter cassette.
    - —One detector tube complete kit, including hand pump and various tubes.
    - —One CGI calibration kit.
    - —One electronic soap bubble calibrator for personal sampling pumps.
    - —One direct reading four-gas instrument for detecting combustible gas/vapor, oxygen and two toxic gases (e.g., carbon monoxide and hydrogen sulfide).

5

Sampling Soils and Sediments

Chapter Objectives

Upon completing this chapter, the student will be able to:

1. **Explain** systems for describing soil, sediment and other geologic materials.
2. **Summarize** procedures for making field observations and taking measurements that must accompany soil and sediment samples.
3. **Describe** the equipment used to sample soil and sediments.
4. **Discuss** sample collection techniques specific to soil and sediment.
5. **Specify** the care and cleaning necessary to maintain the equipment used in sampling soil and sediment, including decontamination techniques.
6. **Describe** the process for selecting containers, preservatives and holding times for soil and sediment samples.
7. **Describe** safety precautions recommended in sampling soil and sediment.

Chapter Sections

5-1 Introduction

Concepts

- The components of soil are minerals, organic matter, water, and air.
- Soils are classified according to grain size.
- Soil, sediment, and rock sampling activities must avoid creating a path for contaminant migration.

Soil and groundwater are connected, and considering one without the other is difficult. Contaminants migrating through soils, sediments and rocks may affect groundwater or surface water, both of which are vital sources of drinking water. Furthermore, as we shall see in Chapter 6, surface water may have hydrologic connections to groundwater. Thus soils and sediments are important for the protection of our groundwater supply. Contaminated soils and sediment may also affect the environment and human health if contaminants are allowed to enter the food chain through absorption by plants, insects, fish or other animals.

To assess soil, sediment or other geologic materials, a technician should be able to support the engineer, site manager or other environmental professional in two site characterization strategies: 1) describing soils and sediments in a consistent manner in order to characterize and document site conditions, and 2) sampling the soils and sediments with appropriate equipment and techniques so that samples are valid and adequate for determining the extent of contamination present at a site.

This chapter will introduce systems for describing soils and sediments as well as the equipment, procedures and safety precautions for sampling them.

Figure 5-1: Sequence of tilted strata along Interstate 70, Jefferson County, Colorado.

Regulatory Authority for the Sampling of Soil, Sediment and Other Geologic Material

EPA is concerned with protecting the environment and human health. Any of the federal regulations that address groundwater contamination also cover contamination of geologic materials. When EPA requests an investigation to determine the extent of contamination, it is in fact requiring the sampling of soils, sediments, surface water and groundwater. Additionally, if contamination is suspected to have entered an area of rock, then sampling of the rock materials may also be required. Defining the extent of contamination means that its horizontal and vertical extent must be determined. The most prominent regulatory programs that require these investigations are the Resource Conservation and Recovery Act (RCRA); the Comprehensive Environmental Response, Compensation, and Liability Act (CERCLA); and the Superfund Amendments and Reauthorization Act (SARA). Included within RCRA and SARA is the federal Underground Storage Tank program, which may require sampling of soils, sediments and rocks in the investigation of a release. Other state regulatory programs may also require such sampling.

Defining Soil and Sediment

Effective and thorough sample collection of soil, sediment or geologic materials is most successful when the individual taking and describing the samples has knowledge of and experience with these types of materials. Although the description and study of soils, sediments and rocks is a vast area of study, basic skills for the entry-level environmental technician should include the ability to communicate with an environmental professional about soil concepts and carry out procedures as directed. The more experienced technician should then become proficient and more independent in applying the standards such as those published by the American Society for Testing and Materials (ASTM) and introduced in this chapter.

For clarity and consistency, each sampling plan should define the terms "soil" and "sediment," since their definition depends largely on one's perspective. In agriculture or biology soil is the material that supports plant growth and microbial activity. An engineer or architect, on the other hand, might consider soil as any material that can be excavated; his or her interest might be more related to the strength and stability of the soil for building purposes than to anything else. Geologists would look at soil from yet a different perspective, taking into consideration its origin, composition and development. In the environmental industry soil may be regarded to have all of these meanings.

Also, depending on one's perspective, sediment may be included in the category of soil. This is especially true if soil is considered to be any material that is present above rock (also called **bedrock**). Many environmental investigations regard soil and sediment to be equivalent terms. Yet other investigations might use the two terms as having distinct meanings; for example, soil could be considered as the material that supports vegetative growth, while sediment would be any material beneath the first few feet, after root systems have disappeared from the profile. Sediment also might be regarded as the **unconsolidated material** present in streams, creeks and rivers.

Chapter 3 distinguished sediment from soil to introduce the differing equipment used for sampling beneath flowing or standing water bodies. In this chapter, however, soil and sediment will be considered equivalent in the sense that the classification systems described here shall apply to both soil and sediment.

In general, we may describe a soil by its components. It is made up of particles of organic matter, minerals, water and air. The organic matter, known as **litter**, is composed of decaying plant materials such as twigs, stems and leaves. When litter decomposes into a residue it is known as **humus**. The mineral material derives mainly from weathered rocks. The mineral material is described – and the soil classified – by the size of its grains or particles. Several classification systems are in use in the environmental industry. Three of the most common ones are:

—The Unified Soil Classification System (USCS), which is described in the American Society for Testing and Materials (ASTM) standards D-2487 and D-2488.

—The American Association of State Highway and Transportation Officials (AASHTO) system (M 145-87).

—The U.S. Department of Agriculture, Soil Conservation Service (ASCS) system.

Box 5-1 ■ Soil Profile

The Earth's surface is composed of a series of layers: soil, **subsoil** and bedrock.

—Soil is the loose material that covers much of the Earth's land surface.

—Subsoil is the material beneath the soil but above bedrock. The subsoil is a shattered or weathered bedrock material, which has not broken down into soil. (**Weathering** is the physical and chemical process that breaks a rock down into soil material.)

—Bedrock is the solid rock material that underlies the soil and subsoil.

A vertical section through a soil, subsoil and bedrock sequence is referred to as a soil profile. Soil scientists refer to the layers as **horizons**. Soil comprises the A and B horizons, subsoil is the C horizon, and bedrock is the D horizon. Soil profiles vary depending on climate because climate influences rainfall and vegetation.

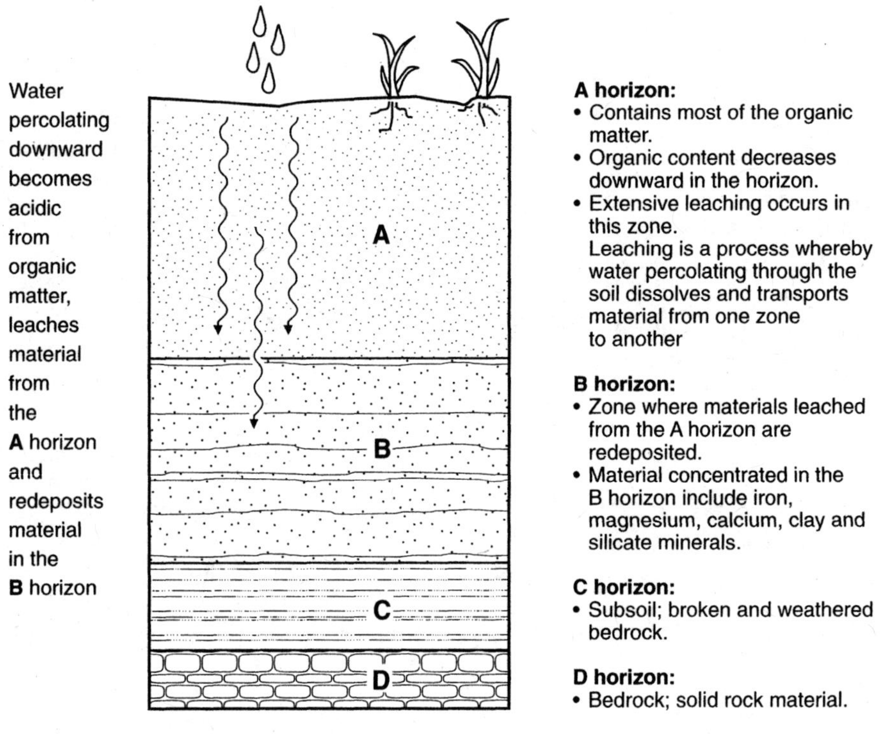

All three of these systems consider the grain size distribution in the soil material. The percentage of each grain size in a particular soil is referred to as its **texture**.

The ASTM Unified Soil Classification System (USCS)

The USCS is a common soil description system in use in the environmental industry. In this system, grains are described as gravel, sand, silt or clay and refer only to grain size without any reference to the grain's mineral composition. The system classifies soils into either coarse or fine grained.

For the fine-grained components of the soils, the USCS introduces the concept of **plasticity** of a soil. Material is considered to have high plasticity if it can be kneaded between the fingers and palms of the hand and then readily rolled into a "worm" –

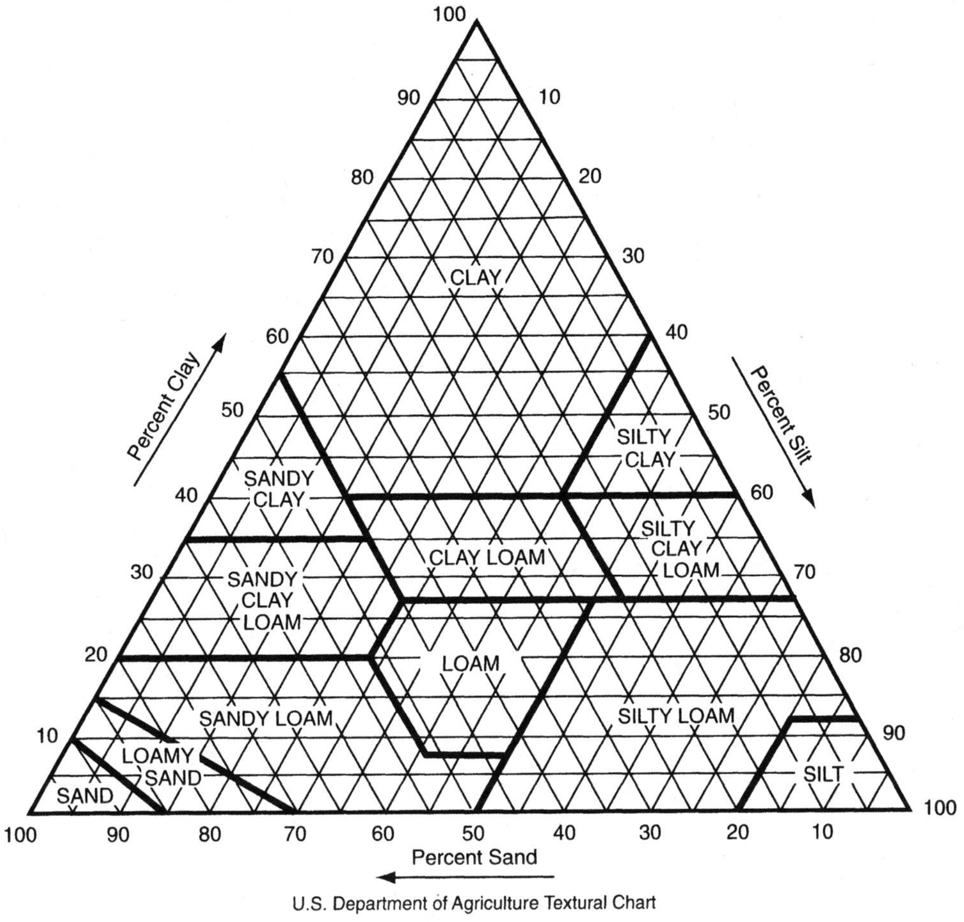

Figure 5-2: Ternary diagram for USDA, Soil Conservation Service Soil Classification System.

that is, a long cylinder shape – which can be bent without crumbling. Conversely, material that crumbles and cannot be rolled or molded into a worm shape is regarded as having low plasticity. Plasticity depends on the water content of a soil which allows the soil to be molded. When fine-grained soil material has low water content it will be more rigid or crumbly. The soil's **liquid limit** is the point at which soil goes from a plastic state to a semiliquid state.

A soil is considered to be coarse-grained if 50 percent or more of it is composed of the coarser materials, such as sand and gravel. With coarse-grained soils, the concept of grading is introduced. If the soil is composed of a range of particle sizes from coarse to fine, including the intermediate sizes, it is called **well-graded**. If a soil does not present a variety of particle sizes and is composed mostly of one or two different sizes of grains, then the soil is **poorly-graded**.

Nonplastic	Cannot be shaped into a worm.
Slightly Plastic	Weak worm shape that crumbles or is easily broken. Also sometimes called low plasticity.
Plastic	Worm shape that resists breaking during moderate handling. Sometimes referred to as medium plasticity.
Very Plastic	Worm shape that may easily be bent and otherwise manipulated before breaking. May also be referred to as high plasticity.

Table 5-1: Plasticity description.

The AASHTO System

The American Association of State Highway and Transportation Officials (AASHTO) developed a soil classification system (AASHTO Designation M 145-87) to standardize the description of soil materials in the design and construction of highways. This system uses a No. 200 sieve to separate soil into two general classifications: Granular Materials, such as stone fragments, gravel and sand, or Silt-Clay Materials, which are soils with a high content of silt and/or clay. AASHTO generally rates a soil made up of granular materials as excellent to good for use as the subgrade under-road construction. Soil analyzed as a silt-clay material is generally rated as fair to poor for subgrade use.

The USDA Soil Classification System (SCS)

The USDA developed a soil classification system based on texture of materials. In this system, percentages of sand, silt and clay-sized grains in a soil are plotted on a **ternary diagram**, which is a three-sided graph (see Figure 5-2). The plotted percentages result in a name for the soil type.

In comparing classification systems, the technician will discover that the sizes designated for gravel, sand, silt and clay particles are not consistent. Some environmental professionals prefer to follow the U.S. Department of Agriculture's definitions of these grain sizes (see Table 5-2). For a more detailed classification system, see ASTM's designation of particle sizes in Table 5-4 later in the chapter.

Gravel	76.2	-	2.0 mm
Sand	2.0	-	0.05 mm
Silt	0.05	-	0.002 mm
Clay	less than 0.002 mm		

Table 5-2: U.S. Department of Agriculture's Textural Soil Classification of Particle Sizes.

Considerations in Characterizing Soils

Above and beyond identifying the soil and/or sediment type by using a soil classification system, the technician also describes the characteristics of the samples collected. For example, when describing a soil's texture and plasticity, the technician should also report its organic litter/humus content and its color. Any other observations about the soil should also be recorded in field notes: comments about odor, staining or discolorations, the presence of construction debris. Construction debris includes bricks, concrete, cinders, nails and so on. It is common for environmental field investigations to be located in areas where industrial or commercial

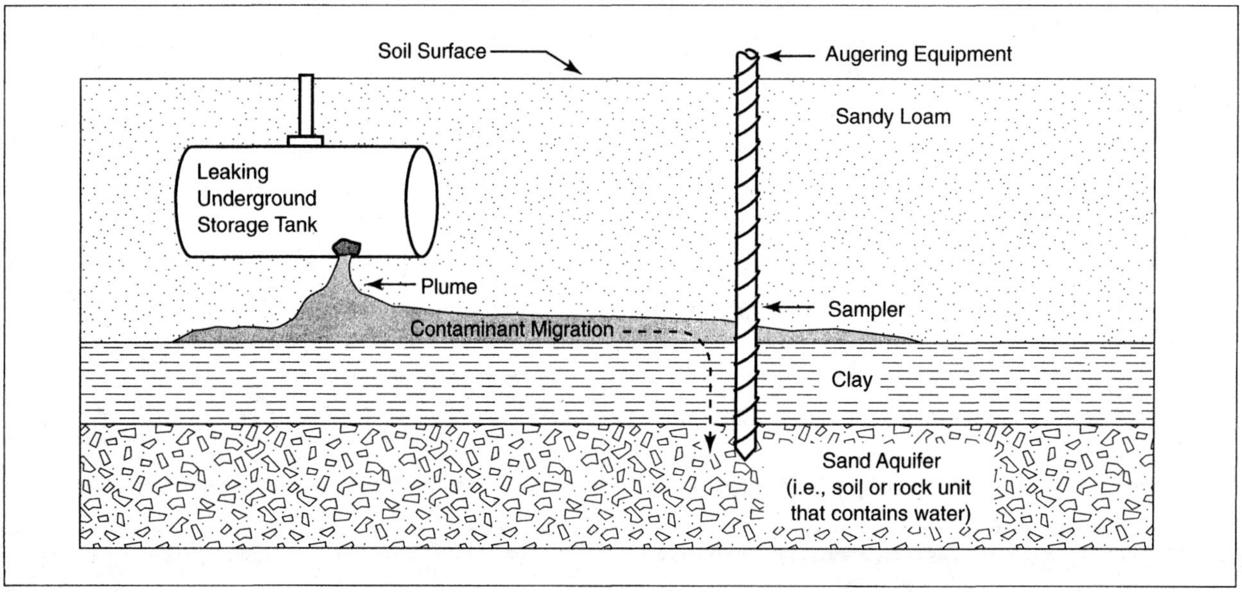

Figure 5-3: A sampling hole through a clay layer may create a conduit for a petroleum product to enter the groundwater.

Box 5-2 ■ Soil Porosity

The ratio of openings (voids) to the total volume of a soil, sediment, or rock is referred to as its porosity. Porosity is expressed as a percentage of void space in a material. When void spaces interconnect, they permit fluids to move through the material. This ability of the material to transmit fluids is called permeability.

Soil – which is typically a mixture of gravel, sand, silt and clay – tends to be very porous (typically 55%), because particles tend to clump together. However, a porous soil, sediment, or rock may not be permeable, because the voids might not interconnect in such a way that a fluid can flow through. For example, clays are porous (40% - 70%) but not permeable. A clay can absorb liquid but cannot readily transmit liquid.

Roots, worm-holes and animal burrows add to the porosity and permeability of soil.

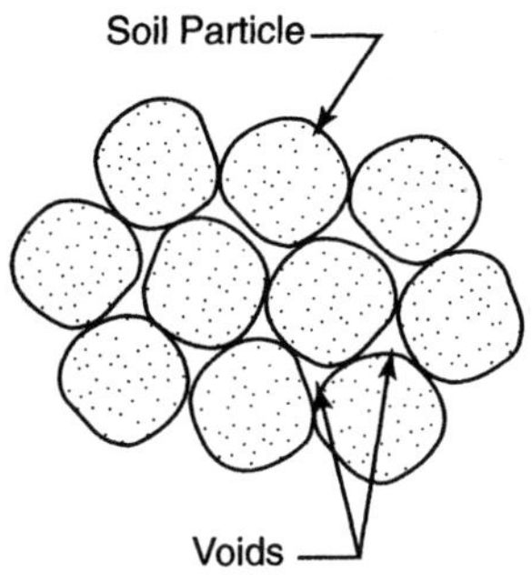

Material	% Porosity
Equal-size spheres (e.g., marbles)	
Loosely packed	48%
Tightly packed	26%
Soils/sediments	
Clay	40-70%
Silt	35-50%
Sand	25-50%
Gravel	25-40%
Rocks	
Limestone	0-20%
Sandstone	5-30%
Shale	0-10%

activity has been present for several decades, so construction debris mixed with soil or sediment is not necessarily an unusual finding.

Some soils are naturally porous and allow migration of fluids. These types of soils include sandy or gravelly materials. Others, such as clays or silty clays, may not allow fluids to move through readily. These physical characteristics are important in terms of the migration of the contaminants through them. When sampling soils during a field investigation, the technician must consider the consequences of sampling through a clay layer, thereby creating a pathway for the migration of the contaminants deeper into the soil profile (see Figure 5-3).

Subsurface Geology: Bedrock

If the sampling plan requires the collection of rock units, special rock-coring equipment may be necessary. The description of the rock types being sampled is likely to be assigned to an individual specially trained in **petrology**, which is the systematic classification of rocks.

The characteristics of some rock types may favor the migration of fluids and therefore of contaminants. The sampling of these rock units is important in defining the extent of contamination. Other rock types are relatively **impermeable** (incapable of transmitting fluids); they stop the migration of contaminants, rather than favoring it. If an impermeable rock type is encountered, it may not be wise to continue sampling through that rock unit, not only because it would not help the investigative process, but also because the technician might inadvertently

Porous and permeable rock types	Generally impermeable rock types if unfractured
Sandstone	Shale
Limestone	Slate
Dolostone	Granite
Gypsum	Gneiss
Conglomerate	Basalt
Any type of rock that has been fractured	

Table 5-3: Rock permeability.

provide a **conduit** or path for the migration of contaminants into deeper, previously uncontaminated sources of groundwater. Caution is recommended whenever the technician is tracking a contamination plume, which is a body of contamination originating from a specific source.

The **porosity** of a rock is the ratio of its volume of empty or void spaces to its total volume. The **permeability** of rock is the ease with which fluids can pass through. Table 5-3 lists some rock types that are generally known for their porosity and permeability. The greater the porosity and permeability of a material, the more fluid can be transmitted through it. Some geologic situations, such as a faulted or fractured zones in the earth's crust, may increase a rock's permeability by several times. **Faults** are ruptures along which movement has occurred, and **fractures** are cracked or broken zones in the earth's crust.

Checking Your Understanding

1. Explain how sampling may provide a conduit for contaminants.
2. Describe the components of soil.
3. Define the terms organic content, plasticity, liquid limit, porous and impermeable.
4. Name the three common soil classification systems.

5-2 Describing Soils and Sediments

Concepts

- Field notes describe soil by such criteria as color, particle size data, particle shape, moisture, consistency, structure, odor and reaction to hydrochloric acid.
- The technician determines if fine-grained soil is clay or silt and if coarse-grained soil is sand or gravel.
- The SCS ternary diagram names soil types according to the relative percentages of sand, silt and clay.

The choice of classification scheme to be used for descriptions should be agreed upon before fieldwork commences. The selection will depend upon the anticipated use of the data. In this section we will discuss and rely mostly on the **Visual-Manual Method** for the Unified Soil Classification System (ASTM Standards D2487 and D2488) and the SCS ternary diagram system.

A special personal protection note: For health and safety reasons, the technician wears protective gloves during the descriptive process to protect against any chemicals that may be present in the soil or sediment. Although thin gloves may be preferable because it is easier to handle and feel the material with them, they might not be sufficiently safe depending on the level of contaminant encountered. Also, gloves should be changed often to minimize both cross-contamination of samples and the possibility of the gloves failing (known as breakthrough), which would allow skin contact with the contaminants. Additional personal protection procedures are listed at the end of this chapter.

Describing Soil by ASTM's USCS Visual-Manual Method

The ASTM standards provide two methods for describing soils: the Laboratory Method, based on putting soil through a sieve device and on running lab tests on plasticity (see Figure 5-4); and the Visual-Manual Method, based on observations using the senses. The paragraphs below introduce observations in the Visual-Manual Method. (Note: For convenience, the rest of the chapter will refer only to soil rather than to soil/sediment). The technician should enter soil descriptions into well-organized field notes applying the criteria listed below.

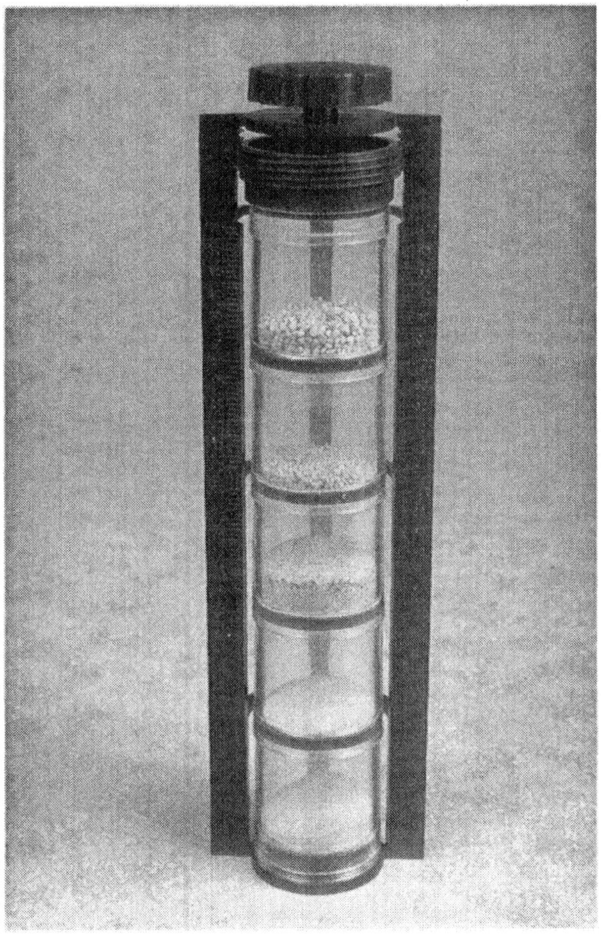

Figure 5-4: The ASTM Laboratory Method is based in part on a sieve test. The sieves, which range in sizes of screen openings, are stacked on a shaker; the top sieve has the largest screen openings, the bottom sieve has the smallest openings. The laboratory procedure involves weighing the sample material before the test and weighing the material retained on each sieve after the test.

Describing Color – Color can be an important element of soil description. Dark brown or black, for example, are associated with organic content. Contaminants may also discolor a sample; soils contaminated with petroleum have an interesting gray-green hue. Discoloration may be useful in tracking a contamination plume through a soil or sediment. Because determining color is subjective, technicians should refer to a color reference chart, such as the one available on the **Geotechnical Gauge** (see Figure 5-5) or the **Munsell Soil Color Chart**. After all, "buff-colored" may mean different things to different people!

Describing Range and Percentages of Particle Sizes – The technician describes the range and percentage of particle sizes observed in the sample. For example, the technician may state in the field notes, "20 percent gravel, 50 percent fine to coarse sand, and 30 percent silt and clay."

Describing Maximum Particle Size – The technician notes the largest grain or particle size observed in the sample. Sand is indicated as fine, medium or coarse grained. If the particle is larger than sand-sized grains, then its size is noted by a measurement, such as "two-centimeter diameter gravel" or "boulders with a maximum dimension 41 centimeters long."

Particle	Size Range (in diameter)	Sieve Size Retaining Particle
Boulders	Greater than 300 mm (12 inches)	Not applicable
Cobbles	75 - 300 mm (3 - 12 inches)	"3-inch"
Coarse Gravel	19 - 75 mm (0.75 - 3 inches)	"3/4-inch"
Fine Gravel	4.75 - 19 mm	No.4
Coarse Sand	2.0 - 4.75 mm	No. 10
Medium Sand	425 µm - 2.0 mm	No.40
Fine Sand	75 - 425 µm	No. 200
Silt	Less than 75 µm*	
Clay	Less than 75 µm*	

* Sieves with openings small enough to separate the smallest particles (silt and clay) are not easily manufactured and are not practical.

Table 5-4: Particle sizes.

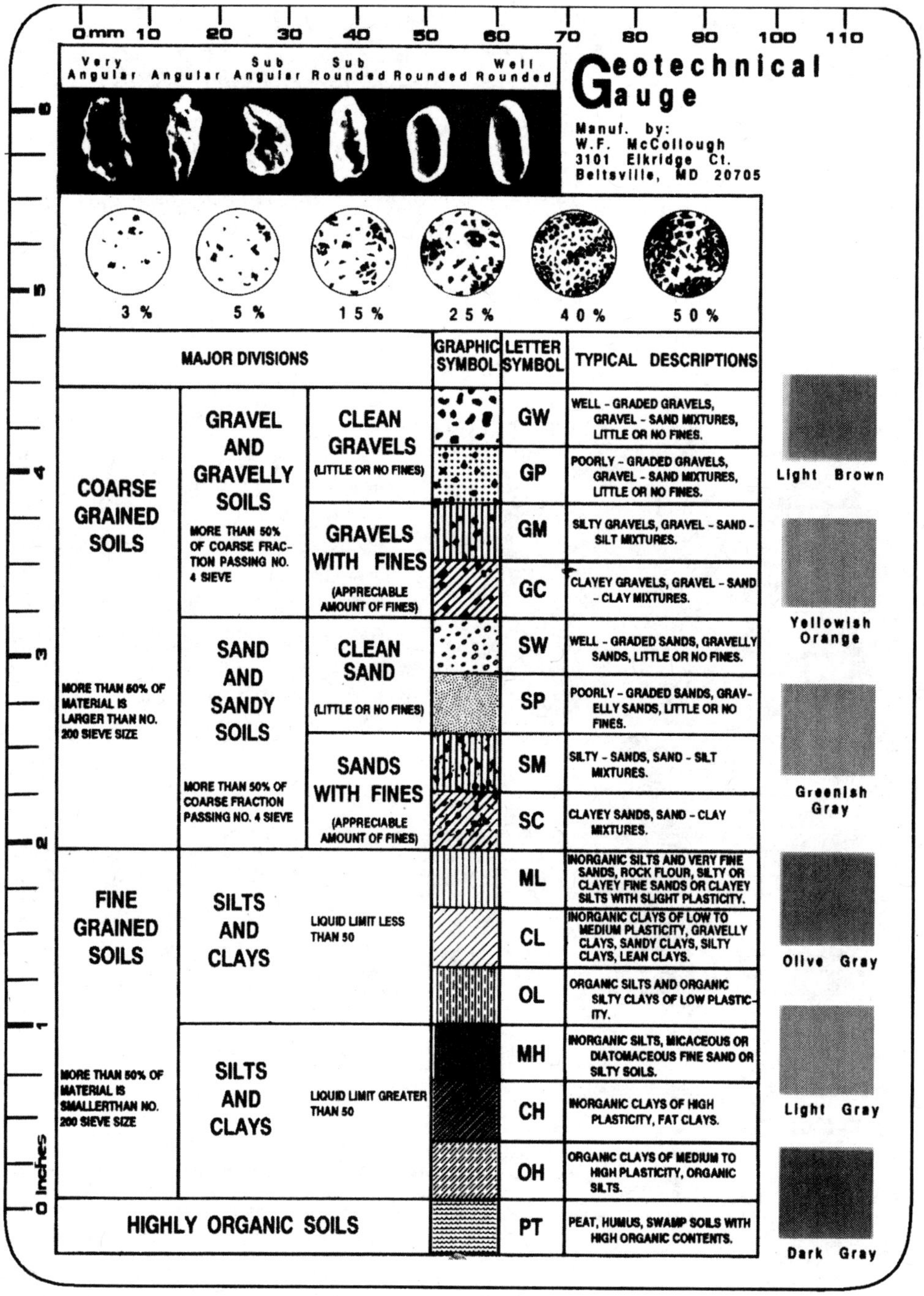

Figure 5-5: The Geotechnical Gauge is an excellent visual aid for field work. Among the reference tools included on the gauge are: soil color chips, a particle size classification chart, a particle percentage estimation chart and a particle shape illustration. The back side (not shown here) includes USCS charts on soil texture. (Reprinted with permission of W.F. McCollough, Beltsville, MD)

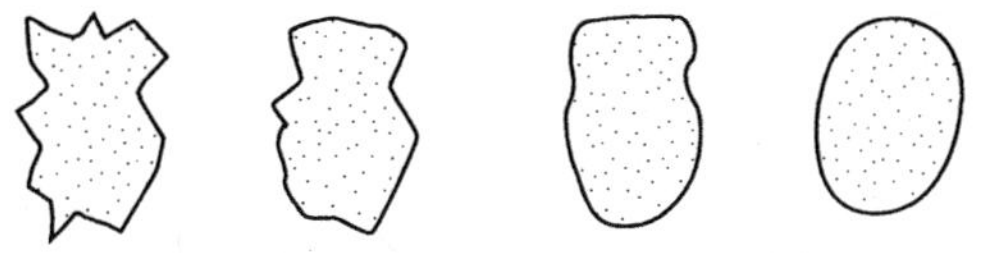

Figure 5-6: Angularity of particles.

Describing Angularity – The surfaces of particles may be quite rounded and smooth; they may have sharp or blunt angles. A range of angularity may even be present among the larger grains of a sample. The technician states the range of angularity in field notes as, for example, "angular to subangular."

Describing Shape – The shape of the grains that make up a sample might be described with terms such as "elongated," "flat," "round" or "prism-like."

Describing Moisture – The moisture condition of the soil might be described as "dry" (dry to the touch or dusty), "moist" (damp, but no visible water), or "wet" (visible water).

Describing Consistency – For fine-grained soils, the consistency of the soil might be described as very soft, soft, firm, hard or very hard.

Describing Structure – If the technician is able to collect an intact, relatively undisturbed sample, he or she may be able to observe and describe the sample's structure. Table 5-6 lists possible structures and their descriptions.

Consistency	Description
Very Hard	Thumb or thumbnail will not indent soil.
Hard	Thumb will not indent soil, but thumbnail will.
Firm	Thumb will indent soil 6 mm (1/4 inch) or less.
Soft	Thumb will indent soil about 25 mm (1 inch).
Very Soft	Thumb will penetrate soil more than 25 mm (1 inch).

Table 5-5: Criteria for describing consistency in soil/ sediments.

Describing Odor – Generally, the technician describes a sample's odor only if it is noticeable without deliberately smelling the sample. Because of the threat of exposing the lungs and other organs to a potentially toxic substance, the technician should definitely not sniff a material. Therefore, monitoring devices rather than noses should screen samples for contamination.

Describing Depth – The technician records the depth at which each sample is collected. Soils change with depth, and these changes may affect the movement of contaminants through the subsurface. Neglecting to record the depth from which a sample has been obtained leaves a particularly serious gap in the data, making it difficult for data users to track the contaminant's movement.

Describing HCl Reaction – If dilute hydrochloric acid (HCl) is available, the technician may place a few drops on the sample and describe its chemical reaction to determine the presence of calcium or other carbonates. Carbonates react by fizzing or bubbling.

Miscellaneous Descriptions – Additional useful comments include observations about the presence of roots or root holes, difficulty in obtaining a sample by the sampling means employed, caving of the trench or hole where the sample is being collected, or any other notable item. It is better to enter any observations, however seemingly unimportant, in the logbook rather than rely on remembering it later. What may be considered unremarkable or common in the field during sampling could become important further on in the investigation.

Structure	Description
Homogeneous	Same appearance throughout.
Stratified	Alternating layers of varying material or color at least 6 mm thick.
Laminated	Alternating layers of varying material or color less than 6 mm thick.
Fissured	Breaks or fractures along planes.
Slickensided	Surface of fractured planes appear polished or glossy.
Blocky	Soil that can be broken into smaller lumps which resist further breakdown.
Lensed	Inclusion of small pockets (or lenses) of different materials.

Table 5-6: Criteria for describing structures in soil/ sediments.

Identifying and Describing the Fine-grained Components of Soil

Since fine-grained materials are very small, it is sometimes difficult to determine what is clay and what is silt. The following is an outline of how to describe fine-grained materials.

1. Separate out enough of the fine-grained material to be able to test it for plasticity. Try to roll a specimen of the soil between your hands into a long worm shape about 1/8-inch in diameter. The sample's plasticity should be classified according to the list in Table 5-1.
2. Determine whether the soil is organic or inorganic (see Table 5-8). Organic soils generally have a dark brown to black color and an earthy odor. They also lighten in color when dried in the air. Organic soils often are nonplastic or slightly plastic. If the soil is organic, the next step is to determine if it is a silt (OL) or a clay (OH). This may be difficult to distinguish, but generally clays tend to stick to sampling equipment and are harder to wash off the equipment. Silts often will slip right off sampling equipment as the equipment is being washed after sampling. The designation OH/OL indicates that the soil could be either type of material or a mixture of these fine-grained materials. If the soil is inorganic, determine if it is lean clay (CL), fat clay (CH), silt (ML), or elastic silt (MH). See the identification criteria in Table 5-7.

Identifying and Describing the Coarse-grained Components of Soil

Coarse-grained soils have sand or gravel components. The sand or gravel might be combined with silt or clay. The following guidelines apply to naming coarse-grained soils:

—A sample is considered coarse-grained if it contains less than 50 percent fine-grained materials. It is considered gravel if the percentage of gravel is greater than the percentage of sand. It is a sand if the percentage of gravel is equal to or less than the percentage of sand.

—The sample is well-graded if it has a wide range of particle sizes and substantial amounts of the intermediate particle sizes. It is poorly graded if it consists mostly of one size (uniformly graded) or if it has a wide range of sizes with some of the intermediate grain sizes missing.

—It is a clean sand or a clean gravel if the percentage of fine-grained material is estimated to be five percent or less.

—The sample is a gravel with fines or a sand with fines if the percentage of fine-grained material is at least 12 percent, but not more than 50 percent. It is a clayey gravel (GC),or a clayey sand (SC), if the fines are clays. It is a silty gravel (GM) or a silty sand (SM) if the fines are silts (see Table 5-8).

—A sample that contains between five and 12 percent fine-grained materials is identified using two of the group symbols. The first group symbol represents a clean sand or clean gravel (SW, SP, GW, GP), and the second symbol represents a sand or gravel with fines (SC, SM, GC, GM).

Soil Name	Soil Symbol	Plasticity	Other
Silt	ML	Nonplastic or slightly plastic	Soil washes easily off equipment.
Elastic Silt	MH	Slightly plastic to plastic	Material is somewhat resistant to being washed off sampling equipment.
Lean Clay	CL	Plastic (medium plasticity)	Material sticks to sampling equipment; hard to wash off.
Fat Clay	CH	Plastic to very plastic (medium to high plasticity)	Soil material sticks to sampling equipment; hard to wash off, must be scraped off.

Table 5-7: Identification of inorganic fine-grained soils.

group symbol plus the words "with clay" or "with silt" to indicate the fines. For example: "poorly-graded sand with silt, SP-SM."

The group name will thus correspond to the first group symbol plus the words "with clay" or "with silt" to indicate the fines. For example: "poorly-graded sand with silt, SP-SM."

—If a material is mostly sand or gravel and contains at least 12 percent but not more than 50 percent of the other coarse-grained constituent, the words "with gravel" or "with sand" need to be added to the group name. For example: "poorly-graded gravel with sand, GP."

Table 5-8 is a soil classification chart, which is adapted from ASTM Standard D 2487. It organizes the criteria for identifying fine-grained and coarse-grained soils.

Identifying Soil Type with the USDA SCS Ternary Diagram

An alternative method of classifying soil is the one developed by the United States Agriculture Department's Soil Conservation Service system. This system employs a ternary diagram. From estimates of the percentages of sand, silt and clay in a sample, the technician can determine the category to which the soil sample belongs by plotting the percentages on the diagram.

A technician may determine, for example, that a soil sample is 10 percent clay, 30 percent silt and 60 percent sand. By plotting these percentages on the ternary diagram the technician identifies the soil's texture as a "sandy loam."

Criteria for Assigning Group Symbols and Group Names Using Laboratory Tests			Soil Classification	
			Group Symbol	Group Name
COARSE-GRAINED SOILS More than 50% retained on No.200 sieve	Gravels More than 50% of coarse fraction retained on No. 4 sieve	Clean Gravels Less than 5% fines	GW	Well-graded gravel
			GP	Poorly graded gravel
		Gravel with Fines More than 12% fines	GM	Silty gravel
			GC	Clayey gravel
	Sands 50% or more of coarse fraction passes No. 4 sieve	Clean Sands Less than 5% fines	SW	Well-graded sand
			SP	Poorly graded sand
		Sands with Fines More than 12% fines	SM	Silty sand
			SC	Clayey sand
FINE-GRAINED SOILS 50% or more passes the No. 200 sieve	Silts and Clays Liquid limit less than 50	Inorganic	CL	Lean clay
			ML	Silt
		Organic	OL	Organic clay Organic silt
	Silts and Clays Liquid limit 50 or more	Inorganic	CH	Fat clay
			MH	Elastic silt
		Organic	OH	Organic clay Organic silt
HIGHLY ORGANIC SOILS	Primarily organic matter, dark in color and organic odor		PT	Peat

Table 5-8: Soil classification chart.

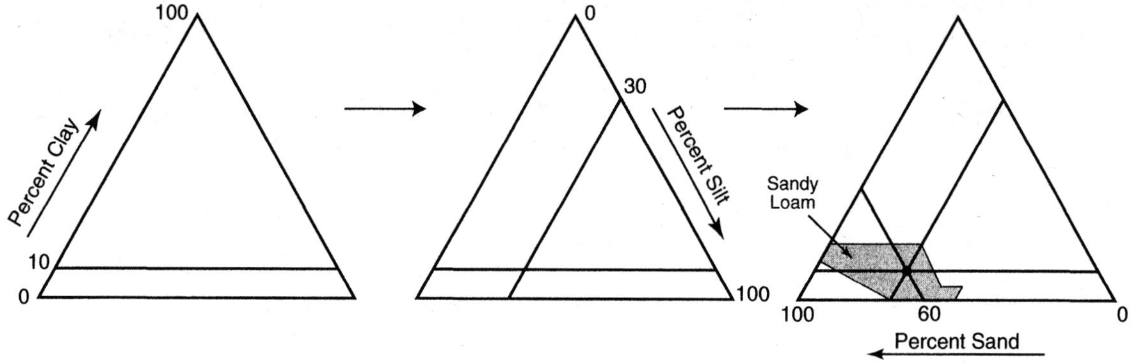

Figure 5-7: Three steps to classify soil by the SCS system. (See Figure 5-2 for complete ternary diagram.)

The ternary diagram is based only on the amounts of sand, silt and clay present in a sample, which means that any gravel or boulders are discounted during the plotting. Let's say, for example, that a soil is estimated to have 10 percent gravel, 65 percent sand, 20 percent silt, and 5 percent clay. If the 10 percent gravel is excluded from the analysis, the sample components add up to 90 percent rather than 100 percent. The percentages of sand, silt and clay can be mathematically corrected as follows to add up to 100 percent:

Each component's percentage is divided by 90 percent or 0.9:

Sand:	65 ÷ 0.9 =	72.2 %
Silt:	20 ÷ 0.9 =	22.2 %
Clay:	5 ÷ 0.9 =	5.6 %
		100.0 %

Now that the sand, silt and clay percentages add up to 100 percent, they can be plotted on the ternary diagram. According to the diagram, this soil fits the classification of a "sandy loam." Because of the 10 percent gravel, however, the soil may be called "sandy loam with some gravel." As the percentage of gravel increases, the soil may be called "sandy loam with gravel," or a "gravelly sandy loam."

Regardless of which soil classification system is selected, technicians may be under time restrictions that inhibit their ability to describe soils in a proper manner. Pressure is on them to get the job done as quickly as possible. Nevertheless, detailed collection of all possible data, including proper soil descriptions, is very important. Not entering observations may be a lost opportunity and may result in higher costs later on. With practice and experience the technician will become efficient and comfortable in describing soil and sediment samples within a given time frame.

Checking Your Understanding

1. List and describe at least eight criteria for describing soil.
2. Differentiate between clay and silt.
3. Differentiate between gravel and sand.
4. Differentiate between the terms "poorly sorted" and "well graded" as they apply to a gravel or sand.

5-3 Sampling Soils, Sediments and Other Geologic Materials

Concepts

- Advance arrangements include arranging for access to the site, contacting utility companies, consulting with the laboratory, and scheduling equipment and operators.
- Sampling contractors keep a truck or boxes fully stocked with equipment and supplies.
- Sampling methods depend on what the technician is trying to sample, sample depth, and whether the sample is grab or composite.
- The protective requirements of OSHA and EPA's Hazardous Waste Operations and Emergency Response standard apply to sampling activities.

Prior to leaving the office for the site, the technician takes care of advance arrangements and equipment preparation; this makes the job onsite easier and it reduces the potential for error.

Logistical Arrangements

The following arrangements are part of the preparation for collecting soil, sediment or geologic samples. Some of these items need to be done several days to a couple of weeks in advance of sampling. Each sampling outing and each site may have special needs or requirements not listed below.

—Call the site owner/operator to arrange for access to the site, to notify him or her when to expect the samplers, and to explain what the samplers plan to do while onsite. Although this is not necessarily the responsibility of the sampling technician (the project manager or project planner may be the appropriate person to make this call), it is to the technician's benefit to ensure this is done. Once onsite, the technician should check in at the appropriate office or other designated location and let the site contact know the samplers are there and what they are doing.

—Call to have the location of underground utilities identified on the site. Depending on local and state requirements, the utility call may need to be made at least 72 hours in advance. Don't call too far in advance since snowfalls and grass mowing can eliminate the markings. If unable to get to the site within the time-frame originally planned, call for the utilities to be located again. Many states have "one-call" utility-locating services, but not all utility companies may take advantage of this convenience. Take notes on which utility companies are contacted for underground utilities to be located. If a utility company is not to be contacted (e.g., water company or electric company), find out why that type of underground utility is not being located. When going to the site for sampling, take the utility locating list along. If evidence shows that a specific kind of underground utility was missed, get clarification from the utility companies involved. Also contact the site owner/operator to determine if they themselves have buried any lines on the site, and where those lines are located. Unlocated or improperly located utility lines may cause accidents or even death.

—Call the laboratory and alert staff to the probable date of sample arrival. This is the time to discuss any concerns about sample holding times and laboratory turn-around times. Many labs will provide sample containers and coolers for transport of the samples, so request these materials as well. Plan how to get the samples to the lab.

—Make arrangements for equipment and/or personnel to be available for the sampling event. This is particularly true if the sampling plan requires the use of a drill rig. In this case, arrangements must be made several weeks to several months in advance. Drill rigs usually have dedicated two-person crews. The rig and the crew are expensive and companies try to schedule them far in advance. Even if the services of a drill rig are not needed, other equipment or personnel may be unavailable if not reserved in advance.

—Make arrangements for travel and accommodations, if necessary.

—Read and understand the Site Safety Plan.

—Check equipment to make sure it is in good working order. Calibrate air monitoring equipment

Box 5-3 ■ General Equipment/Supply Needs

Each site and sampling outing may have special equipment or supply requirements. A few items, however, are likely to be needed on every project. A technician may think of a few other items indispensable in fieldwork, so this list is certainly not all-inclusive.

Plans/Procedures

— a copy of the site-specific safety plan
— a copy of the sampling plan
— Standard Operating Procedures for sampling

Documentation Supplies

— logbook and pen that will write on wet paper
— permanent, fine-tipped marker
— chain-of-custody forms and seals
— other forms
— microcassette recorder, tapes and batteries
— spare batteries for all equipment using batteries
— camera and film
— graph paper
— ruler (English and metric)
— conversion table(s)
— clipboard
— calculator

Clothing, PPE and Personal Supplies

— rain gear and warm clothing
— boots
— gloves (may need several different types for different chemicals encountered)
— Tyvek™ suits
— duct tape
— respirator with appropriate cartridges
— safety glasses/goggles
— hearing protection (earplugs, etc.)
— hard hat
— hat to shield from the sun
— sunscreen and insect repellent (note that these may contain organic materials that could contaminate samples)
— drinking water
— first aid kit
— business card (technician's business card or identification card as well as supervisor's/ project manager's card)
— money
— flashlight
— mobile phone and/or radio(s)
— area map(s)
— local phone book
— watch

Decon Supplies and Equipment

— plastic waste bags
— paper towels
— distilled and/or deionized water
— potable water (for decontamination if such water is unavailable onsite)
— nonphosphate detergent
— decon solutions/chemicals
— brushes for decon
— buckets for decon
— sprayers dedicated to decon solutions
— aluminum foil
— roll(s) of plastic sheeting
— 55-gallon drums (DOT-approved, as discussed in Chapter 7)

Sample Storage Supplies and Equipment

— sampling table (folding)
— plastic zip-top bags
— sample jars
— sample labels
— packing tape (clear)
— cooler
— ice or cooling source
— thermometer for cooler (max-min thermometer)

Sampling Equipment

— toolbox and tools
— bucket, bowl, spoon and knife (all stainless steel)
— shovel
— tape measurer and/or measuring wheel
— compass
— health and safety screening equipment (which might include combustible gas meter, oxygen meter, toxic gas meters, colorimetric tubes, photoionization detector, flame ionization detector, and/or a radiological screening instrument)
— AND all the site-specific sampling equipment!

Whew! What a list! Sometimes companies will simply dedicate a van or truck to sampling so the vehicle can be kept stocked and ready at all times. These supplied vehicles still need to be checked and restocked on a regular basis. If all supplies must be loaded for each sampling outing, large plastic chests with lids, or other big containers that can be packed and loaded easily, are recommended.

Specific Soil Sampling Equipment and Procedures

The choice of the sampling methods depends on what the technician is trying to sample (physical aspects of the material and the analyses to be performed); the sample depth (shallow vs. deep); and the type of sample (grab vs. composite).

Surface Soil Samples Collected with Spade, Shovel or Scoop

Soil samples that need to be collected from within the top foot of soil may be adequately obtained using a spade, a shovel or a scoop. The technician should not use a tool that has been plated as this may adversely affect some sample results. Rather, stainless steel tools are used for digging whenever possible. The ten steps listed below are general and are therefore used for collecting samples not only by spade, shovel or scoop but also by any type of soil sampling equipment introduced in this chapter:

General Steps

1. Take field notes in the field logbook. Strive for completeness in the notes. Include sample locations, soil descriptions, procedures used, health and safety screening data, conversations with site visitors or the client, site sketches, measurements, weather, etc. Notetaking is an ongoing process during sampling and other work done at the site.
2. Peel back the sod, if present. Save the sod for restoration of the hole after sampling is complete. Remove the top layer of soil or debris.
3. When collecting samples with a spade, shovel or scoop, use a clean, stainless steel tool to remove and set aside or discard a thin layer of soil from the area that came into contact with the tool used for digging. It is best to put this discarded soil on plastic sheeting (placed nearby for convenience) to avoid spreading around any potential contaminants and to make cleanup easier.
4. Pack sampling containers. If volatile organic analysis is to be performed, transfer a portion of the sample directly into an appropriate labeled sample container with a stainless steel spoon or clean sampling gloves. Pack the container to the top, making sure there is no headspace present in the jar. Tightly secure the top of the container. Immediately label the sample. For other types of grab samples place a portion of the soil into a clean stainless steel bowl or bucket and mix thoroughly to obtain a homogeneous sample. If the sample appears to be homogeneous, skip mixing the sample and transfer the sample directly into the sample containers. Place the sample into the proper containers with the appropriate chemical preservatives (if any) and secure the lids tightly. Label the containers before the sample is added or immediately after the sample container is filled and closed. If compositing, place samples from all sampling locations or intervals into a stainless steel container and mix thoroughly to homogenize the material. After mixing is complete, put an aliquot into the appropriate container and tightly secure the lid. Immediately label the container.
5. Complete the chain-of-custody form as samples are collected.
6. Put all samples into a cooler chilled by ice (or an alternative) to 4 degrees Celsius (4° C equals 39.2° F).
7. In many states, shallow holes can simply be backfilled with the removed soil material, eliminating the need for containerization and disposal of some of the soil. Follow the sampling plan's directions. (The planner will have checked state regulations to determine the proper handling of the soil.) Otherwise, place both the soil and the plastic sheeting in labeled drums. Seal the drums and situate them in a secured spot on the site. These investigation-derived wastes will need to wait for laboratory analytical results before they can be appropriately disposed of.
8. Clean and restore the sampling locations. Add clean topsoil or other acceptable material back into the holes to fill them to ground level.
9. Decontaminate equipment and place decontamination fluids in an appropriate container such as a lined 55-gallon drum. Label and seal drum before leaving site (see item 7 above).
10. Before leaving the site, map sampling points in field notes and review the field notes for completeness.

Soil or Sediment Collected with a Hand Auger

A hand auger is a manually operated stainless steel shaft with a T-shaped handle at the top connected to a turning, cutting device at the working end of the tool. Judging from its appearance, one may think the hand auger looks like some type of post-hole digging tool. Several types of hand augers are available depending on what type of material is to be sampled and what kind of sample is to be obtained.

An auger may be of the bucket type, which has a cylindrical receptacle at the bottom that gathers and holds the soil. Some bucket augers, such as a mud auger or sand auger, may be designed to work better with certain types of soils or sediments. Continuous flight hand augers are spiral-shaped and look like a corkscrew. The continuous flight auger can dig a hole down to a certain level at which a thin-wall tube sampler can be pushed down into the soil for the sample collection. An advantage of using the thin-walled tube sampler with the auger is that this arrangement minimizes the disturbance of the sample.

Figure 5-8: Different auger buckets are used for different types of soil.

Manual sampling with a hand auger is usually limited to depths of 20 feet or less. The depth at which the auger can effectively be used depends on the type of material as well as the physical capabilities of the people operating the device. In some materials, the hole will not stay open; sands and gravels will cave in. Encountering groundwater, or water present in the soil, can also cause holes to collapse. Gravels themselves are difficult, if not impossible, to sample with a manual auger. Other materials, such as construction debris, may also prevent the auger from penetrating to the depth desired. As one penetrates deeper into the ground, the physical labor required for sampling increases to the point where power equipment becomes necessary. In general, hand augers are effective sampling devices only where the soil is easily excavated and the resulting hole, also called a boring or borehole, will not cave in or collapse.

Steps to Review

1. Take field notes in the field logbook as described earlier in the General Steps.
2. Peel back the sod, if present. Save the sod for restoration of the hole after sampling is complete. Clear the area of surface debris and the first 3 to 6 inches of soil, if desired.
3. Assemble the clean auger device: attach T-handle and auger to the appropriate ends of the shaft.
4. Plan to use clean augers and tube samplers when collecting samples between holes or at different depth intervals in the same hole.
5. Begin augering by applying downward pressure on the auger and rotating the auger in a circle. Periodically remove the auger and deposit whatever soil has accumulated from the augering activity onto plastic sheeting spread on the ground near the hole. The plastic sheet prevents contaminated soil from contacting the surrounding ground. The sheeting also makes cleanup easier. If necessary, attach more shaft extensions to the auger as the hole gets deeper.

6. Upon reaching the desired sampling depth, carefully remove the auger from the hole.
   —When sampling directly from the auger (i.e., from the bucket of the bucket auger or from the flight of the continuous flight auger), collect the sample from the auger and handle in the appropriate fashion.
   —When using a thin-walled tube sampler, remove the auger from the end of the shaft and replace it with a clean, thin-walled tube sampler. Carefully lower the tube sampler into the borehole, gradually forcing the tube sampler into the soil. Avoid scraping the borehole sides or hammering/vibrating the sampler and shaft, which could cause the boring walls to collapse on top of the sampler device. Remove the tube sampler from the borehole and detach the tube sampler from the shaft. Remove the soil core from the sampler. Discard the top inch or so of the core sample, as this is likely to be material that was knocked to the bottom of the hole from the walls of the borehole.

Figure 5-9: Slide hammer.

7. Pack sampling containers as described earlier in the General Steps.
8. Complete chain-of-custody form as samples are collected.
9. Put all samples into the cooler.
10. If allowed by state regulations, backfill shallow holes as described in the General Steps.
11. Clean and restore the sampling locations as described in the General Steps. Depending on the depth of the hole and if groundwater is encountered, add some bentonite (a type of clay) back into the hole to seal it and prevent contamination from traveling down the hole to groundwater. Check state requirements for "hole abandonment" to know what is appropriate in the local situation.
12. Decontaminate as described in the General Steps.
13. Complete field notes as described in the General Steps.

Soil or Sediment Collected with a Slide Hammer

A **slide hammer** is another manual tool for collecting relatively shallow soil samples. The tool itself is a weight on a slide located on top of a shaft, which is connected to a stainless steel core barrel that collects the sample. The slide hammer is used to drive the core barrel into the ground. Extensions can be added to the shaft to enable samples to be collected at depth. The core barrel accommodates stainless steel liners or plastic liners for ease in removing the soil sample from the device. If a stainless steel liner or plastic liner is used in the sample collection, a relatively undisturbed sample may be collected.

Steps to Review

1. Take field notes in the field logbook as described in the General Steps.
2. Peel back sod and spread plastic sheeting on the ground as described in the General Steps.
3. Assemble the clean slide hammer device: attach slide hammer to one end of the rod and the clean core barrel to the other end.
4. Use the hammer to beat the sampler into the ground to the desired depth. Attach additional rods as required. Count the number of hammer blows needed to reach the desired depth and record this information in the logbook. (The number of hammer blows is valuable geotechnical information related to the strength and type of material being sampled.)
5. Once the desired sampling depth is reached, rock the slide hammer sampler back and forth and then lift the sampler out of the hole. If it does not lift easily out of the hole, use the slide hammer and hammer in reverse (upward) to beat the device back out of the sample hole.
6. Unscrew the core barrel from the sampler and slide the sampling liner out of the core barrel. If desired, seal the ends of the sample liner and label it for transport to the lab. If a sample liner is not used, scrape soil out of the core barrel using a stainless steel spoon or knife. Handle the sample in the appropriate fashion (see item 7 below).
7. Pack, label, and store sampling containers as described in the General Steps.
8. Complete chain-of-custody form as samples are collected.
9. Put all samples into a cooler.
10. Backfill shallow holes or containerize waste soil according to local regulations and store onsite as described in the General Steps.
11. Clean and restore the sampling locations according to local regulations, and seal hole with bentonite as described in the General Steps.
12. Decontaminate as described in the General Steps.
13. Complete field notes as described in the General Steps.

Soil or Sediment Collected with a Split Spoon Sampler or Continuous Sampler

Split spoons and **continuous samplers** are sampling devices driven into the ground through a borehole prepared by a drill rig. These devices penetrate to great depths. As pointed out earlier, the drill rig is usually assigned a dedicated two-person crew. The crew is generally not responsible for the soil descriptions, sampling, mapping, or any other part of the investigation.

The split spoon sampler is a stainless steel tube that opens into two halves along the long axis of the device. It is often two inches in diameter (outer diameter) and two feet in length. The split spoon sampler is attached to the end of a rod on the drill rig and lowered to the bottom of a borehole. The borehole has previously been drilled to the desired depth using a **hollow stem auger** through which the split spoon is lowered. Once at the bottom of the borehole, the split spoon is hammered into the ground using a weight (commonly 140 pounds) on the drill rig. The technician should record in the field log: 1) the number of **hammer blows** (i.e., a weight dropped from a height of 30 inches) that it takes to drive the split spoon into the ground, and 2) the number of hammer blows required to move the split spoon every six inches. For a 24-inch long split spoon, for example, a series of four numbers should be recorded in field notes – one for every six inches.

The continuous sampler is another stainless steel tube that opens along the long axis into two halves. It may have various diameters (two- and three-inch diameters are common) and also varying lengths, although most commonly it is five feet in length (the same length as the hollow stem auger sections). The continuous sampler is placed at the end of the rod on the drill rig and lowered to the bottom of the borehole through the hollow stem auger; the drill rig pushes the sampler into the ground.

Both the split spoon sampler and the continuous sampler are retrieved from the borehole and opened longitudinally, exposing the sample within. Both devices may be equipped with stainless steel liners or plastic liners.

The split spoon sampler and the continuous sampler are most effectively used with hollow stem augers, since the augers can be left in the ground. This supports the hole and prevents the walls from collapsing, allowing the sampling device to be lowered to the bottom of the boring to obtain undisturbed soil beneath the hole.

Steps to Review

1. Take field notes as described in the General Steps.
2. Peel back sod as described in the General Steps. Clear the area of surface debris and remove the first few inches of soil.
3. Near operations around the drill rig, but not in the way, choose a spot to set up 55-gallon drums to contain soil cuttings that come out of the borehole as drilling progresses. Spread plastic sheeting by the borehole to prevent soil cuttings from contaminating the surrounding ground. If the plastic sheeting interferes with the safe operation of the drill rig, then minimize the spread of the soil as much as possible and, at regular intervals during the drilling process, transfer soil cuttings to the drums prepared for that purpose.
4. Assemble the clean split spoon sampler or continuous sampler device; you may include a liner. Attach the sampler to the end of the drill rig's rod. The drill crew may perform this activity.
5. The drill crew uses the drill rig to hammer or push the sampler into the ground to the desired depth. If using the split spoon sampler, count and record the number of hammer blows needed to penetrate to the desired depth.
6. Once the desired sampling depth is reached, the drillers will lift the sampler out of the hole using the drill rig.
7. The drillers remove the sampling device from the drill rod. They unscrew the ends of the split spoon or continuous sampler and split the sampler open. If a sample liner is used, its ends may be sealed, and the liner may be labeled and prepared for transport to the lab. If a sample liner is not used, or if the liner will not be sealed and sent to the lab, pack the sampling containers as described earlier.
8. Complete chain-of-custody form as samples are collected.
9. Put all samples into a cooler.
10. Backfill shallow holes or containerize waste soil according to local regulations and store onsite as described in the General Steps.
11. Clean and restore the sampling locations according to local regulations and seal hole with bentonite as described in the General Steps.
12. Decontaminate as described in the General Steps.
13. Complete field notes as described in the General Steps.

Obtaining Soil Samples by Direct Push Methods

Direct push uses techniques that push tools into the ground to remove the soil without the use of drilling. One type of direct push system is known as Geoprobe®. The first Geoprobe® was built in 1988 for EPA. It is a hydraulically powered machine that combines the weight of the vehicle the Geoprobe® is attached to with percussion of up to 2,000 blows per minute to drive sampling devices into the ground. The Geoprobe® is typically used for investigations to depths of 60 feet, but some intrepid field teams have used the equipment to depths exceeding 100 feet. The device's maximum depth at any site is limited by soil type and bedrock.

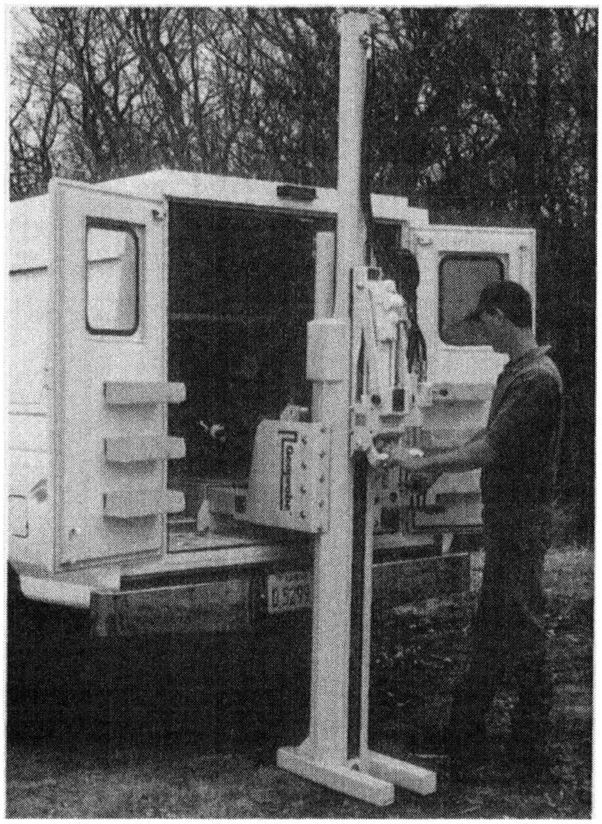

Figure 5-10: Since the Geoprobe® is pushed directly into the soil, it does not produce soil cuttings; this reduces the necessity of handling investigation-derived wastes.

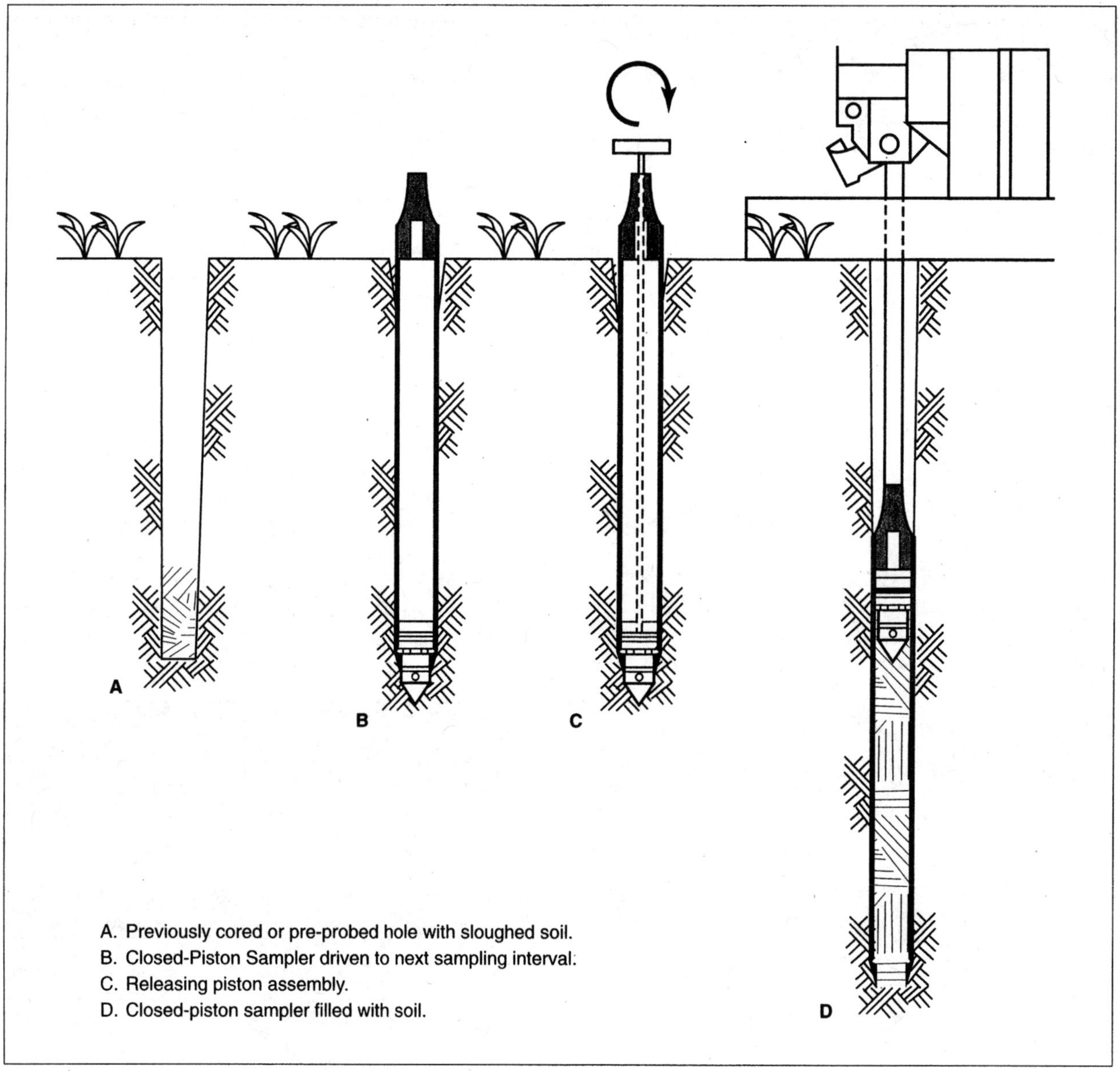

Figure 5-11: Phases of Macro-Core® Closed-Piston soil sampling. (Diagram from Geoprobe Macro-Core® Soil Sampler SOP Technical Bulletin 95-8500, Sept. 1996 rev.)

The Geoprobe® and other direct-push devices are being used with increasing frequency in environmental investigations for the following reasons:

—The technique's reduced costs in comparison with traditional drilling methods

—Speed with which sampling can be accomplished

—Minimal disturbance of the site

—No production of soil cuttings

The Geoprobe® is used most commonly to collect **discrete interval** soil samples using a **large bore sampler** that is two feet (60.96 cm) in length and 1.375 inches (3.49 cm) in diameter. This sampler may be used with brass, stainless steel, Teflon® or CAB (cellulose acetate butyrate) liners. Another sampling device, the Macro-Core®, allows continuous soil cores to be recovered. The Macro-Core® may also be used with a liner. Collection procedures with a large bore sampler or Macro-Core® sampler are outlined below. However, this outline is not a substitute for both hands-on training in using the

Geoprobe® system and/or the operating procedures supplied by the equipment manufacturer.

Steps to Review

1. Take field notes as described in the General Steps.
2. Clear the area of sod and surface debris as described in the General Steps.
3. Assemble the clean large bore sampler or Macro-Core® sampler device including the liner. Make sure the sampler is correctly and tightly assembled. Attach the sampling device onto the leading end of the Geoprobe® probe rod.
4. Drive the sampler into the ground.

   —For the large bore (discrete interval) sampler, which obtains a sample 22 inches (55.88 cm) long by one inch (2.54 cm) in diameter (volume of sample: 283 cm^3), follow this procedure:

   -Drive the sampler to the top of the desired sampling interval, attaching additional probe rods as needed. A "stop-pin" located at the top of the sampler prevents soil from entering the large bore sampler while it is being driven to the desired depth.

   -Remove the drive cap from the top of the probe rods. Using a series of extension rods linked in series to the appropriate length, lower the extension rods down the center of the probe rods to the stop pin at the top of the sampler.

   -Rotate the extension rod until the threads on the rod engage the stop pin threads on the top of the sampler. Continue rotating until the stop pin is removed from the top of the sampler. Remove the extension rods and stop pin from the probe rods.

   -Replace the drive cap onto the top probe rod, attaching an additional probe rod if necessary.

   -Mark the probe rod at a point 24 inches above the ground surface. Drive the sampler no more than 24 inches. With the stop-pin removed, the point at the base of the sampler is displaced into the sampler as the soil sample fills the large bore sampler.

   -Retract the probe rods from the hole. Remove sampler from the hole.

   -Remove the sample from the large bore sampler. If using appropriate sample liners, the ends of the sample liner may be sealed, labeled and prepared for transport to the lab. If the liner is not be sealed and sent to the lab, pack the sampling containers as described in the General Steps. A utility knife or special extrusion device may be necessary.

   —For the Macro-Core® (continuous) sampler, which obtains a sample 43 inches (109.22 cm) long by 1.5 inches (3.81 cm) in diameter (volume of sample: 1,245 cm^3) follow this procedure:

   -Position the sampler beneath the driving mechanism of the Geoprobe® and drive into the ground its full length to collect the first continuous sample.

   -Remove the sampler by attaching a pull cap to the drive head and retracting the sampler device using the powered controls.

   -To collect more continuous samples below the first one, connect probe rods together to the length required to allow the bottom of the sampler to reach the next sampling interval. Lower the Macro-Core® sampler into the same hole as the previous sample and use the Geoprobe® controls to apply downward pressure and the percussion hammer as necessary to drive the probe four feet to collect the continuous sample.

   -Remove the sampler by reversing the controls to pull the sampler up to the top of the hole. Once the top of the sampler has appeared above the ground surface, attach a pull cap to the Geoprobe® drive head and then remove the sampler the rest of the way using the powered controls.

   -Remove the sample liner from the Macro-Core® sampler. The ends of the sample liner may be sealed, labeled and prepared for transport to the lab. If liner will not be sealed and sent to the lab, pack the sampling containers as described in the General Steps. A utility knife or special extrusion device may be necessary.

5. Pack sampling containers as described earlier.
6. Complete chain-of-custody form as samples are collected.

7. Put all samples into a cooler chilled with ice (to four degrees Celsius).
8. Backfill shallow holes or containerize waste soil according to local regulations and store onsite as described in the General Steps.
9. Clean and restore the sampling locations according to local regulations and seal hole with bentonite as described in the General Steps.
10. Decontaminate as described in the General Steps.
11. Complete field notes as described in the General Steps.

Drilling to Obtain Rock Samples

A drilling rig is required to obtain samples or cores of bedrock. Different drilling methods are available; chips and/or cores of the rock being drilled may be obtained, depending on the method and set-up. It is up to the project manager or project designer to determine which is the most appropriate method for a particular project. In most cases, any sampling of the rock material would not be the responsibility of the field technician; it would most likely be performed by a geologist onsite. To increase awareness of activities that may be required on a project, however, a brief description of drilling methods is given here.

—Rotary Drilling: A drill bit on the end of the drill pipe is rotated against the rock material to form the bore hole. As the drill bit and pipe are rotated, circulation fluid is pumped into the hole to flush cuttings to the surface and to cool and lubricate the drill bit.

—**Percussion Drilling**: The drill bit is hammered or beat on the rock to force the rock to break. **Cable-tool drilling**, which is a type of percussion drilling, is slow but has the advantage of not requiring very much drilling fluid to carry the material out of the hole. With **air percussion drilling**, air powers a down-hole hammer and carries the cuttings to the surface of the hole. A similar percussion drilling method is the air-operated **casing hammer** which uses a second tool to carry rock cuttings to the surface. Both air percussion drilling methods require in-line petroleum lubrication of internal parts, which presents the danger of introducing petroleum to the hole, potentially compromising an investigation for hydrocarbons.

Samples collected as chips are put into containers – bags or bottles – and labeled according to their collection point, depth and related information. Samples collected as cores are labeled and placed into special core boxes or tubes.

Any material generated during an investigation is considered investigation-derived waste and must be handled accordingly. In addition to any material that comes out of the borehole, such as water or rock chips, all material added to the borehole, such as drilling fluids, must be recovered. The investigation-derived wastes must be containerized, labeled, analyzed and disposed of in accordance with local, state and federal requirements.

Sediment Collected with a Dredge

Sometimes a sample is needed from a submerged area. The water may be relatively shallow (only a few inches) or many feet deep. It may be a static water body like a pond or a lake, or flowing water like a river or a stream. The sediment beneath the water may be collected directly by use of a hand-held device such as a shovel or auger, or indirectly using a dredge. The method used to collect the sample depends on the depth of water at the sampling location and the physical characteristics of the sediment to be sampled. The procedures for collecting soil samples using a shovel, a thin-walled tube sampler, and an auger have already been described above; we will devote this section to describing how to collect sediment samples using a dredge.

The Ekman and the Ponar dredges are two commonly used dredge types. Both dredges are box-like devices with jaws at the bottom that shut when triggered, trapping sediment. Part of the skill involved in using a dredge is the technician's ability to lower the dredge through the water to a position above the desired sampling point and then to let the dredge free-fall through the water into the sediment without the jaws of the dredge closing before impact. The free-fall is intended to allow the dredge to penetrate the sediment layer.

Steps to Review

1. Take field notes as described in the General Steps.
2. Assemble the dredge. Attach a sturdy nylon or steel cable to the dredge in the manner indicated

by the manufacturer, ensuring that the dredge is in the open position.

3. Slowly lower the sampler to a point above the sediment surface where the sample is to be collected.

4. Drop the sampler sharply into the sediment. Release the dredge's jaw release mechanism by whatever method the manufacturer indicates. This may mean depressing a button on the upper end of the extended cable or pulling up on a line. Even if you think the impact of the dredge as it hits the sediment has triggered the dredge's release mechanism, be sure to manually release this mechanism so that the sediment stays in the dredge as it is raised to the water's surface.

5. Raise the dredge to the surface of the water and slowly decant (pour off without disturbing sediment) any free liquid through the top of the sampler. Be careful to retain fine sediments.

6. Open the dredge and allow the sediment to drop into a stainless steel bucket. If too little sediment results, repeat steps 2 through 6 until enough has been obtained.

7. Pack sampling containers as described in the General Steps.

8. Complete chain-of-custody form as samples are collected.

9. Put all samples into a cooler chilled with ice (to four degrees Celsius) or alternative refrigerant.

10. Excess sediment removed from the sampling locations should be containerized in a drum as described in the General Steps.

11. Decontaminate as described in the General Steps.

12. Complete field notes as described in the General Steps.

Field Screening of Soils and Sediments

Air monitoring instruments such as the photoionization detector (PID) or the flame-ionization detector (FID) may be used to screen or test samples for indications of possible contamination. Other devices such as pH meters may also be used to screen samples for contamination if those devices can reasonably be expected to detect indications of the contaminants of interest. Field screening techniques allow the technician to make a judgment concerning the extent of contamination; this is especially important when the investigation goal is to determine the full horizontal and vertical extent of contamination. Data from field screening may also be used as criteria for choosing the samples to be sent to the laboratory for analysis.

The readings on a PID are not specific to a chemical (see Chapter 4). All chemicals present that have ionization potentials lower than the lamp strength can be detected by the PID. Dust, moisture and condensation may cause a PID to give readings at values less than what would be expected with a dust-free, dry lamp. FIDs can only detect organic compounds. Thus, readings obtained from a PID, a FID or any air monitoring instrument must be interpreted with caution. As with all instruments, the manufacturer's instructions should be read, understood and followed.

Field Screening Headspace with a PID

Using a PID to screen in the field is a common way to check soil for possible contamination. **Headspace** is the space above the sample surface within a closed container. It is the space in which accumulation of volatile gases, organic or inorganic, may occur. Headspace gases must be screened the same day the samples are collected. Before and after using the PID, the technician must calibrate the instrument with an appropriate calibration gas according to the manufacturer's instructions. The technician records the calibration data and PID lamp strength in the field logbook.

1. Immediately upon removing soil or sediment from the sampling device, transfer the material to a clean glass jar (8 fluid ounce size or larger), filling the jar halfway. Quickly cover the top of the jar with one or two sheets of clean aluminum foil. Screw the cap on the jar and seal tightly. Label the jar to indicate the contents.

2. Allow 10 minutes for headspace development. This development period allows time for the gases to rise to the top of jar. Shake the jar for 10 to 15 seconds at both the beginning and the end of the headspace development period. If ambient temperature is less than 32°F, move headspace sample jars to a heated vehicle or

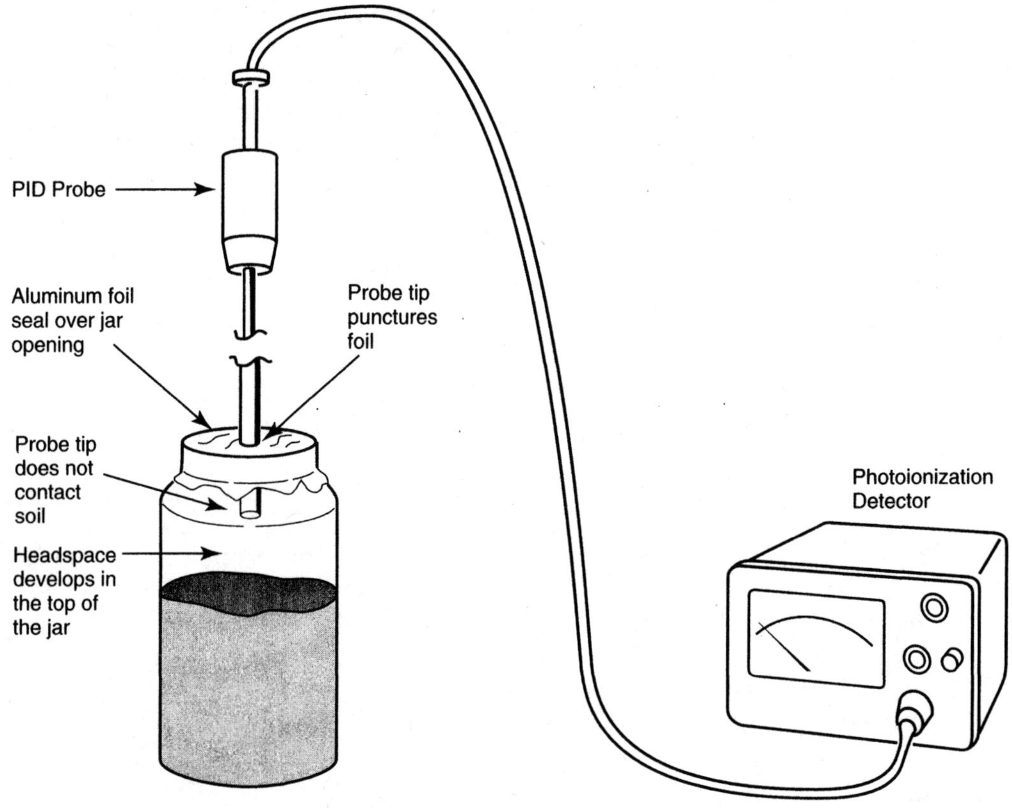

Figure 5-12: Field screening headspace with a PID. The jar containing the soil sample shows detail of probe tip entering jar through foil but not contacting the soil itself.

building for headspace development and measurement. The volatilization of the gases, and thus headspace development, is dependent on the temperature and surface area of the sample and the vapor pressure of the contaminant to be measured.

3. When headspace development is complete, carefully remove the screw-on lid to the jar and expose the foil seal. Quickly but carefully puncture the foil seal with the instrument sampling probe to a point about one half of the headspace depth. Be careful not to allow water droplets or soil particles to enter the probe. Immediately look at the meter response.
4. Record the highest meter response. Maximum response usually occurs within two to five seconds. Erratic meter response may be due to high contaminant vapor concentration or moisture. Erratic meter response data should be noted and probably discounted.
5. After headspace readings have been made, dispose of soil field screening samples according to state and federal regulations.

Equipment Decontamination and Cleaning

Decontamination of field sampling equipment is very important and necessary to evaluate accurately the presence or absence of chemical constituents at low concentrations. Poor decontamination efforts can spread contaminants between sampling locations and between sites, expose workers to contaminants on sampling equipment, and reduce data quality and reliability. In general, decontamination should be performed before and after each sample is collected.

Some of the equipment – split spoon sampler, hand auger bucket, dredge, stainless steel bowl – comes into direct contact with the sample matrix itself during sampling. We will call such equipment **sample contacting equipment**. Other equipment used in the sampling activities may come into contact with soil or sediment, but does not directly contact the sample itself. This equipment includes augers, drilling rods, cables, and also equipment handles. We will call this type of equipment **sample non-contacting equipment**. ASTM's standard D5088-90 indicates that, in general, two different decontamination protocols may be used for equipment, depending on whether it is classified as sample contacting equipment or non-sample contacting equipment. We will cover both practices in this section. The site sampling plan should specify what decontamination protocols are to be followed; it should stipulate objectives, desired precision and accuracy, materials and equipment, step-by-step procedures for decontamination, and procedures for collecting QA/QC samples.

Equipment and material for decontamination may vary according to the QA/QC needs of a sampling effort, the contaminants of concern, and the size of the equipment being used for sampling. In general, decontamination is a wet process. Water, detergent, solvents and even acids may be used to clean equipment thoroughly. Equipment decontamination may be summarized as a series of steps:

1. Brush/scrape soil and sediment off equipment.
2. Wash with a nonphosphate detergent solution.
3. Rinse with **control water** (control water is **potable water**, preferably from a municipal water system of known chemical composition).
4. Rinse with desorbing agents. These include acids for removing inorganic constituents or solvents for removing organic constituents. The acid should be 10 percent nitric acid or a hydrochloric acid solution made with deionized water. If the sampling equipment is low-carbon steel, only one percent acid solution should be used. The solvents may be pesticide-grade isopropanol, acetone or methanol.
5. Rinse with organic-free, reagent-grade deionized water.
6. Air-dry.
7. Wrap in plastic or aluminum foil.

Sample Contacting Equipment	Sample Non-Contacting Equipment
Split Spoon Sampler	Augers
Continuous Sampler	Drilling Rods
Hand Auger Bucket	Cables
Dredge	Equipment
Stainless Steel Bowl	
Trowel	

Table 5-9: Examples of samples contacting and non-contacting equipment.

The use of desorbing agents may generate waste material that must be handled and disposed of appropriately. To avoid this, some project administrators or managers modify or delete the desorbing agent step from the sampling plan when the project allows it. At a minimum, sample equipment should be washed with a detergent solution and rinsed with control water.

To minimize generation of waste material when desorbing agents are used, the technician may apply a rinse of acid solution or solvent using a stainless steel sprayer. Spraying and rinsing are performed over a bucket or other basin dedicated to that decontamination fluid. Additionally, each sprayer is dedicated to use with only one type of decontamination solution. Sprayers should be clearly labeled to prevent confusion and to maintain the quality of decontamination solutions.

QA/QC samples (such as equipment blanks, rinsate blanks and field blanks) may need to be collected. As outlined in Chapter 2, these samples are used to determine the effectiveness of decontamination. If equipment has not been properly decontaminated between the collection of samples, the quality and accuracy of the data obtained from those samples may be compromised. Decontamination fluids are handled like water samples (see Chapter 6).

The process of decontamination should be arranged in an organized and linear fashion. With the completion of each step in the decontamination process, the equipment proceeds to the next step without delay and without being carried great distances. In many cases the technician will not have the luxury of having several pieces of the same sampling equipment available. Rather, as soon as a piece of equipment is decontaminated, it is used to take another sample. It is therefore important to the timely completion of the sampling activities for technicians to decontaminate efficiently.

Decontamination Procedure for Sample Contacting Equipment

1. Brush and scrape soil or sediment off the equipment.
2. Using a nonphosphate detergent solution, wash the equipment by hand with a scrub brush.
3. Rinse with control water.
4. Rinse with an inorganic desorbing agent such as 10 percent nitric acid or hydrochloric acid solution (may be deleted if samples will not be analyzed for inorganic chemicals).
5. Rinse with control water.
6. Rinse with an organic desorbing agent (solvent) such as pesticide-grade isopropanol, acetone or methanol.
7. Rinse with organic-free, reagent-grade deionized water.
8. Allow equipment to air-dry.
9. Wrap equipment in plastic or aluminum foil to avoid contact of the sample equipment with potentially contaminated surfaces.
10. Place decontamination fluids in lined 55-gallon drums or other appropriate container. Label and seal the drums, and store them in an appropriate, secured spot on the site. These investigation-derived wastes will need to wait for laboratory analytical results before they can be appropriately disposed of.

Decontamination Procedure for Sample Non-Contacting Equipment

1. Brush and scrape soil or sediment off the equipment.
2. Wash equipment with a power washer or steam cleaning machine and nonphosphate detergent solution, or wash by hand with a scrub brush and nonphosphate detergent solution.
3. Rinse with control water.

In some cases, all or some of steps 4 through 10 of the procedure for sample contacting equipment may need to be performed for non-sample contacting equipment as well. The choice is determined by sampling and decontamination objectives, desired precision and accuracy as well as the QA/QC procedures outlined in the sampling plan.

Selection of Containers, Preservatives and Holding Times for Geologic Materials

Appropriate containers, preservatives and holding times are determined by the analytical methods chosen. Details should be outlined in the site sampling plan. Preparation for going onsite includes

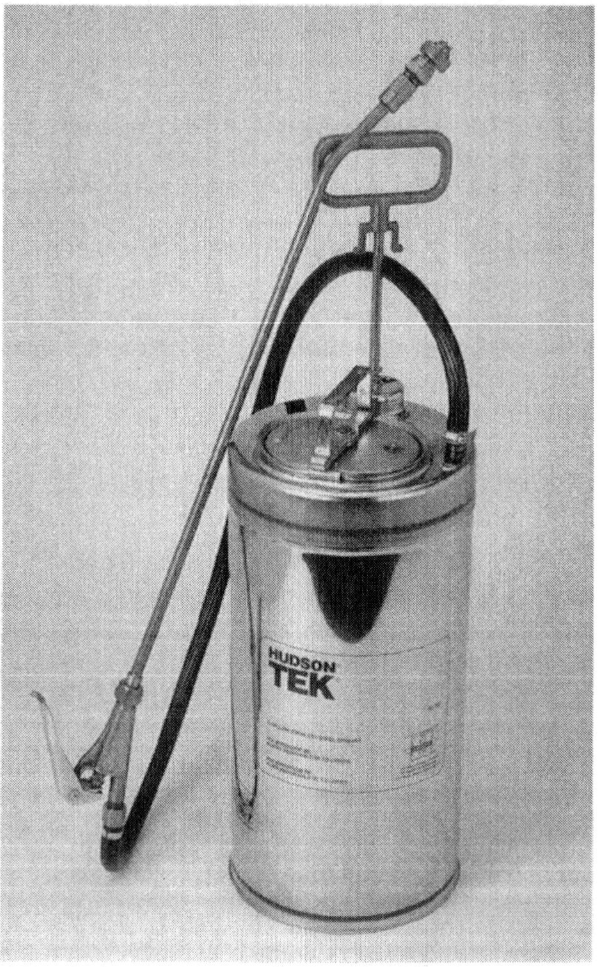

Figure 5-13: Each sprayer should be dedicated to use with only one type of decontamination solution.

Box 5-4 ■ Underground Utility Standard Color Designations

Red	Electrical cables and conduits
Yellow	Gas, Oil and Chemical Lines
Blue	Water Systems
Green	Sewer and Storm Drains
Orange	Communications (TV, phone, fiber optics, telegraph)

Utility locations are marked on the ground with paint, stakes, flags or signage using these colors. What color would the sign to the right be?

consulting the site sampling plan to be sure to take enough of the right type of containers with the appropriate preservatives for the sampling activities. In addition, the technician should keep in mind the holding times for the samples and make advance plans to deliver the samples to the laboratory as soon as possible. When holding times are exceeded, analytical results may be inaccurate or altogether unreliable.

Containers

Glass jars are generally used as sample containers for soil and sediment analysis. These jars may have volumes of 4, 6, 8, 16, or 32 fluid ounces. Sometimes a project-analytical methodology or a regulatory authority may specify that 40 or 120 milliliter glass vials be used for sample containers. For soils and sediments, wide-mouth jars are normally used to allow material to be packed in the jar with relative ease. Glass jars resist most chemicals and protect against permeation of materials into or out of the jar. The lids of the jars are usually made of a **thermoset** material that has a Teflon® lining for chemical resistance. The lids of glass vials may also feature a Teflon® silicone septum. The septum cap allows liquids to be injected into and withdrawn from the sample container. Many laboratories will provide clean sample containers as a customer service. Chips of rock may also be collected in glass jars; a core of rock may be placed in plastic or waxed cardboard until further study or analysis of the sample is undertaken.

Preservatives

Typically soil and sediment samples are chilled to 4°C by placing the sample container into a cooler with ice. This is often the only preservative technique used for the samples. Whenever an analytical method requires chemical preservatives, it is best to add the preservatives to the sample containers before leaving the office and entering the field. Many laboratories will provide sample containers with appropriate preservatives in the containers.

Holding Times

Holding time is the maximum length of time a sample may be held after collection before it degrades and loses integrity as a valid measurement of chemical contamination. Preservation extends the holding time of a sample, but it only provides for a delay in the inevitable changes that will occur to the sample once it is removed from its place of origin. Holding times vary depending on the analytical method. Some analyses have short holding times (Chromium VI at 24 hours or less), while others hold longer (metals, with the exception of Chromium VI and Mercury, at six months). Some samples need to go through an **extraction** process within a certain period, after which they have a longer holding time to complete the rest of the analysis (phenols at seven days prior to extraction and 40 days after extraction). The technician needs to be conscious of the holding time of a sample; samples need to be sent to the laboratory for analysis as soon as possible after collection.

Safety Precautions Recommended in Sampling Geologic Materials

The protective requirements of OSHA and EPA's HAZWOPER standard (29 CFR 1910.120) are relevant to sampling activities for soil, sediment and other geologic material. The sampling technician and his or her employer are responsible for knowing and complying with the requirements of these and all other pertinent OSHA general industry or construction standards. The technician is also responsible for reading, understanding, and complying with the procedures written in the site safety plan.

Among the safety procedures the technician should observe are:

—Always be certain that all underground utilities have been located before doing any sampling of soil, sediment or other geologic materials. Make note of the overhead utility lines as well. Review where utilities are located and be mindful of them while working.

—Use appropriate devices for real-time monitoring of the air around the work area. Be certain of the proper functioning of monitoring devices through maintenance and calibration before and after each day's sampling activities. Devices with audible and visible alarms are preferred.

—Be aware of a potentially explosive or dangerous atmosphere, and be ready to leave the area if necessary.

—Select and use the appropriate level of PPE. Always wear appropriate chemical resistant gloves while sampling.

—Take precautions against accessibility problems, such as uneven terrain and water hazards.

—Powered equipment such as drill rigs may present their own special hazards. Cables may snap violently, drill rods may break, moving parts may also break under strain. Use caution when working with large equipment.

—When working in public areas, secure the work area with barricades, caution tape, etc. Wear orange vests or other bright-colored clothing. Use the field support vehicle as an additional barrier whenever possible. Fieldwork does not always occur in deserted areas; sometimes the work is at or near a busy commercial facility. Maintain awareness of the surroundings, do not permit passersby to wander into the work area, and be alert to absent-minded motorists who are not accustomed to your presence in an area!

—Use the buddy system. Also, check in with appropriate personnel by radio or phone on a pre-arranged or regular basis.

—Do not eat, drink or smoke while performing sampling activities without first going through personal decontamination.

—Do not use matches or lighters onsite.

—Follow SOPs for equipment.

—Use intrinsically safe, explosion-proof electrical devices.

—Use nonsparking tools.

—Do not perform hot work in hazardous atmospheres. (Hot work produces heat or a static electricity ignition source; examples of hot work include: burning, welding, riveting, drilling, abrasive blasting and space heating.)

Checking Your Understanding

1. List and describe ten general steps for collecting a soil or sediment sample.
2. Describe the equipment used in sampling soil and sediment.

3. Discuss sampling techniques specific to soil and sediment.
4. Specify the care and cleaning necessary to maintain the equipment used in sampling soil and sediment, including decontamination techniques.
5. Describe the process for selecting containers, preservatives and holding times for soil and sediment.
6. Describe safety precautions recommended in sampling soil and sediment.

Summary

As part of site characterization activities, the technician should be able to 1) use one or more established soil classification systems to describe in field notes the soils and sediments encountered onsite, and 2) sample the soils and sediments with appropriate equipment and techniques.

Soil – which is comprised of minerals, organic matter, water and air – is classified according to grain size: fine-grained soils are clay or silt and coarse-grained soils are sand or gravel. Soil is described according to several criteria: color, particle size data, particle shape, moisture, consistency, structure, odor as well as reaction to hydrochloric acid.

Before going onsite, the technician makes advance arrangements, reviews site safety procedures, and prepares sampling equipment and supplies.

While onsite, the technician follows the sampling plan for collecting and storing samples, repairs the site, decontaminates equipment, and properly handles investigation-derived wastes. During sampling, the technician must be aware of avoiding the creation of paths that could cause or increase contaminant migration.

During all sampling-related activities, the protective requirements of OSHA and EPA's Hazardous Waste Operations and Emergency Response standard apply.

Contaminants migrating through soil, sediments and rock units may affect groundwater or surface water. Chapter 6 will discuss methods and equipment for sampling groundwater and surface water.

Application and Critical Thinking Activities

1. Explain the situation in Figure 5-3. For example:
   a. Describe the apparent path of the plume.
   b. Assuming that a split spoon sampler is inside the hollow stem auger, describe what you would expect the core of soil to look like upon opening the sampler.
   c. Predict the consequence of continuing to drill through the clay layer.
   d. If you were a technician in this situation, what would you do?
2. Acquire from the American Society for Testing and Materials (ASTM) the following standards: D 2487, which is the Standard Classification of Soils for Engineering Purposes (Unified Soil Classification System), and D 2488, which is the Standard Practice for Description and Identification of Soils (Visual-Manual Procedure).

   Given several soil samples, use the ASTM standards to describe the soils and identify each soil's components. Respond as though you were writing field notes in a log:
   a. Describe the soil according to the ASTM criteria.
   b. Identify the fine-grained components.
   c. Identify the coarse-grained components.
3. With the data from the previous question, plot each soil's sand, silt and clay components on the SCS ternary diagram to determine a name for each soil. Respond as though you were writing field notes in a log.
4. Obtain a manual auger and arrange for permission to dig at least one hole (e.g., at school or in your own yard). Before actually digging, review safety procedures (e.g., call to locate the underground utilities). Pack and label sample containers and repair the site as described in the General Steps. Document your activities with field notes.
5. Suppose you have contacted utility companies before going onsite to sample. When you arrive, you notice that the power company has marked a four-inch gas main with spray paint and flags. According to your sampling plan, three of the targets are located on or close to the utility markings. What do you do? Suppose also that you suspect the backfill around the gas main is a conduit for contaminants at the site. What do you do?

6

Sampling Water

Chapter Objectives

Upon completing this chapter, the student will be able to:

1. **Summarize** procedures for making field observations and taking measurements to accompany water samples.
2. **Describe** the equipment used to sample water.
3. **Describe** the instruments, including their calibration, that are used to sample water.
4. **Discuss** sample collection techniques specific to water.
5. **Specify** the care necessary to maintain the equipment used in sampling water.
6. **Indicate** the process for selecting containers, preservatives and holding times for water samples.
7. **Describe** safety precautions recommended in sampling water.

Chapter Sections

6-1 Purposes of Water Sampling

Concepts

- Drinking water, wastewaters, waters used for recreation, industry and navigation, as well as waters associated with site investigations, are all subject to routine water sampling.
- Drinking water is sampled for microbiological organisms, inorganic chemicals, volatile organic compounds as well as synthetic organic chemicals.
- Effluents from industrial wastewater and from publicly owned treatment works (POTW) are sampled for six "conventional pollutants" and more than 60 toxic pollutants.

As introduced in earlier chapters, the project planner for a site investigation develops a written sampling plan, which sets down the procedures and sampling techniques for collecting investigative and quality control samples. Although water sampling techniques are applied to both water-based and nonwater-based liquids, the focus of this chapter is on sampling water-based materials. Chapter 7 will detail how the techniques are used for nonaqueous materials, such as those found in industrial processes or storage tanks.

Water samples are routinely collected for several broad purposes, including:

Figure 6-1: Water is one of our most precious resources. We protect it by setting up and implementing regulatory standards for groundwater and surface waters.

1. To determine if drinking water meets federal or local quality standards;
2. To determine if wastewater meets federal or local **pretreatment** and effluent discharge standards;
3. To determine if water quality meets standards for recreation, the propagation of fish and other wildlife as well as for agriculture, industry and navigation.
4. To determine the extent and nature of groundwater and surface water contamination during site investigations.

Drinking water is sampled for microbiological organisms, inorganic chemicals (IOC), volatile organic compounds (VOC), and synthetic organic chemicals (SOC).

—Microbiological organisms typically sampled for include **coliforms**, such as the bacterium Escherichia coli (E. coli) and **protozoans** (single-celled organisms such as Giardia lamblia, which causes intestinal infections in humans).

—Inorganic chemicals are those that do not contain carbon. Water is often tested for lead, which is known to cause neurological disorders in humans. The use of nitrogen fertilizers has led to an increase in nitrates and nitrites in water. Nitrates in water have been linked to congenital malformations, cardiovascular disorders, hypertension and methemoglobinemia. Methemoglobinemia is a condition that limits the supply of oxygen to tissues, which is particularly damaging to infants.

—Volatile organic compounds are carbon-containing compounds that readily evaporate. Volatile compounds may be manufactured as, for example, chloroform or trichloroethylene (TCE); they may also occur in nature in natural gases such as methane or propane, or in the volatile oils of certain pine trees (i.e. turpentine).

—Synthetic organic chemicals (SOC) are carbon-containing compounds that are manufactured. They may or may not be volatile. Examples include polychlorinated biphenyls (PCBs), herbicides and pesticides, and many industrial solvents and sealants.

Effluents from industrial wastewater and from publicly owned treatment works (POTW) are sampled for six conventional pollutants and more than 60 toxic pollutants. The conventional pollutants are:

—BOD (biochemical oxygen demand), which is the amount of dissolved oxygen needed to break down organic matter in the water;

—TSS (total suspended solids), that is, particulate matter, which in excess can fill a body of water with sludge;

—pH (potential of hydrogen), which is a measure of acidity or basicity; a pH below five or above nine has damaging effects on most living organisms;

—Fecal coliform (bacteria from feces); exposure may cause illness in humans;

—Oil (liquid organic materials that are not soluble in water, such as industrial oils and lubricants);

—Grease (solid organic materials that are not soluble in water, usually associated with food processing and preparation, especially animal rendering).

The toxic pollutants include inorganic chemicals such as cadmium, chromium and lead; volatile organic chemicals such as ethyl benzene and the solvent naphthalene; and synthetic organic compounds such as the banned pesticide DDT and the pesticide Dieldrin.

EPA water standards and their associated requirements are published in the Code of Federal Regulations, Title 40, Subchapter D – Water Programs. Operators of drinking water plants and POTWs are generally certified by their states. Operation of water and wastewater plants requires specialized training, which includes the procedures for sample collection, among many other responsibilities. These specialized, certified plant operations are not within the scope of this book.

Regulatory Framework

Water sampling is performed to carry out the requirements of many federal acts, but especially the following:

Safe Drinking Water Act of 1974 (SDWA) – Under the Safe Drinking Water Act and its later amendments, EPA sets standards for Maximum Contaminant Levels (MCLs) of toxic and carcinogenic compounds in drinking water and Secondary Maximum Contaminant Levels (SMCLs) for substances that are of aesthetic concern (taste, color, odor) but are not health hazards. The SDWA also provides for assigning the **sole source aquifer** designation, which provides protection to aquifers that are especially valuable because they are the only source of drinking water in an area.

Clean Water Act of 1977 (CWA) – The Clean Water Act is an amendment to the Federal Water Pollution Control Act of 1972. The two most prominent goals of the CWA are 1) to make water quality sufficient for fish, shellfish, wildlife as well as for recreational purposes, and 2) to prohibit the discharge of toxic pollutants. Amendments in 1987 strengthened requirements for discharge permits, including the discharge of stormwater.

Resource Conservation and Recovery Act of 1976 (RCRA) – RCRA provides a management system for hazardous and solid waste. With the Hazardous and Solid Waste Amendments (HSWA) of 1984, RCRA also regulates underground storage tanks (USTs) and medical wastes. Under RCRA, facilities that generate, treat, store or dispose of hazardous wastes must sample groundwater and surface water. Facilities that use USTs to store hazardous materials or petroleum must also sample water when the UST system has leaked or when the system is being taken out of use.

Toxic Substance Control Act of 1976 (TSCA) – TSCA gave EPA the mandate to regulate individual toxic substances in 1976. Although this is generally applied during manufacturing, distribution and use of the substances, EPA may use the law to control toxic substances discharged into surface or groundwater.

Comprehensive Environmental Response, Compensation and Liability Act of 1980 (CERCLA) – CERCLA is also known as Superfund, which is a fund approved by Congress for urgent actions in response to abandoned or uncontrolled hazardous waste sites. EPA attempts to recover from the responsible parties the Superfund money spent during their sampling and cleanup activities. In 1986, the Superfund Amendments and Reauthorization Act (SARA) expanded CERCLA and provided for response action deadlines, inclusion of federal facilities, emergency planning and community right-to-know.

The focus of this chapter will be on the methods used to collect groundwater and surface water samples for the purpose of determining the extent and nature of contamination in site investigation and characterization. Section 6-2 provides a brief overview of the hydrologic cycle and aquifers to help the student understand the structure and use of groundwater monitoring wells. The next three sections describe the equipment and procedures for sampling groundwater, surface waters, drinking water and wastewater. Section 6-6 discusses preservatives and storage considerations for water samples. Section 6-7 reviews water sampling safety procedures.

Checking Your Understanding

1. List four broad purposes for collecting water samples.
2. Give an example for each of four kinds of water contaminants:
   a. Microbiological organisms
   b. Inorganic chemicals
   c. Volatile organic chemicals
   d. Synthetic organic chemicals
3. List the six conventional water pollutants; list several examples of toxic pollutants.
4. Identify five federal acts that relate to water sampling.

6-2 The Hydrologic Cycle, Aquifers and Wells

Concepts

- The term hydrologic cycle refers to the processes by which water constantly moves above, on, and below the Earth's surface.
- Wherever groundwater occurs, soil and rock units may be classified as either aquifers or confining units.
- Monitoring wells are used to sample groundwater.

The Hydrologic Cycle

The term hydrologic cycle refers to a series of processes involving the constant movement of atmospheric water, surface water and groundwater above, on, and below the Earth's surface. A convenient place to start describing the hydrologic cycle is with the process known as **evapotranspiration**. Evapotranspiration entails the release of water vapor to the air by plant **transpiration** and by evaporation from moist surfaces including the land surface, lakes and the ocean. The water vapor in the air condenses to form clouds that return water to the land surface or to the ocean in the form of precipitation. Precipitation includes rain, dew, snow and hail.

Precipitation infiltrates the soil by the force of gravity after wetting the land surface and the vegetation. The rate of infiltration varies depending

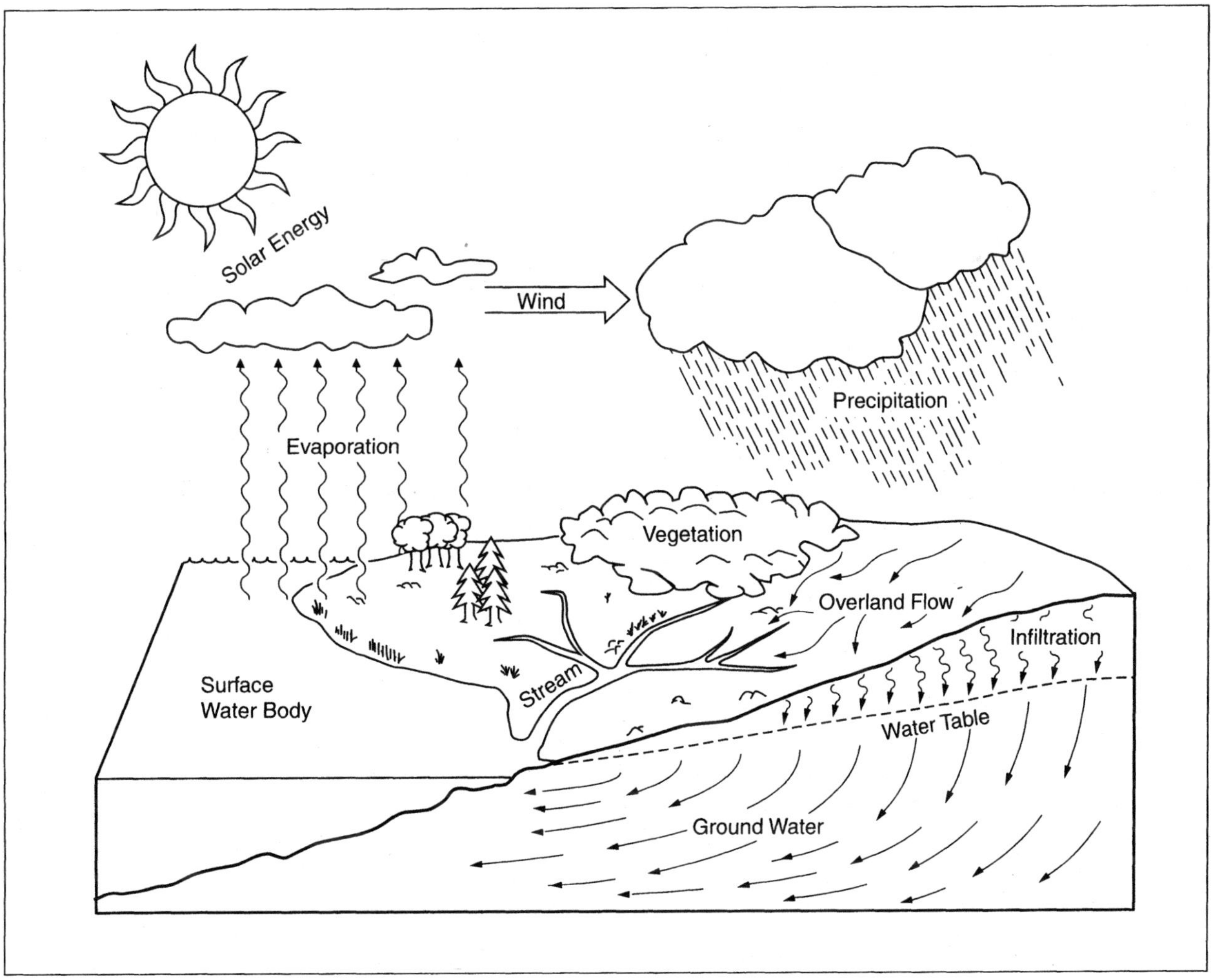

Figure 6-2: Illustration of the hydrologic cycle.

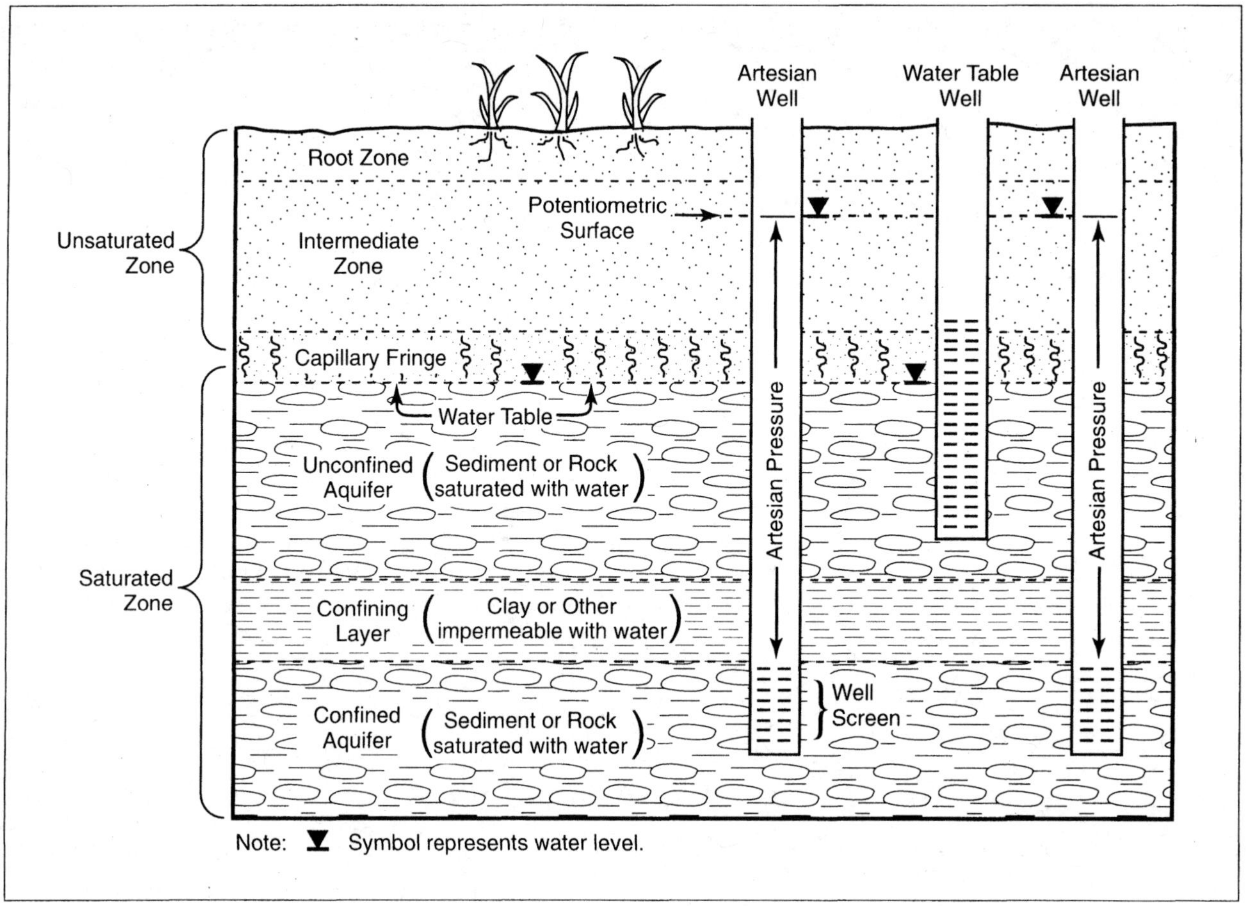

Figure 6-3: Subsurface zones.

on the amount of precipitation and the soil surface itself. If the surface is paved, no water will be able to infiltrate or penetrate the surface. The infiltration rate of clay-rich and silty soils may allow precipitation to penetrate and wet the soil beneath the surface at a rate of a few millimeters per hour. Sandy soils may allow the water to penetrate as much as 300 millimeters per hour. Since there is a limit as to how much water the soil can hold, once the rate of precipitation exceeds the rate of infiltration, **overland flow** occurs; that is, the water flows directly over the land surface.

When water infiltrates the soil, it first enters the soil zone or root zone, which is the zone where plant roots are found. The excess water then percolates slowly down and across the intermediate zone to the zone of saturation, or saturated zone. The intermediate zone is below the root zone and above the zone of saturation. The zone of saturation is where the spaces in soil or rock are filled with water. Just above the zone of saturation is the capillary fringe; this is where water is drawn upward from the zone of saturation by **capillary forces**, which are attributed to molecular attraction and surface tension between soil particles and water. The three zones – soil (root) zone, intermediate zone and capillary fringe – are known as the unsaturated or **vadose zone**. The top of the zone of saturation is the **water table**. At the water table surface, the pressure of the water filling the spaces in the soil or rock equals the atmospheric pressure. The water below the water table surface is known as groundwater.

Besides moving downward, water in both the unsaturated and saturated zones moves laterally. This may result in sites of groundwater discharge such as springs on hillsides. The water may also enter streams, lakes, or oceans from below the land surface. Water that reaches streams or other surface water bodies, both by overland flow and by groundwater discharge, moves to larger streams and rivers, and then to the lakes or oceans, where it is evaporated in a continuation of the hydrologic cycle.

Aquifers and Confining Units

Wherever groundwater occurs, soil and rock units may be classified as either aquifers or confining units. An **aquifer** is a soil or sediment (such as sand or gravel) or a porous rock unit (such as a sandstone) that will yield water in a usable quantity to a well or a spring. This means the sediment or rock must contain open spaces or pores that are filled with water; the pores must be interconnected and large enough so that water can move through them.

Aquifers may be confined or unconfined. A **confining unit** is a sediment (like clay) or a rock unit (like shale) that does not readily permit water to move through it, and that restricts the movement of water either into or out of aquifers. Water is said to be confined if it completely fills an aquifer that is overlain by a confining unit. Such aquifers are referred to as confined aquifers. Where water only partly fills an aquifer, the upper surface of the saturated zone (the water table) is free to move upward and closer to the land surface, or downward as the amount of water in the aquifer decreases. The water in such aquifers is said to be unconfined, and the aquifer is referred to as an unconfined aquifer (also called water-table aquifer).

Anatomy of a Typical Monitoring Well

A well is any excavation that is drilled, cored, bored, driven or otherwise constructed to obtain water. In this chapter we will be focusing on special sampling wells known as monitoring wells, which are used to sample groundwater during site characterization. Monitoring wells are sampled periodically; they are not used for drinking or potable water purposes.

Wells are generally constructed of a casing, also called a **riser**, and a screen. The casing is a piece of solid pipe connected to a screen, which is also made of pipe, but with holes or slots cut into it. An end cap is connected to the bottom of the screen to seal the end off from the sediment or rock beneath it. The casing and screen may be constructed of several different types of materials, including stainless steel, rigid polyvinyl chloride (PVC) pipe, or Teflon®. The diameter of the well casing and screen may vary, but the most common for groundwater monitoring purposes is a two-inch inside diameter (abbreviated to i.d.). The well is inserted into a large-diameter hole that has been bored or drilled to the desired depth. The space between the well (riser + screen) and the borehole's wall is known as the **annular space**. Other important components of a groundwater monitoring well are the well cap, the protective steel outer casing, the lock, the grout, the annular seal and the filter pack or sand pack.

The well cap or expanding plug is a watertight cap that fits snugly into the top of the well casing and prevents water that may have condensed inside the protective outer casing from seeping into the well. The expanding plug may have a hole to accommodate a padlock for securing the well opening. Some wells are completed at-grade, that is, they are almost flush with the ground so that vehicle traffic is not impaired by the well location. The tops of **at-grade wells** will be enclosed in a vault-type outer casing, also called a **manway**. The manway is cemented into the ground and small enough to not be burdensome, but large enough to allow the technician's hands to enter the space and perform tasks such as opening the well and locking the plug. The cover of the at-grade enclosure is made of metal and may be bolted down securely to the enclosure. For wells completed above-grade (sticking up above the ground), the protective steel outer casing is a steel pipe with a locking cap that fits over

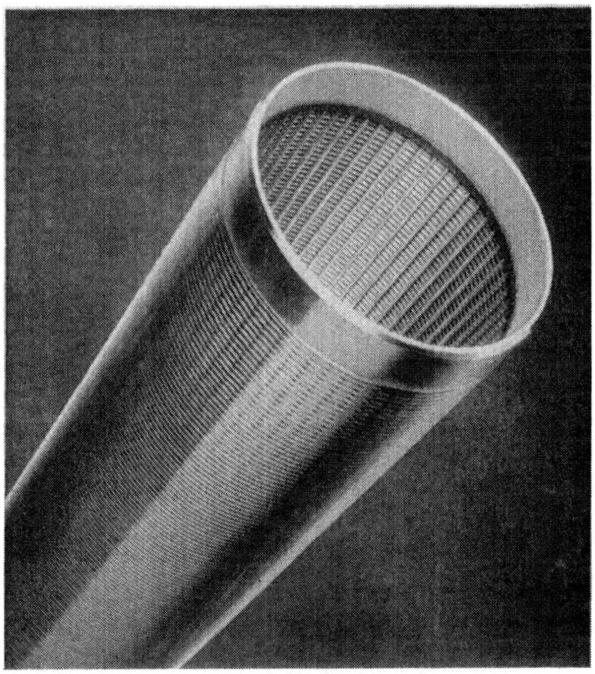

Figure 6-4: The screen area at the bottom of a groundwater monitoring well is built in such a way as to allow the water from the aquifer to flow through it and into the well.

the outside of the well casing and the well cap. It extends to a depth below the local frost-line and is cemented into place with concrete. The outer casing or the well's expanding plug should be locked with a padlock of a material that does not easily rust, such as brass. Above-grade wells may be further protected by guard posts constructed of steel or large-diameter wood that are sunk into the ground below the frost-line and held in place with concrete.

A filter pack or sand pack is inserted around the well screen and into the hole bored for the well. As the name suggests, it is usually composed of sand, pebble or small gravel-sized materials. The material for the filter pack is specifically chosen in each situation so that it does not restrict the flow of water into the well. The filter pack also prevents fine-grained materials from entering the well and causing either the groundwater in the well to be **turbid** (containing suspended fine particles) or the well itself to fill with the fine-grained materials. In some cases a filter pack may not be required because of the characteristics of the subsurface materials where the well will be installed. This decision is made by the project's geologist, hydrogeologist or planner/manager. The filter pack is placed in the bottom of the borehole around the screen and usually fills the borehole to a point two feet or less above the well screen.

Figure 6-6: Well plug.

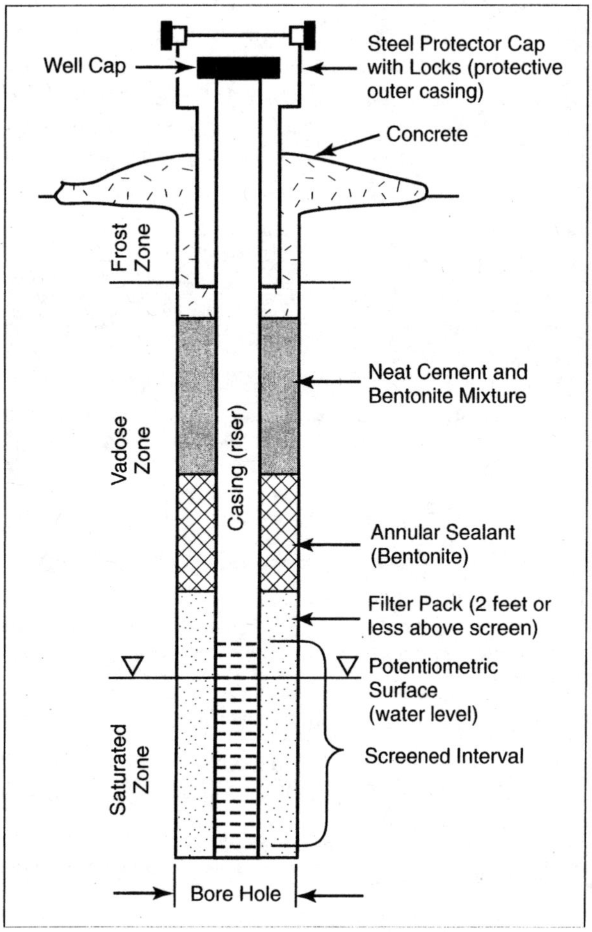

Figure 6-5: Structure of a monitoring well.

Above the filter pack, an annular seal is constructed of a material that will not allow water from the ground surface to percolate directly down the borehole and reach the groundwater. The annular

Figure 6-7: The manway that protects the top of an at-grade monitoring well provides the technician with access to the well cap.

seal material is usually **bentonite**, a clay derived from volcanic ash that swells when water is added. Bentonite comes in chips, pellets or formed into a **slurry**. When water is added to bentonite, the bentonite increases by 10 to 15 times its dry volume. The amount of annular sealant needed in the borehole may vary but it is usually a minimum of one or two feet thick.

Above the annular seal, grout is added around the well casing in the borehole. Grout provides additional protection against downward migration of contaminants; it also increases the strength of the well; it may be composed of neat cement, bentonite, or a mixture of the two. Neat cement is a specific mixture of Portland cement with water (94 pounds of Portland cement with 6 to 7 gallons of water). Portland cement is a cement produced by finely grinding a specific mixture of limestone and shale (or equivalent raw materials), heating the mixture until it is almost fused in a rotary kiln, and then finely grinding the resulting material again. The top of the borehole, including the protective steel outer casing, is sealed in concrete for stability and durability.

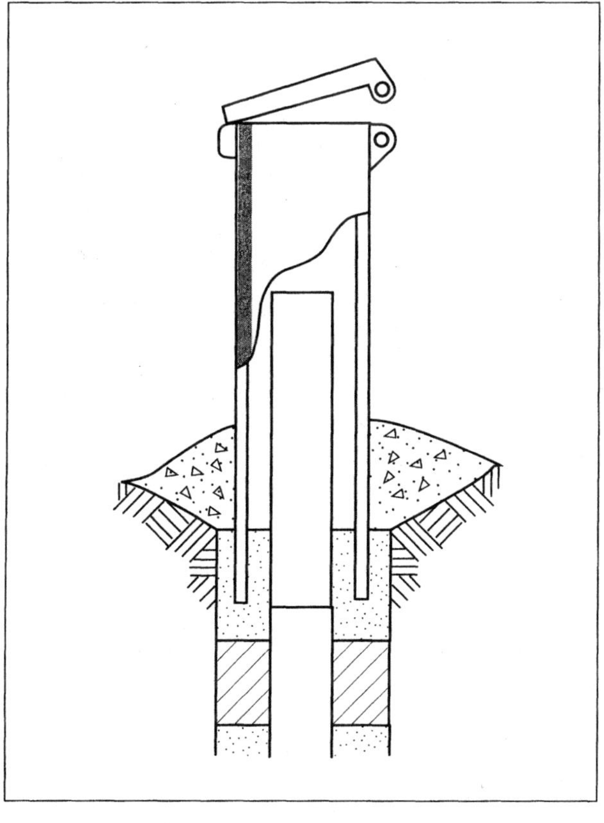

Figure 6-8: Steel outer casing protects an above-grade well.

Wells are usually constructed with the use of a drill rig. Some direct push technology systems such as Geoprobe® (see Chapter 5) also allow for the construction of small-diameter wells in the hole generated by the probe. These wells are similar to the wells described above, with minor modifications resulting from the small size of the well (0.65 to one inch in diameter) or from technology used in the direct push methodology. Direct push systems may also be used to sample groundwater without the installation of a permanent sampling point such as a monitoring well. For sampling groundwater using a direct push technology, the technician should follow the manufacturer's instructions. These instructions will be similar to the ones we shall discuss for monitoring wells, but with some modifications.

Types of Wells

A well placed in such a way that the well screen extends a few feet into the saturated zone of the unconfined aquifer is called a **water-table well**. The water level in this type of well indicates the position of the water table in the surrounding aquifer.

A well drilled into a confined aquifer is referred to as an **artesian well**. The water level in an artesian well stands at some height above the top of the aquifer but not necessarily above the land surface. If the water level in the artesian well stands above the land surface, the well is called a flowing artesian well. The **potentiometric surface**, sometimes referred to as the piezometric surface, is the elevation or level that groundwater will rise to in wells that tap an aquifer. The position of the potentiometric surface in wells that tap a confined aquifer is dependent on the pressure of the water in the aquifer.

Checking Your Understanding

1. Describe the movement of water throughout the hydrologic cycle.
2. Define:
   a. Zone of saturation
   b. Vadose zone
   c. Water table
   d. Capillary fringe
   e. Confining layer

3. Define and differentiate aquifers and confining units.
4. Define and differentiate water-table well and artesian well.
5. Define potentiometric surface or piezometric surface.
6. Sketch and label the parts and dimensions of a typical monitoring well.

6-3 Sampling Groundwater

Concepts

- Sampling from a monitoring well involves developing the well, purging the well, and measuring the well's water level.
- Well development restores aquifer characteristics in the vicinity of the well so that groundwater samples will be representative of the aquifer.
- During development, water is considered stabilized when it is visually free of suspended materials and when temperature, conductivity, and pH measurements remain relatively constant.
- Purging removes the stagnant water to allow water representative of the aquifer to enter the well for sampling.

Each time a well is sampled, specific preparations are required before the groundwater sampling device is lowered to collect the sample. These steps include:

—Opening the well, making initial observations and, particularly, monitoring for potentially hazardous conditions.

—Measuring the water level.

—Developing/purging the well, which involves removing stagnant water or water that contains sediments.

—Measuring pH, conductivity, and temperature as specified in the sampling plan.

Opening the Well

When opening the well, a technician makes observations and records data in a field log. As a well's protective casing is opened, the technician notes the condition of the well. Is the padlock holding up to the elements? Are protective posts still in place? Is there any noticeable corrosion or damage to the well's protective casing? Is there evidence of tampering?

It is recommended that when first taking the well cap off the well, the technician monitor with a photoionization detector (PID) and other gas meters, such as a combustible gas indicator (CGI), to determine if hazardous gases are present. The technician should check the area around the outside of the well riser as well as the inside. As the well's expanding plug (well cap) is removed, the technician notes any sounds or other observations. Is there a hissing or popping sound? Is the plug in tight? Is the plug discolored or deteriorated?

Measuring Water Level

The technician then takes measurements that will be used to calculate the amount of water to be removed before sampling. The technician measures the distance from the top of the well to the water surface and to the bottom of the well (note: if the plug was on tight, the water level may change to **equilibrate** with current conditions). Measurements made with a steel measuring tape weighted at the end are not very exact, considering that typically values are expected to be accurate to the hundredth of a foot. A battery-operated electric water-level meter is essential for measuring the water levels in a monitoring well during site characterization. The water-level meter should have a buzzer or light as well as batteries that can be changed in the field. The meter's measuring tape or line is marked in graduations to the hundredth (0.01) of a foot. The measuring tape may be of various lengths up to 2,000 feet, although 100- and 200-foot lengths are common. Metric tapes are available and are usually marked with graduations as small as 1 millimeter (mm). Some water-level meters have both metric and English measurements on each side of a flat tape. The procedure for using a water level meter is as follows:

1. Turn the meter on and push the test button (many meters have a test button to verify that it is working). If the buzzer or light does not come on when the test button is pushed, check the batteries or get the equipment serviced.
2. Drop the sensor end down the well or borehole. Keep reeling out more tape until the buzzer or light goes on. Then reel it in until the buzzer or light goes off. The point at which the buzzer or light first comes on is where the sensor first encounters water, i.e., the water level.
3. Determine the distance to the water level from the top of the well casing as precisely as possible by reading the number on the measuring tape at the measuring point on the well. The top of the well casing will be surveyed to determine its elevation above sea level or some other established datum. This surveyed point should be marked on the well casing for use as the measuring point in measuring water level. In most cases, the marked point on the well casing will be on the north side of the well. Sometimes the outer protective steel casing will be surveyed instead of the well casing (riser). Whichever point is surveyed, it should be marked for water level measuring use and should always be used. This measurement is referred to as the **top of casing (TOC)**. In the case of measuring depth to water in a borehole before the well is installed, measure to the top of the borehole or to the top of the auger, if it is in the borehole.
4. Raise and lower the water level probe again and check your measurement for the depth to water. Record this measurement in your field notes.
5. To measure to the bottom of the well, turn off the meter switch. Carefully reel the tape out until the tape goes limp. Measure this depth from the marked measuring point on the well casing (twice for accuracy). This is the bottom of the well or borehole. Record this measurement in your field notes.
6. Reel in the tape while wiping off water from the well on a paper towel or sorbent pad.
7. Decontaminate the tape and probe.

Some special probes are capable of measuring water and hydrocarbons in a well or borehole. Since oil and water do not mix, hydrocarbons may sometimes be present at the bottom of a column of water or may be found floating on top of water,

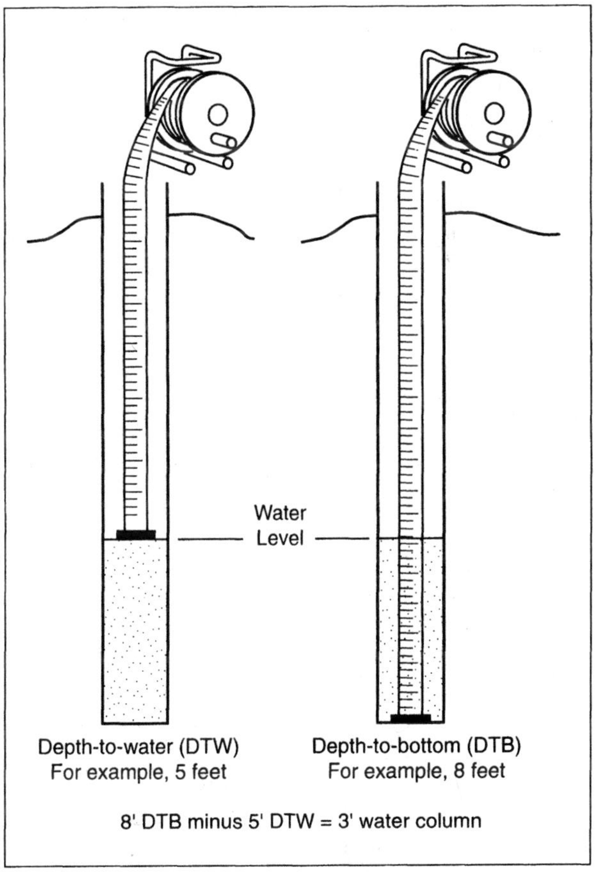

Figure 6-9: Measuring water level.

Box 6-1 ■ Calculating Borehole Volume

Part of the well development process requires that a minimum of three borehole volumes of water be removed. The technician needs to be able to calculate how much water represents a borehole volume. This volume can be calculated by using the following formula:

$$V = \pi r^2 l \text{ (cf)}$$

Where V = volume of borehole (in gallons)
π = 3.14
r = radius of the borehole (in feet)
l = length (in feet) of the water column in the borehole/well.*
cf = 7.48 gallons/ft^3, the conversion factor for changing cubic feet to gallons.

Once the volume of the borehole has been calculated, the volume is multiplied by three to determine the amount that must be removed.

*Depth-to-bottom of borehole (i.e., well) measurement minus the depth-to-water measurement.

depending on whether the hydrocarbons are more or less dense than the water. The hydrocarbon layers are often referred to as **nonaqueous phase liquids (NAPLs)**. When NAPLs are encountered in groundwater in measurable layers, the hydrocarbon layer may also be referred to as **free product**. The meters capable of measuring hydrocarbon layers in water are commonly referred to as **oil/water meters** or **interface meters**. Oil/water meters will have two buzzer and/or light signals to indicate when water is encountered or when hydrocarbons are encountered. The signals for hydrocarbons or water will depend on the instrument's design. A common design has a flashing light and an oscillating on/off tone when water is encountered and a steady light and tone when hydrocarbons are encountered. The process for measuring the depth to water and the depth to hydrocarbons is much the same as outlined above. The technician must record the top and bottom measurements to determine the thickness of the hydrocarbon layer (free product) and the thickness of the water encountered, in addition to the depth to the bottom of the well.

Development of a Monitoring Well

When a borehole is drilled and a monitoring well installed, the sides of the borehole are smoothed, caked, smeared or otherwise modified. This is a process that may disrupt aquifer characteristics such as the rate at which water flows through the aquifer. Well development is a process that restores the aquifer characteristics in the vicinity of the well, so that groundwater samples from the well will be representative of the aquifer. The fine sediments that cake the sides of the borehole in the screened portion of the well are washed into the well and removed during development. If the development process is not performed, these sediments will become suspended in the water, resulting in nonrepresentative samples and silting up of the well.

The process of well development involves taking water out of the well for a period of time to remove the fine sediments that are present. The water may be removed in a manner that agitates it, encouraging fine sediments to be removed. Although wells are sometimes developed using a surge process, which moves or pumps water into and out of the well, it is not a good idea to force air or pump water into any well that will be used for groundwater sampling. The addition of water may introduce other chemicals into the aquifer and thus alter the chemical characteristics of the aquifer being sampled. Air forced into the well and then into the aquifer may damage the well and alter the chemistry of the aquifer.

During development, water is removed from the monitoring well until it **stabilizes**; that is, until the water removed is visually free of suspended materials or sediments and the temperature, conductivity and pH measurements of the water remain relatively constant. These measurements need to be made to verify that these **parameters**, or factors, have stabilized. The amount of water removed must also be measured and put into a container such as a labeled, lined 55-gallon drum and kept until laboratory analysis of this water is complete. A minimum of three borehole volumes of water should be removed. Analysis of **development water** may

Box 6-2 ■ Well Development Data

(to be included in field notes or on a separate form referenced in field notes)

Well Development Data

Date: ______________________ Name of Technician(s): ______________________

Site/Project Name: ______________________

Well Number: ______________________ Well Location: ______________________

Depth to Bottom of Well

Initial Measurement & Time of Reading: ______________________

Ending Measurement & Time of Reading: ______________________

Depth to Water

Initial Measurement & Time of Reading: ______________________

Ending Measurement & Time of Reading: ______________________

Description of Development Method Used: ______________________

____

____

____

____

Time Start: ______________________ Time End: ______________________

Total Volume of Water Removed (in gallons): ______________________

Field Measurements

Date	Time	Volume Removed	Temperature (°C)	Conductivity (mS or μS)	pH	Comments

result in its being classified as hazardous (i.e., investigation-derived waste), and the water therefore must be properly disposed of according to federal, state and local regulations.

While removing water from the well the technician should, at regular intervals, put a sample of the water into a clean glass container and take temperature, conductivity and pH measurements. The interval between the measurements may be every one-half or one borehole volume. It is best to take at least three measurements during the well development process. A sample form for recording well development information is shown in Box 6-2. Instead of using a specific form, the technician may record data by setting up similar columns in the field log.

Because the temperature of the water in the well will change rapidly once the water is out of the well, this value should be measured first. Many instruments offer the feature of measuring all three parameters at the same time. This may be convenient but, if the meter malfunctions, no data can be collected. This is a good argument for using a separate instrument for each of the parameters. Since designs of instruments vary, the technician should read and follow manufacturers' instructions for proper use and maintenance. Guidelines for making temperature, conductivity and pH measurements individually with separate instruments are presented in Box 6-3, Box 6-4 and Box 6-5.

In summary, a well is considered to be developed when:

— A minimum of three borehole volumes of water has been removed;

— There is no visible sediment suspended in the water;

— The intervals of measurements meet these criteria:
  - Temperature does not vary more than ±0.5°C
  - Conductivity does not vary more than ±10%
  - pH does not vary more than ± 0.1 units.

Purging a Monitoring Well Before Sampling

Each time a well is to be sampled, the technician must first purge it, which means again removing volumes of water. Purging removes the stagnant water in the well and allows water representative of the aquifer to enter the well for sampling. A rule of thumb is that three well volumes should be removed to assure representative water is available for sampling. There are exceptions to this rule of thumb:

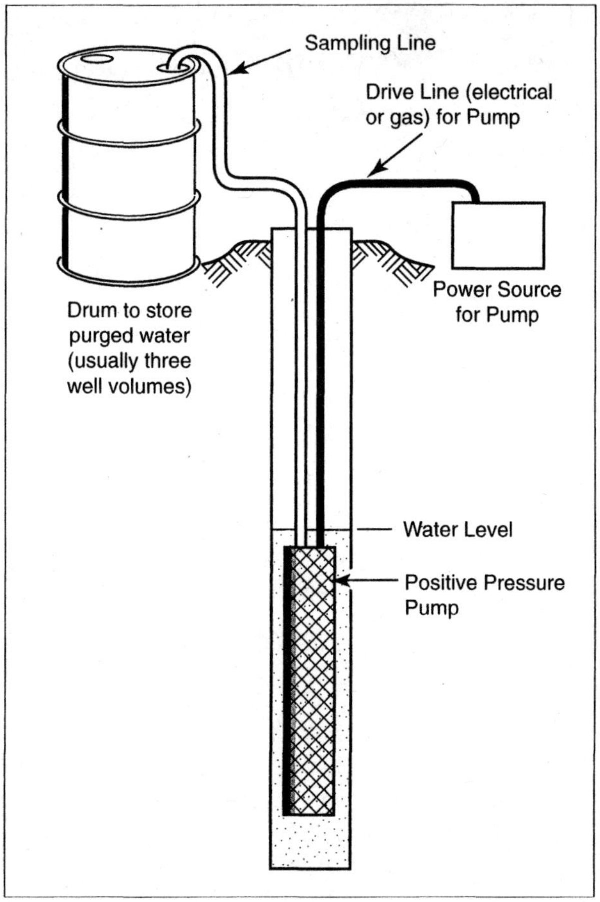

Figure 6-10: Well is purged by bailer or pump (as shown here); purged water is stored in a drum for proper disposal.

Box 6-3 ■ Measuring Temperature with a Thermometer

Materials

—Thermometer accurate to ±0.1° Celsius

—Insulated flask

—Stopper for insulated flask with a hole to hold thermometer

Procedure

1. Rinse clean insulated flask with well water.
2. Fill flask with well water.
3. Insert stopper with thermometer.
4. Wait one to two minutes for thermometer reading to stabilize (watch thermometer reading closely).
5. Measure and record temperature.
6. After measurements are complete, decontaminate the thermometer, flask and stopper.

The accuracy of thermometers used in the field may be checked against readings taken by a precision thermometer certified by the National Bureau of Standards (NSB). Thermometers should be checked for accuracy semi-annually.

Electronic temperature meters are available. The technician should follow the manufacturer's instructions for calibrating and using the meter.

Box 6-4 ■ Measuring Conductivity

Conductivity is a measure of the ability of an aqueous solution to carry an electric current. This is dependent on the presence of ions and the temperature of the solution. Electrolytic conductivity increases with an increase in temperature. Conductivity meters may measure conductivity only or conductivity along with other functions, such as temperature.

The Readout

The technician will record in the log the readout of the conductivity meter. Conductivity is reported on the instrument in any one of the following measurement units:

— μmhos (micro-mho) – the mathematical symbol for mho is Ω;
— mmhos (milli-mho);
— mS (milli-Siemens);
— μS (micro-Siemens).

Siemens (pronounced sēmunz) is a unit of electrical conductance in the International System, equal to one ampere per volt. Mho (pronounced mō) equals one Siemens. Conductance is mathematically reciprocal to resistance; resistance is measured in ohms. Mho is the backward spelling of ohm.

The technical meanings of Siemens, mho and ohm are beyond the scope of this text. The technician's role is to:

1. Calibrate the meter according to the manufacturer's instructions.
2. Perform the measurement.
3. Record in the log or on a data chart the readings and the measurement unit.
4. In the case of well development, repeat the conductivity measurement, as well as the temperature and pH measurements, each time water is removed until the measurements stabilize.
5. Decontaminate and store the instrument according to the manufacturer's instructions.

Materials

The following materials are needed:

— Temperature-compensating conductivity meter and probe
— Large beaker or glass jar or bucket
— Small glass beaker
— Conductivity calibration solution; for example, 0.01 M KCl (potassium chloride)
— Deionized water in a squeeze bottle

Conductivity Calibration Procedure

Always record calibration information in the field notes.

1. Turn meter on; wait for it to warm up.
2. Rinse probe with deionized water; wipe gently to dry. Then hold the probe's sensor in the air while checking the meter readout. The meter should read 0.0 μmho or 0.0 μS.
3. Immerse the probe into a small glass beaker containing 0.01 M KCl solution.
4. Measure the temperature of the 0.01 M KCl calibration solution and then read the conductivity. At 25°C, the conductivity meter should show a reading of 1413 μmho or 1413 μS. (For other conductivity calibration solutions, check the label for the expected conductivity reading.) Some models of conductivity meters correct for temperature or their manufacturer provides a temperature conversion chart. If not, the technician should convert the conductivity for the current water temperature to the conductivity of the calibration solution at 25°C:

$$\text{Conductivity at } 25°\text{C} = \frac{\text{measured conductivity}}{0.0191(t - 25) + 1}$$

where "t" equals temperature.

If the meter is not reading the expected conductivity value, the technician should check the instrument's batteries, use fresh calibration solution, and inspect the probe for defects. If the batteries check out and fresh solution is used, but the meter still doesn't calibrate, the probe or its sensor should be replaced or the meter checked by the manufacturer.

Measurement Procedure

1. Put well water into large beaker, jar, or bucket. Immerse probe into water to allow it to adjust to the water's temperature. (Refresh this water as needed to maintain temperature consistency.)
2. At each point in time when this measurement is needed, put well water to be measured into a large beaker, jar, or bucket. This can be the same container used in step #1 above or a different one. Take and record measurement.
3. After measurements are complete, decontaminate the containers used to hold the well water and the conductivity meter probe. Most manufacturers require the conductivity meter's probe (electrode) to be stored in deionized water. Be sure to follow the manufacturer's instructions for the use and storage of the instrument.

Box 6-5 ■ Measuring pH

pH is a measure of a solution's acidity.

pH can be measured using pH indicator paper with a color scale or a pH meter. The technician immerses the pH indicator paper into the well water and compares it with the color scale provided with the paper. The closest match of the color to the scale indicates the approximate pH value of the solution. Most pH indicator papers are divided into colors correlating to pH divisions of 0.5 to 1 pH unit.

A pH meter uses electrodes on a probe that is immersed into the well water being sampled. Measurements with a pH meter can usually be made to the nearest 0.1 pH unit.

Materials

—pH meter with probe
—pH buffer solutions: 4.00, 7.00, 10.00
—deionized water in a squeeze bottle
—battery-powered magnetic stirrer
—4 or 5 beakers
—wipes

Calibration Procedure

1. Turn meter on; wait for it to warm up. If the meter has a dial or button to compensate for temperature, adjust it according to the manufacturer's instructions. Some meters automatically compensate for the temperature.
2. Check meter's batteries and verify that pH = 7.00 on the meter without the probe attached.
3. Attach probe.
4. Rinse probe with deionized water, wipe gently, and immerse into a beaker containing pH 7.00 buffer solution. The beaker should be placed on a magnetic stirrer that minimizes the change in the pH of the solution from solids settling out of solution and the transfer of the solution's molecules into the air. Check the meter's pH readout. If necessary adjust the meter's calibration knob until the reading is 7.00.
5. Remove probe from 7.00 solution and rinse with deionized water. Wipe and immerse into a pH 4.00 buffer solution in a beaker placed on the magnetic stirrer. Check the meter's pH read out. If the reading is not 4.00, adjust the meter's "slope" with the appropriate knob or button as specified by the manufacturer.
6. Remove the probe from the 4.00 solution and rinse with deionized water. Wipe and immerse into a pH 10.00 buffer solution in a beaker placed on the magnetic stirrer. Read the pH value indicated by the meter. It should be at 10.00 or very close. Do not adjust the meter's calibration or slope knobs at this point; the instrument should be calibrated. If the reading is not 10.00, the instrument needs repairs or the probe should be replaced.
7. Remove the probe from the 10.00 solution and rinse with deionized water. Wipe the meter gently and place into a pH 7.00 buffer solution while waiting to take a reading of the pH of the well water.

Measurement Procedure

1. Put well water into a clean beaker, and place it on the magnetic stirrer. Rinse the pH meter's probe with deionized water and wipe gently. Immerse the probe into the well water sample and wait a minute or until the sample's pH reading remains steady.
2. Record measurement.
3. After each sample's pH reading, rinse the probe with deionized water and return the probe to the pH 7.00 buffer solution. Use a clean beaker for each sample.
4. After measurements are complete, decontaminate the containers used to hold the well water and the pH meter's probe. Most manufacturers require the electrodes on a pH meter's probe to be stored in a pH 7.00 buffer solution, not in deionized water. Be sure to follow the manufacturer's instructions for the use and storage of the instrument.

In case the pH meter malfunctions, it is wise to include a supply of pH indicator paper with the sampling supplies. The pH paper is not as accurate as a properly calibrated and working pH meter, but it is preferable to obtaining no data at all.

Exception 1: Removing three well volumes causes the well to go dry. If this happens, make note of how much water was removed and let the well sit for a few hours. Later, check the well and measure how much water is in the well. The water found in the well is representative of the aquifer.

Exception 2: The rate at which water is removed is faster than the rate at which water enters the well, and water from the aquifer is observed cascading down the sides of the well. This cascading action may cause chemicals present in the groundwater to volatilize and escape. This should be avoided. The technician must carefully remove an adequate volume of groundwater at a rate that ensures the water present in the well is representative of the aquifer.

Many sampling plans will specify that three well volumes should be purged before sampling the well water. Water removal during purging, like water removal during well development, may be based upon borehole volume. This means that the well volume would be considered to include the sand pack outside of the well screen.

Some project planners, however, consider a well volume to be based on the radius of the well screen and riser only. The volume is calculated the same way but using the radius of the well (i.e., riser and screen) rather than the one of the borehole. The table below provides a shortcut for the well volume calculation using a simple formula; it is based on the length of the water column in the well and the diameter of the well riser and screen (the well diameter).

Suppose, for example, a well is two inches in diameter and has a seven-foot water column. Using Table 6-1 the well volume is calculated as follows: Well volume = 7×0.1632 or about 1.14 gallons.

Some sampling plans will specify that purging continue until well water parameters such as pH, temperature, conductivity, or other chemical parameters have stabilized. The technician should always follow the purging guidelines provided in the sampling plan. Purged water is considered to be investigation-derived waste and may need to be stored in a labeled container until sampled and/or properly disposed of. This should also be indicated in the sampling plan.

The method of purging should be specified in the sampling plan as well. Purging may be accomplished using a bailer or a pump, which are described below. Whatever device is chosen, the technician must always purge with clean (decontaminated) equipment.

Well Diameter	v (volume in gal/ft)
1 inch	0.0408
2 inches	0.1632
3 inches	0.3672
4 inches	0.6528
6 inches	1.4688

Table 6-1: Calculating well volume.
Well volume (in gallons) = lv; where l = length of water column (in feet) and v = volume in gallons per linear foot as determined from the table.

Purging and Sampling Devices: Bailers and Pumps

Bailers

Bailers are the simplest, most commonly used devices for removing water from a well. In general, a bailer is a long tube with a ball check-valve at the bottom. Some bailers are designed with a check-valve at the top of the bailer. The top check-valve allows water samples to be collected from a specific point in the well. Bailers, which are available in various lengths and diameters, may be constructed of stainless steel, Teflon®, polyvinyl chloride (PVC) or polyethylene. A line is attached to the top of the bailer and it is lowered into the well to remove water. (The bailer is illustrated in Chapter 3, Figure 3-8.)

Advantages to using bailers include:

—Manual device: no power source needed.

—Readily portable.

—Inexpensive (may be disposed of after one use).

—Easy to decontaminate.

—May be dedicated to a specific well by a hanging device inside the well (thus, diminishing the possibility of cross-contamination).

—Rapid, simple method for removing small volumes of purge water.

—Stagnant water is removed first.

—Minimal escape of volatile organic compounds while water sample is in the bailer.

—Can capture a sample of **immiscible** materials (i.e., hydrocarbons) from the top of the water column in the well.

Disadvantages to using bailers include:

—Can be time-consuming to use, especially for large, deep wells.

—Labor intensive.

—When water is transferred from the bailer to the sample container, a loss of volatiles or addition of atmospheric oxygen may occur because of water turbulence. This disadvantage has been addressed by several manufacturers of bailers by providing controlled-flow, bottom-emptying attachments for their bailers.

—Some bailers are very lightweight and take a long time to sink into the water, increasing the time needed to complete purging and/or sampling activities. Some manufacturers have added weights to the lightweight bailers to solve this problem.

The technician typically follows this procedure when using a bailer:

1. Lay clean plastic sheeting around the well to prevent contamination of the bailer line with materials on the ground.
2. Attach a clean line securely to the bailer top so that it is not inadvertently lost in the well. Be sure the free end of the line at the surface is tied securely to the well or other object at the surface. The line should be strong enough to lift the bailer full of water repeatedly during purging and sampling. The line may be of stainless steel or nylon.
3. Lower the bailer into the well until it is completely submerged.
4. Lift the bailer out of the well carefully, minimizing or avoiding any scraping of the bailer on the sides of the well. Make sure the line falls onto a clean area of the plastic sheet and does not touch the ground. As the bailer is lifted, the check valve will close, trapping water inside the bailer.
5. Empty the water from the bailer into a bucket or container until the well is purged. The amount of purge water removed from the well can be gauged by the volume in the bucket, if the bucket is marked with volume levels, or by counting the number of full bailers obtained from the well if the bailer's volume is known. This purge water should be handled in the appropriate manner for disposal. Handling and disposal of purge waters should be outlined in the sampling plan.
6. After purging is complete, water samples from the bailer can be placed into the appropriate sample containers. Carefully and gently pour water from the bailer into the sample containers or use special bottom-emptying devices available with the bailer. It is common practice to use the same bailer for purging and sampling; however, always follow the instructions outlined in the sampling plan concerning use of the same bailer for purging and sampling. Document purging and sampling activities in the field notes.
7. Decontaminate and/or dispose of the bailer, sampling line, plastic sheeting and bucket as specified in the sampling plan.

Pumps

Several types of pumps are available. The most common sampling and/or purging pumps are non-gas contact bladder pumps, submersible pumps, suction pumps and inertia pumps.

Non-gas contact bladder pumps, also referred to as gas-operated squeeze pumps, are positive-displacement pumps that use a pulse of gas to push the water sample up out of the well. The gas does not contact the water sample which is collected by a flexible Teflon® bladder inside the rigid pump housing. The pump alternately fills and discharges. During the fill cycle, the bottom check valve opens, allowing the bladder to fill with water while an upper check valve prevents liquid in the discharge line from dropping back into the bladder. During the discharge cycle, the bottom check valve closes while compressed gas injected into the pump housing squeezes the bladder, unseating the upper check valve and forcing water up the discharge line. Positive pressure is maintained on the pump at all times. Teflon® tubing is recommended for use as the water discharge and air lines.

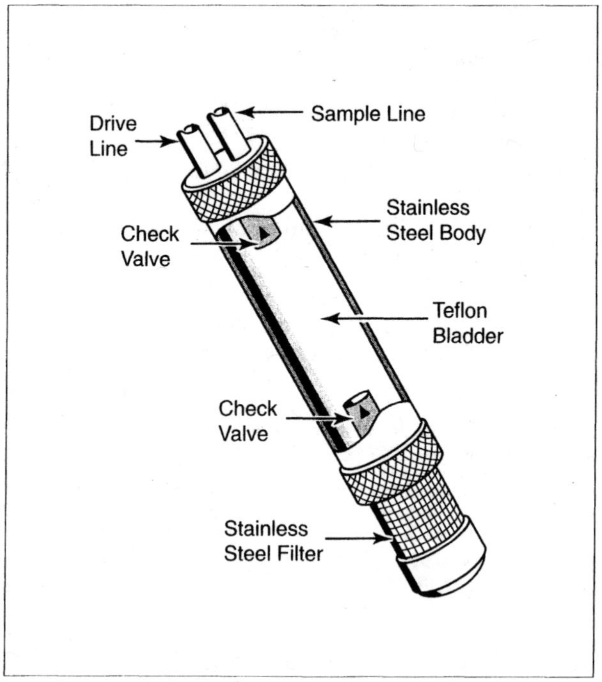

Figure 6-11: Non-gas contact bladder pump.

Advantages to using bladder pumps:

—Maintains integrity of sample.

—Easy to use.

—Can be permanently installed in a well.

—Electronic controller on the pump equipment allows water pumping rate to be varied as needed.

Disadvantages to using bladder pumps:

—Requires a power source and compressed air or gas supply.

—Difficult to clean.

—Wells deeper than 100 feet have decreased pumping rates. The bladder pump cannot be used in wells deeper than 250 feet.

—Supply of bottled air or nitrogen and/or compressor may be difficult to obtain and cumbersome. Compressor power needs to be increased with deeper wells. Relatively large amounts of gas are needed for lifting water from deeper wells.

—Expensive, especially for bladder pumps dedicated to wells.

The technician typically follows this procedure for using the bladder pump:

1. Lay plastic sheeting around the well to prevent contamination of pump, hoses or lines.
2. Assemble the bladder pump apparatus according to the manufacturer's instructions.
3. Lower the pump assembly into the monitoring well. Place the pump deep enough into the well so that pumping does not cause the water level in well to drop below the pump intake (running a pump without water may damage the pump). Adjust pumping rate and pump placement in the well while purging if necessary.
4. If the pump has a flow meter to measure the volume of water pumped, make sure it is attached and in ready position.
5. Turn on the power supply and purge the well of the desired volume of water or until field chemical parameters (temperature, pH, conductivity) have stabilized.
6. Collect purge water in a bucket or container. Handle purge waters as specified in the site sampling plan.
7. After purging is complete, adjust flow rate of pump for sample collection. Allow well to recharge, if necessary. Pump flow rates for sample collection will be much lower than the flow rate for purging.
8. Fill sample containers as indicated in the sampling plan. Make notes in the field notebook on purging and sampling activities.
9. Disassemble the pump apparatus after use. Decontaminate the Teflon® tubing and bladder or replace with new dedicated tubing and bladder. Decontaminate pails or buckets used to collect the purge water. Dispose of plastic sheeting.

Submersible pumps are inserted into the well to pump water to the surface. Newer submersible pumps are constructed of relatively inert materials that do not contribute trace levels of contaminants to the sample, such as stainless steel and Teflon®. The newer submersible pumps also have electronic controls for adjusting the rate of water pumped for sample collection.

Advantages to using submersible pumps:

—Portable.

—Generally reliable, do not require priming.

—Relatively high pumping rate (depending on size of pump and its depth in the well). Some pumps are capable of variable pumping rates.

—Greatest advantage may be for purging wells before sampling with another device such as a bailer (if pump has a check valve to prevent purged water from flowing back into the well).

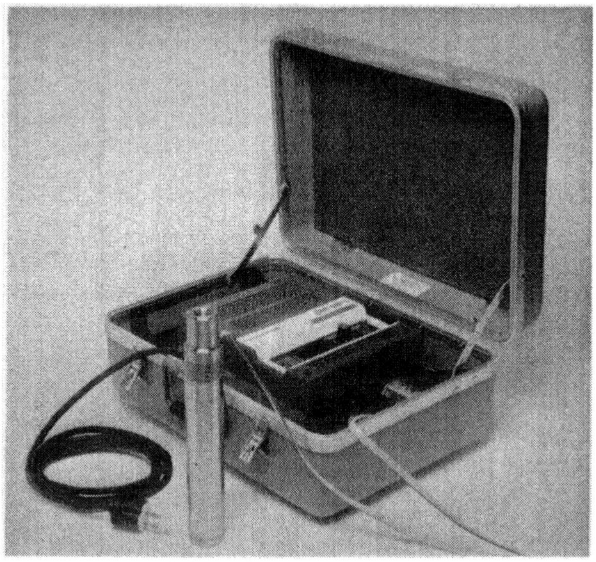

Figure 6-12a: Submersible pump.

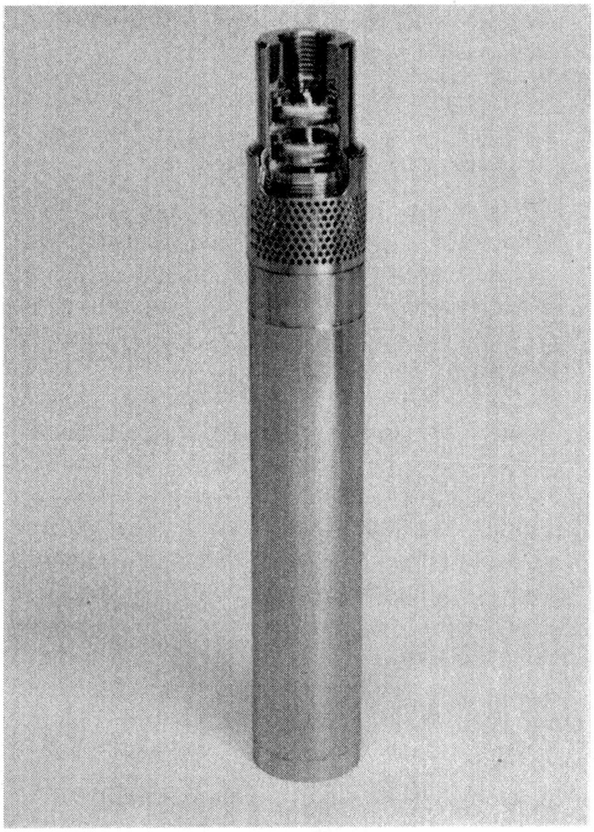

Figure 6-12b: Close-up of submersible pump.

—Some pumps are constructed of inert or nearly inert materials.

Disadvantages to using submersible pumps:

—Heavy and cumbersome; may be difficult to use.

—Power source required.

—Expensive.

—May be damaged by silt or sediment, necessitating frequent repairs.

—May be impractical in shallow wells or wells with low water yields.

—Depending on the pump construction materials (plastic, rubber or metal), trace organics/inorganics may leach from the pump and affect the accuracy of measuring those materials.

—Pump may generate water samples that contain suspended sediments which may result in higher metal concentrations.

—Decontamination may be difficult.

The technician typically follows this procedure for using a submersible pump:

To purge a well with a submersible pump, follow the steps indicated earlier for the bladder pump. If the submersible pump is to be used for sample collection, follow the additional bladder pump sample collection steps. If the submersible pump is not to be used for the actual sample collection, remove the pump from the well, decontaminate it according to the site sampling plan, and follow the sampling steps listed in the sampling plan for the specified sampling device (e.g., a bailer).

A *suction pump* is a surface-mounted pump that creates a vacuum; the vacuum draws water through a sampling tube and the pumping system. Suction pumps may be used for purging water from shallow wells (less than 25 feet deep). All suction-pumping systems, except for the special suction pump known as the peristaltic pump, are unacceptable for sample collection because the water may be exposed to metals and lubricants. Additionally, the movement of the water in the pump may cause **degassing** of the water, which causes a loss of volatile organics. The theoretical suction limit for suction pumps is 34 feet. Thus, if groundwater is located deeper than 34 feet in the well, suction pumps won't work. Realistically, suction pumps are not effective with water at depths greater than 25 feet.

Peristaltic pumps move the water within the sampling tube without allowing the water to contact the pump. Degassing may still occur with peristaltic pumps; the flexible sampling tube may be constructed of a material that leaches chemicals into the water. The pumping rate on peristaltic pumps can be adjusted.

Advantages to using suction pumps:

—Portable.

—Inexpensive.

Disadvantages to using suction pumps:

—Only useful in wells up to 25 feet deep.

—Vacuum may cause loss of dissolved gases and volatile organics.

—Pump must be primed and vacuum may be difficult to maintain.

—May result in pH modification.

—Most types cannot be used for sample collection (some peristaltic pumps excepted).

For purging wells, the technician follows the procedures for bladder pumps and submersible

Box 6-6 ■ Well Sampling Data

(to be included in field notes or on a separate form referenced in field notes)

Well Sampling Data

Date: ______________________ Name of Technician(s): ______________________

Site/Project Name: ______________________

Well Number: ______________________ Well Location: ______________________

Well Elevation: ______________________ (Surveyed measuring point at top of casing, TOC)

PURGING: (All measurements are made from the measuring point at TOC)

Initial Measurements Time Initial Measurements Made: ______________

Depth to Water (DTW): ______________________

Depth to Bottom of Well (DTB): ______________________

Length of water column in well (L): ______________ (DTB – DTW = L)

Calculation of volume of water in well:

Ending Measurements Time Ending Measurements Made: ______________

Depth to Water (DTW): ______________________

Depth to Bottom of Well (DTB): ______________________

Description of Purging Method Used (include pump or device used): ______________________

____

____

Total Volume of Water Purged (in gallons): ______________________

Field Measurements (Purging)

Date	Time	Volume Removed	Temperature (°C)	Conductivity (mS or μS)	pH	Comments

SAMPLE COLLECTION:

Depth to Water (measured from top of casing)

Initial Measurement & Time of Reading: ______________

Ending Measurement & Time of Reading: ______________

Description of Sampling Device & Method Used: ______________________

____

____

Sample Collection Data

Sample ID	Time	Pumping Rate	Field Chemical Parameters (i.e., pH, temperature conductivity)	Lab Analysis to be Performed	Number of Containers	Types of Containers	Preserv-atives	Appearance & Other Comments

pumps. Suction pumps are not, however, recommended for collecting samples. If a pump is desired for sampling, bladder pumps are preferred.

Inertia pumps are simple tube or pipe devices with a check valve located at the bottom of the pump. The tube is manually raised and lowered several times, which forces water to move upwards in the tube. Although inertia pumps may be used to purge a well, the agitation and lack of flow control of these devices make them unacceptable for use as sampling devices.

Advantages to using inertia pumps:

—Portable.

—Inexpensive.

Disadvantages to using inertia pumps:

—Only useful in wells less than 70 feet deep.

—May be time-consuming to use.

—Labor intensive.

—May only be effective in narrow-diameter wells (i.e., two-inch diameter).

The technician should always follow the instructions specified in the sampling plan for the choice and use of purging and sampling devices. Purging and sampling activities should be documented in field notes; tables set up in the field notebook will assist in thorough notes and data collection. A suggested format for recording well sampling data is shown in Box 6-6.

Physical Tests on Wells and Aquifers

Site characterization includes describing the aquifer characteristics of a site by performing **slug tests** and **pumping tests**. Performance of these tests requires training and experience beyond the scope of this book, but a brief description of these tests is included for the student's information.

Slug tests are used to measure the hydraulic conductivity of the aquifer at a singular sampling point of a well. Hydraulic conductivity is the rate at which a fluid (groundwater) can move through the permeable medium (the aquifer). Slug tests are performed by instantaneously removing a known volume, or a "slug", of water from a well, which will reduce the water level by a known amount. Then the rate of recovery is measured, that is, the time it takes for water to rise to its initial level before the slug was removed.

Pumping tests measure hydraulic characteristics of the aquifer as well as the rate at which a well will yield water. These tests are performed by pumping a well for a long period of time and observing the change in the water level in the well and/or other surrounding wells.

Checking Your Understanding

1. Describe the data gathering associated with opening a monitoring well.
2. Explain why a well is developed and when its water is considered stabilized.
3. Explain why a monitoring well is purged before sampling.
4. Calculate the volume of a bore hole that is eight inches in diameter by nine feet deep. The depth-to-water measurement is three feet.
5. Describe how the technician takes measurements for temperature, conductivity and pH.
6. Calculate well volume for a well using the following information:

   casing + screen = 3" i.d.
   DTW = 21'
   DTB = 30'
7. Using the data from the previous question, calculate how much water needs to be removed from the well if the sampling plan indicates three well volumes will be purged before sampling.
8. Explain why the bladder pump may be preferred over the various submersible pumps.
9. How is development water or purge water handled after it is removed from the well?

6-4 Sampling Surface Waters

Concepts

- When working around surface water bodies, the technician wears a personal flotation device and appropriate protective clothing for safety.
- During sampling activities, disturbing underlying sediments may contaminate water samples.
- Devices for collecting samples at depth include the bomb sampler, the Kemmerer bottle and the subsurface grab sampler.
- The direct method of sampling the surface requires collection activities that begin at the furthest downstream target.
- Samples may be collected at the water's surface using a dip sampler or by directly collecting water in a sample container.

Surface waters include lakes, rivers, creeks, streams, ponds, surface impoundments and drainage or irrigation ditches. These settings often present challenges to sampling. The technician may need to use a boat, lean off a bridge, or wade in the water to get to the sampling point. When working around surface water bodies, the technician should always wear a personal flotation device (PFD) and other appropriate protective clothing.

Surface waters can be categorized by their uses:

—Potable water supply

—Fish, shellfish and wildlife maintenance and propagation

—Recreation (swimming and boating)

—Agricultural or industrial

—Navigation

Samples may be collected from the water's depth as well as from the surface. The sampling position (deep vs. shallow) depends on the chemical contamination suspected. If the chemical is more likely to be found at or near the bottom of the water body because of the density of the chemical, then it makes sense to sample the water "at depth," not near the water's surface. The choice of sampling devices will also depend on the characteristics of the water body: shallow or deep? still or flowing? broad or narrow?

When sampling surface waters, the technician must take care not to disturb sediments. The sediments may have contamination or metals associated with them that the free flowing water normally does not have. It is wise to follow the practice of sampling at the furthest downstream location first, since sample collection procedures often disturb sediment or otherwise alter downstream water quality characteristics.

The technician should always decontaminate sampling devices between samples and put samples into a cooler with ice as soon as possible after collection.

Sampling Devices

Devices for collecting samples at depth include the bomb sampler, the Kemmerer bottle and the subsurface grab sampler. Bailers and pumps may also be used to sample surface waters, although they are more commonly used in groundwater sampling.

The devices used to collect samples of surface waters at depth can, in some instances, also be used to collect water from the top or near the top of the water's surface. However, dip samplers or the "direct method" of sample collection are often the best techniques to use for samples at the surface.

Bomb Sampler

The bomb sampler is a device for obtaining discrete samples at a specific depth or interval in the water.

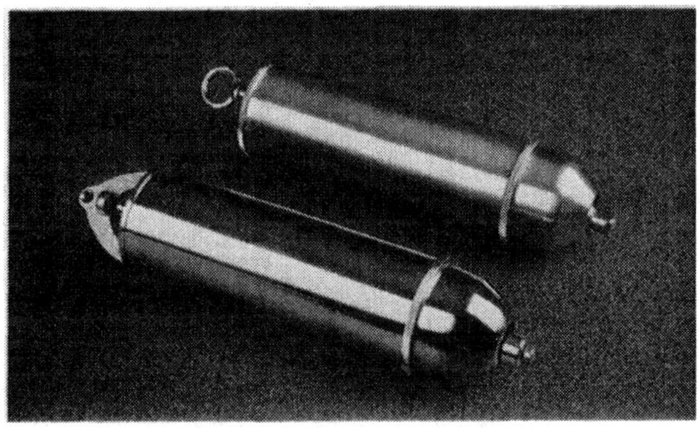

Figure 6-13: Bomb samplers.

The sampler is somewhat cylindrical in shape and may be constructed of stainless steel, Teflon® or nickel-plated brass. This device has two separate cables, one for lowering and raising the sampler in the water and the other to open a plunger for sample collection.

The technician typically follows this procedure when using the bomb sampler:

1. Assemble a clean bomb sampler device. Make sure device is closed.
2. Lower the bomb sampler into the water with the suspension cable. Keep the cable attached to the trigger of the plunger relaxed. When the desired sampling depth is reached, open the sampler by pulling the trigger cable taut. This allows the reservoir chamber to fill with liquid.
3. Allow the trigger cable to go slack; this should close the bomb sampler device. Retrieve the sampler with the suspension cable used to lower the device.
4. Position the sampler over the appropriate sample container. Transfer the sample to the container by pulling on the trigger cable to open the sampler.

Kemmerer Bottle

Like the bomb sampler, the Kemmerer bottle is a device for obtaining discrete samples at a specific depth or interval in the water. Kemmerer bottles may be constructed of stainless steel and/or Teflon®. The sampler has a cylindrical body with stoppers at the top and bottom of the device. The stoppers are held by a rod which extends lengthwise through and parallel to the center of the cylinder body.

The technician typically follows this procedure when using the Kemmerer bottle:

1. Assemble a clean Kemmerer bottle so that the stopper end pieces are away from the cylinder body of the bottle; in other words, arrange the device in the "open" position. This will allow water to pass through the sampling device.
2. Lower the Kemmerer bottle into the water with the suspension cable. When the desired sampling

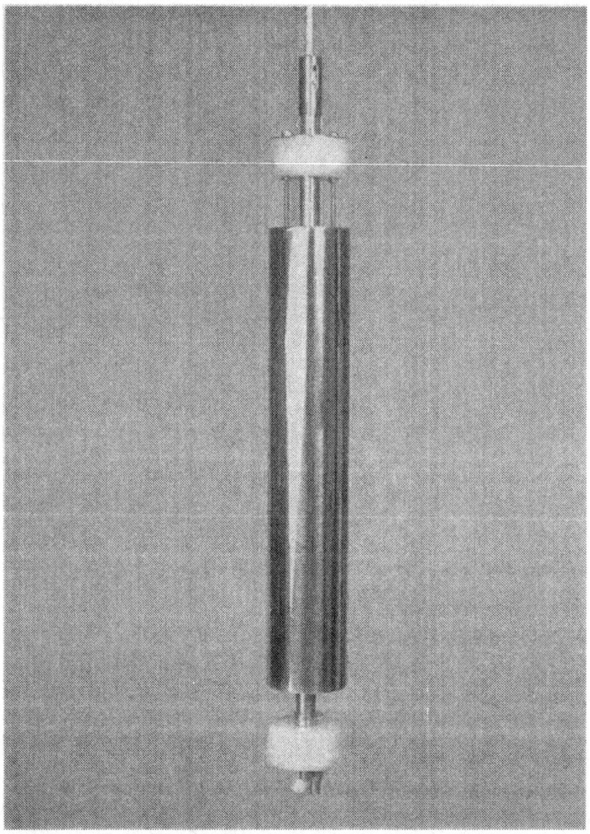

Figure 6-14: Kemmerer bottle.

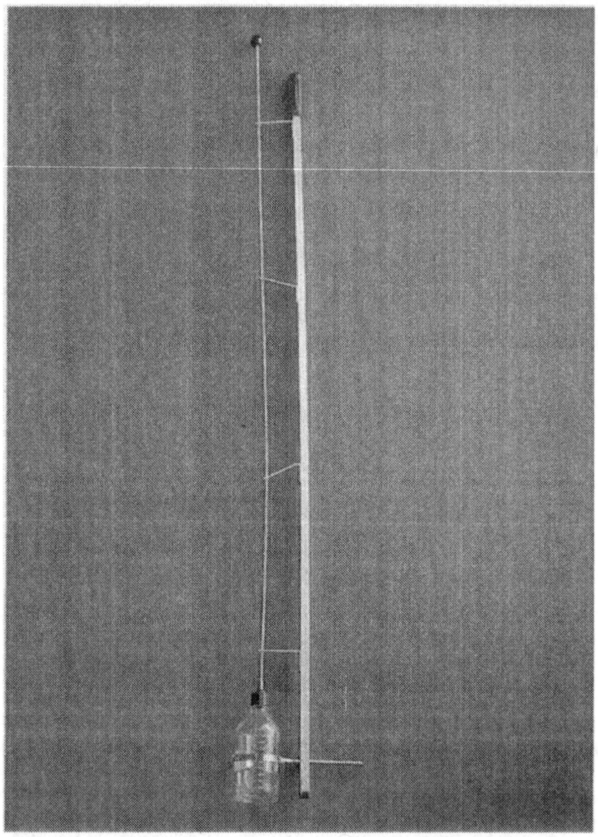

Figure 6-15: Subsurface grab sampler.

depth is reached, "send down the messenger"; this will close the sampler, trapping a sample of the liquid inside.

3. Retrieve the sampler with the suspension cable used to lower the device.
4. A drain or valve at the bottom of the bottle is used to transfer the sample into sample containers. Before transferring the sample, discharge a small amount of the water collected to clear any sediment or contamination from the valve.

Subsurface Grab Sampler

Subsurface grab samplers are also designed to collect samples at specific depths. In one design, a capped sample container is clamped or attached to the end of the grab sampler apparatus. The container can be opened at the desired depth to collect the sample and then closed before retrieval.

The technician typically follows this procedure when using the subsurface grab sampler:

1. Clamp the sample container onto the end of the sampler's aluminum tubing. Affix the cap opening device to the container's cap.
2. Submerge the closed container to the desired sampling depth.
3. Operate the cap opening device to unscrew the lid on the bottle. Allow liquid to enter the bottle.
4. Re-close the cap on the grab sampler.
5. Lift the sampler out of the water.

Other designs are available for subsurface grab samplers. Always read and understand the instructions for using the particular subsurface grab sampler you are operating.

Figure 6-16: Dip sampler.

Dip Sampler

Dip samplers, which are also known as pond samplers, consist of a plastic beaker attached at the end of a long pole. Some samplers may have adjustable-length poles. These devices are intended to collect samples at the water's surface. In some cases, a stainless steel, "soup ladle"-type dipper may be the most appropriate sampler to use.

The technician typically follows this procedure when using the dip sampler:

1. Assemble the clean sampler according to the manufacturer's instructions.
2. Extend the pond sampler to the sample location.
3. Gently lower the sampler at approximately a 45° angle into the water, just at the water's surface.
4. Allow the water to fill the sampler.
5. Rotate the sampler into an upright position.
6. Retrieve the sampler and transfer the sample into the appropriate sample container with as little agitation as possible.

Direct Method

The direct method involves the technician filling a hand-held sample container by submerging it directly into the water. When utilizing this method of sample collection, it is very important for the technician to use the appropriate protective clothing (gloves, waders or other boots, personal flotation devices, etc.) as well as the proper technique for collecting the sample. If sampling in a stream or other flowing water, the technician should always start the sampling process at the sample collection point furthest downstream. This will prevent sediment stirred up during the sampling process from contaminating the samples. Sample containers to which the laboratory has already added a preservative are not appropriate for this sampling method.

The technician typically follows this procedure when using the direct method:

1. Access the sampling point from a downstream position.
2. Open the sample container and point it upstream while standing downstream. If sampling is being conducted from a boat, avoid collecting the sample in the boat's wake.
3. Hold the sample container at a 45° angle and lower it gently into the water. (Alternatively, the sampling plan may call for the sample container to be inverted before it is lowered into the water and gently rotated upright underwater before the container is lifted out of the water.)
4. Allow the sample container to fill. Avoid collecting debris on the water's surface.
5. Gently lift the container out of the water. Attach the lid to the container securely. Wipe the container off with a paper towel or other disposable cloth.

Checking Your Understanding

1. Explain why the technician should avoid disturbing sediments during surface water sampling.
2. Describe the features of surface and subsurface water samplers and the procedures for their use:
   a. Bomb sampler
   b. Kemmerer bottle
   c. Subsurface grab sampler
   d. Dip (pond) sampler
3. Describe the precautions to be followed when using the direct method of taking a sample from the water surface.

6-5 Sampling Drinking Water and Wastewater

Concepts

- Deteriorating lead, copper or asbestos pipes in a drinking water supply system may be a contamination source.
- Drinking water may be sampled directly from a potable well or from a tap.
- Wastewater influent samples are for reference and comparison; effluent samples indicate the quality of water discharged.
- Dissolved oxygen can be analyzed in the field electronically or in the laboratory by the Winkler method.

Drinking water and wastewater sampling may be required for site characterization at hazardous waste sites. Drinking water sources at cleanup sites are sampled to determine if contamination is entering the water in supply lines. The water supply system itself may be a source of contamination if the piping is constructed of lead, copper or asbestos and is deteriorating. Wastewater is a potential source of contamination to surface water and groundwater. Wastewater may need to be sampled to determine if it is causing contamination at the site, to verify that a site's **NPDES (National Pollution Discharge Elimination System) permit** is being followed, or in the case of a special treatment system designed to treat contaminated water at the site, to verify that the wastewater from the system meets minimum treatment standards.

Sampling Drinking Water

Water may need to be sampled directly from a potable well. In many cases, potable wells may be sampled in the same way that monitoring wells are sampled. The well will need to be purged until fresh water from the aquifer is obtained.

Often water is not sampled from the water source (the well), but directly from the water distribution system (the tap). The sampling plan should specify where and how to take the samples. Some samples, such as those for lead and copper analyses, are referred to as **first draw samples** and need to be collected after the water has been sitting undisturbed in the plumbing system for a number of hours (usually six hours or more). Other samples, such as those for volatile organic compounds (VOCs), require that the system be flushed for a period of time.

This section of the chapter is an overview intended for technicians who will collect samples for site characterization purposes. The specific training of public water system operators whose responsibilities include compliance with the Safe Drinking Water Act is beyond the scope of this book. Information about operator training is available through state health and environmental departments as well as through the National Environmental Training Association (NETA).

Sampling Wastewater

Both the **influents** and effluents from wastewater generating processes are sampled. Influent samples are taken for reference and comparison purposes. Effluent samples are indicators of the quality of discharged wastewater. Effluent samples are collected downstream of the process that generated wastewater but before the water enters the sanitary sewer, storm sewer, or other discharge point. Wastewater samples are best taken from discharge areas where good mixing occurs. The equipment and procedures

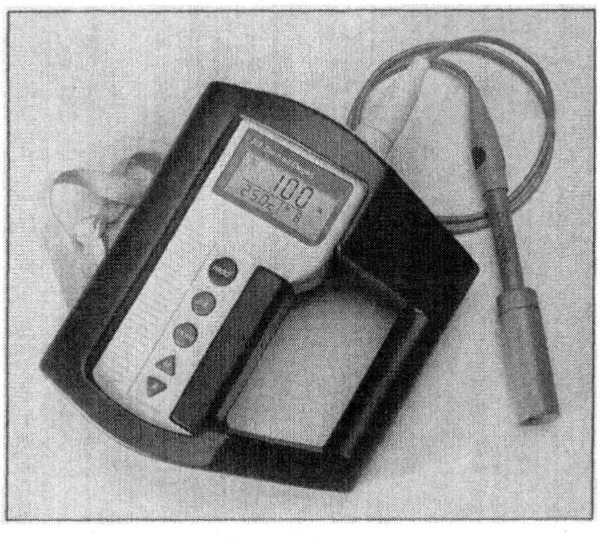

Figure 6-17: Dissolved oxygen meter.

Box 6-7 ■ Collecting Drinking Water Samples

Samples to be analyzed for copper and lead contamination are taken from the first draw of the water through a tap. Most other samples are taken from a flushed system in which the water line has been cleared for several minutes so that sampled water will come from the main water supply.

1. Study the sampling plan and laboratory instructions carefully before sampling.
2. Organize equipment: sample containers; rack to keep containers in an upright position; ice chest chilled with frozen cold packs; tools for removing aerators or other tap attachments; cloths for cleaning the sampling area; a butane torch if required by sampling plan.
3. Beware of introducing contaminants when collecting microbiological samples. Coliforms may be carried in dust, which may contaminate an open sample container during water sample collection.
4. It is important that the selected site be physically clean and free from visible dirt and that taps be located in an area protected from wind. Bacteria are easily carried on minute fecal and decay particles. At dirty, exposed locations it is difficult to collect an uncontaminated sample. The total coliform analysis is simply a presence/absence test; if just one coliform bacterium gains entrance to the sample container, the test will indicate a contaminated water supply.
5. If sampling from a tap, remove attachments such as a hose, aerator or strainer, and wipe dust or other materials off the tap. If required by the sampling plan, expose the tap opening to a butane flame to burn off remaining organisms.
6. Avoid substituting one type of container for another, even if they look the same; for example, one plastic container may seem to be the same as another yet be of a different composition that could potentially interfere with a sample.
7. Remove a sample container cap just before – and replace it immediately after – collecting a sample. Because contaminants are easily transferred from a contaminated cap to the water being sampled, determine how best to protect a cap from contamination while it is off the container; for example, avoid laying a cap on the ground or on other potentially contaminated surfaces, avoid touching a cap's inner surface, and do not place a cap in a pocket.
8. Determine from the sampling plan instructions if the sample is from "first draw" or from flushed line. Maintain the flow to about the diameter of a pencil so that the water will easily enter the container without touching its outer edges. Avoid letting the water from the tap flow down the outside of the container. Keep flow constant throughout the collection. Fill containers according to instructions from the lab or sampling plan; do not overfill, underfill, or rinse and refill. If sampling for VOCs, follow the procedure described in Section 6-6.
9. Complete the documentation. The sample label should include the following: ID number(s); sample type (raw, plant tap, entry point, or distribution); exact location of sample site; date and time sample was collected; collector's name; type of sample (grab or composite); and type of analysis to be conducted.
10. If a preservative is packaged separately and must be added to the sample after collection, follow directions carefully. Some preservatives are hazardous, such as nitric acid (HNO_3). Store samples upright in a rack in the ice chest. To chill samples, use frozen ice packs. Avoid dry ice as it may solidly freeze the samples, and avoid ice since melting causes potential contamination and packing problems. Adhere to holding times, especially in coliform analysis.
11. Perform field measurements (e.g., pH, temperature, conductivity) as directed by the sampling plan. Measure temperature with a handheld thermometer or a combined temperature/pH electrode and meter. Hold a thermometer in the water stream until the temperature reading stabilizes, or remove the faucet aerator and run enough water to flush the line before filling a sample bottle to slightly overflowing. Cap the sample with a cap that allows insertion of the thermometer. Record the temperature once the thermometer stabilizes.

Calibrate the pH probe as specified by the equipment manufacturer. After taking the pH reading, rinse the electrode with deionized water and return it to its holding bottle.

Field analyze suspended materials with a turbidimeter. Follow manufacturer's instructions for calibration and use.

Field analyze chlorine level as instructed by sampling plan (e.g., by colorimetric method or amperometric titration method) within 15 minutes of collecting the sample. If not to be analyzed immediately, keep chlorine samples still and away from bright light.

Source: Cynthia Root, "Operator Training Guide on Total Coliform Rule," Phoenix, NETA and CETC (under a grant from EPA), 1993

described for collecting samples from surface waters are applicable to wastewater sampling.

The NPDES permit, which is a regulatory permit under the Clean Water Act, will specify the parameters to be analyzed in the water. Dissolved oxygen (DO) is a key parameter commonly measured in evaluating wastewater process control and water pollution. DO is analyzed either in the field by using an oxygen-sensitive membrane electrode method, or in the laboratory by the Winkler method of iodometric titration.

The Winkler method, though the most precise and reliable way to measure DO content in water, is not convenient for field work. Samples to be analyzed by the Winkler method have a holding time of eight hours and involve collecting a water sample in a special glass bottle with a glass stopper. The Winkler method also requires that manganese sulfate and alkali-iodide-azide reagents be added to the sample after collection and just before titration (see Chapter 8).

It is far more convenient to measure DO level in the field by using an oxygen-sensitive membrane electrode. Since DO meters are sensitive to temperature, corrections must be made to compensate; many meters do this automatically. Some manufacturers specify that the meter needs to be calibrated at four-hour intervals during field work and also at the end of the working day. The calibration should be done at least once daily. While the instrument is taking a DO reading, there needs to be a flow of the solution under the membrane of the electrode; therefore, the technician either moves the probe around in the solution during readings or uses a magnetic stirrer. DO is measured and reported either as a percent by volume or as mg/L (ppm).

Checking Your Understanding

1. Distinguish first draw samples from flushed line samples.
2. Describe the equipment, procedure and considerations associated with collecting drinking water samples.
3. Describe the field observations and measurements associated with sampling drinking water.
4. Explain the rationale for clean techniques when sampling for coliforms.
5. Explain the purpose of collecting both influent and effluent samples.

6-6 Storing Water Samples

Concepts

- VOC samples, which are always grab samples, are collected first to minimize the loss of volatile compounds and are accompanied to the lab by trip blanks.
- Water samples containing highly toxic hexavalent chromium are sampled and handled differently from other metal samples and have a holding time of only 24 hours.
- Samples containing dissolved and suspended metals are field- or lab-filtered to capture particles larger than 0.45 microns (μ) in size.
- Inorganic, nonmetallic parameters include nitrogen and phosphorus, which are plant nutrients; cyanide; bacteriological organisms; radiological materials; various minerals such as chloride and calcium; total organic carbon and halogen; and chemical and biochemical oxygen demand.

Since time during field work is precious and limited, it is best to be well-organized before going onsite. Studying the sampling plan and all of its technicalities is important in achieving sampling goals for the site investigation. The sampling plan will tell the technician what kinds of samples to collect and where and how to collect them. This includes specifications for QA/QC samples (See Chapter 2). The Chapter 5 list of equipment required for sampling soils may be expanded for water sampling activities. Before going to the field, sample containers may be prepared with the specified preservative and partially completed labels may be attached. Information that can go on labels prior to sampling includes ID number, sample location and analysis requested.

Sample containers, preservatives, analytical methods and holding times should be specified in the sampling plan (see Appendix 6, Sampling Usage Guides). The sampling plan should be followed in detail; any deviations from the plan should be documented in the field notes. In almost all cases, and unless otherwise specified in the sampling plan, samples need to be chilled to 4°C. Chilling preserves the sample's integrity by slowing down chemical and biological processes that could change the sample's chemistry. Other preservation methods include adding chemicals such as acids or bases to the sample container before the sample is collected to slow down or stop chemical reactions that could affect the accuracy of analyses. These preservatives include:

—hydrochloric acid (HCl)

—nitric acid (HNO_3)

—sulfuric acid (H_2SO_4)

—sodium hydroxide (NaOH)

—sodium thiosulfate ($Na_2S_2O_3$)

—sodium sulfate (Na_2SO_4)

—ascorbic acid ($C_6H_8O_6$)

Specially prepared sample containers, to which preservative has already been added, are often provided by the laboratory. These pre-preserved sample containers are not appropriate for use in the direct method of surface water sample collection, since the preservative may wash out as the container is lowered directly into the water.

Sample containers are chosen with the contaminant in mind. For example, samples that will undergo analysis for organic molecules will be collected in a glass container, never in plastics. Plastics are formed from organic molecules and could affect the outcome of the sample analysis. In addition, organic molecules can permeate plastic containers, potentially increasing or decreasing contaminant levels in the sample. Samples collected for metal analysis are collected in plastic containers, because silica could leach out of a glass container, thus affecting the lab results.

Accurate and precise chemical analysis requires an adequately large sample. This may require the collection of more than one container of the sample. The number of containers to be collected and/or the quantity of sample needed should be specified in the sampling plan.

As samples are collected, label information should be completed, chain-of-custody forms filled out, and field notes taken in order to detail sampling activities and data collected. Waterproof ink should be used. Whenever possible, two people should work together to collect samples: one person to operate the sampling device, the other to handle the sampling containers.

When collecting samples and filling sample containers, the technician should always wear gloves specified by the glove manufacturer as appropriate for the chemicals and materials being sampled. Gloves should be changed often to prevent cross-contamination.

Caps must not be removed from sample containers until the technician is ready to sample. The technician should not touch the inside of the sample container or the cap, even if he or she has gloves on. Unless the sampling plan indicates otherwise, samples are transferred directly from the sampling device to the sample container and placed as soon as possible into a cooler chilled with ice to 4°C.

Storing Volatile Organic Compounds (VOCs)

Volatile organic compounds (VOCs) are present in gasoline, kerosene, organic solvents, paints, degreasers and trihalomethanes. When the sampling plan calls for a variety of chemical analyses, VOC samples are collected first to minimize the loss of the volatile compounds. VOC samples are always collected as grab samples. They should not be composited, because the compositing process would cause a loss of VOCs from the sample.

Sample containers used for VOCs are 40 milliliter (mL) borosilicate glass vials with a screw cap. The screw cap contains a Teflon®-lined silicon septum that can be punctured by the laboratory to withdraw an aliquot of the water for analysis. These septums should be kept clean and should not be written on. Generally, three 40 mL containers are collected for each sample. VOC samples are generally preserved by chilling to 4°C, but some sampling plans may require a chemical preservative. The containers may either come from the laboratory pre-preserved or they may need to be prepared with preservative by the sample technician. The holding time for VOC samples is seven to 14 days, depending on the analytical method.

A trip blank is required for each cooler containing VOC samples The trip blank is a VOC sample container that has been filled with deionized water, usually by the laboratory, and is kept with the VOC sample vials throughout the trip. It is intended to detect any contamination of samples that may result from organic compound vapors permeating the septum of the sample container during sampling or sample transportation.

The technician typically follows this procedure when sampling for VOCs:

1. Remove the screw cap from the vial. Do not touch the septum or inside of the cap.
2. If the sample is to be collected from a tap, remove the aerator from the spout. Turn on the cold water faucet to flush the system of water that may have been sitting in the piping. Let the water run until its temperature stabilizes. Decrease the water flow to a thin stream to fill the sample vials.
3. While filling the vial from the tap or sampling device, avoid agitation of the water. Fill the vial completely, letting a convex **meniscus** form at the upper surface of the water.
4. Place the screw cap for the vial over the meniscus and close securely. Do not delay in performing this step as volatile compounds may escape from the sample.
5. Invert the sample and tap it gently to check for air bubbles. If air bubbles are present, discard the sample and collect another. You may try carefully removing the cap and adding a few more drops of water from the sampling device to obtain a sample without bubbles. However, this process releases VOCs and may cause the preservative to be diluted. It is best if you do not reuse the sample container unless preservative is not required by the sampling plan or you add preservative to the vial (generally you will not have preservative with you in the field unless it is in pre-prepared sample containers.) Never reuse vials between two separate samples.
6. Fill three vials or the quantity required by the sampling plan.
7. Complete the sample labels and make sure they are securely attached to the vials. Place the vials in a foam holder and/or wrap them securely to prevent breakage during shipment.
8. Immediately place samples in a cooler with ice and chill to 4°C. A trip blank should be present in the cooler. It is best to segregate VOC samples in a separate cooler away from other samples to avoid cross-contamination.

Storing Extractable Organics

Fuel oils, coal tars, phenols, plasticizers, dye residues, wood preservatives, greases, pesticides, polychlorinated biphenyls (PCBs) and dioxins are some of the semivolatile chemicals or chemical products that are put through an extraction process before analysis. The extraction process is necessary because these chemical compounds have low vapor pressures (they are semivolatile) and will not readily change to the vapor state for analysis like VOCs. A solvent such as methylene chloride or hexane is used to extract the semivolatile compounds. In some analyses, the sample needs to be extracted within a certain period of time (most often within seven days) after sample collection. The extraction then must also be analyzed within a set period of time after the extraction process, which may vary from seven to 40 days.

Samples for extractable organic analyses are collected in one-liter amber-colored borosilicate glass bottles with Teflon®-lined screw caps. The number of bottles to be filled with the sample varies according to the requirements of the specific analysis, the analytical method, and the laboratory's requirements, but should be specified in the sampling plan. The bottles should be filled with the sample water until they are almost full. As with all samples, the cap should be securely screwed on the sample bottle. Some types of samples, such as those for greases, PCBs and pesticides, may need to be preserved with the addition of sulfuric acid (H_2SO_4) or sodium hydroxide (NaOH). Other types of samples may need other chemical preservatives. The preservative will be specified by the analytical method and the sampling plan. Additionally, the samples should be placed in a cooler with ice and chilled to 4°C.

Storing Metals

Analysis for metals in water samples may include total metals, dissolved metals, suspended metals, and the highly toxic hexavalent chromium (Cr^{+6}). Hexavalent chromium samples are handled differently from other water samples.

Samples to be analyzed for metals are collected in polyethylene bottles with polypropylene caps (polyethylene and polypropylene are specific types of plastics). The bottles should be filled with the sample until they are almost full, and the lid screwed on tightly. The quantity of sample required for analysis depends on the method and laboratory. As always, the technician must follow the sampling plan.

Total Metals

Samples to be analyzed for total metals are preserved with nitric acid (HNO_3) and have a holding time of six months. They do not need to be chilled in a cooler.

Dissolved Metals

Samples to be analyzed for dissolved metals should be filtered in the field during sample collection before the sample is placed into the plastic bottle containing the preservative, usually nitric acid (HNO_3). In some cases, hydrochloric acid (HCl) may be the preservative specified in the sampling plan. The filters consist of a mat of fibers or a thin membrane designed to capture particles that are larger than 0.45 microns (μ) in size. The technician filters the sample by using a pressure or vacuum filtration apparatus; a filter may also be placed directly in-line with a groundwater sampling pump or bailer, which reduces the need to clean the filtration apparatus. It also minimizes the potential of exposing the sample to the atmosphere and additional contaminants. Before filtering water samples, it is recommended that the filter first be washed or flushed with deionized water. The technician should use a new filter for each sample. Samples that have been field filtered have a holding

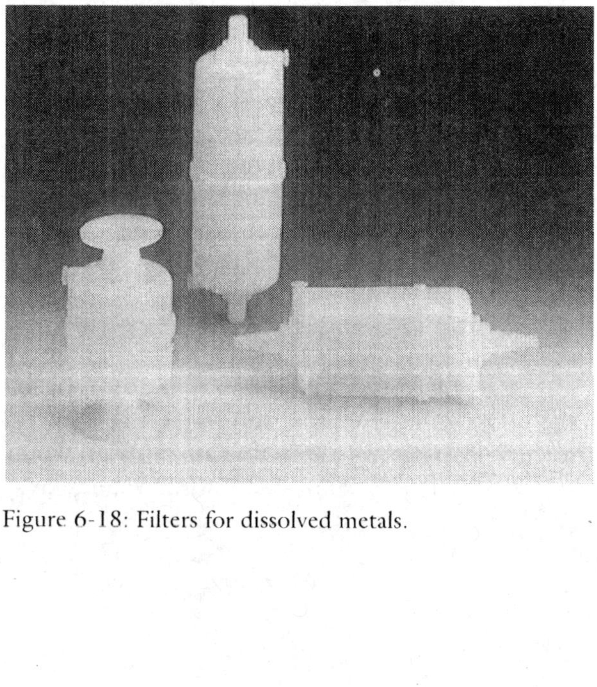

Figure 6-18: Filters for dissolved metals.

time of six months and do not need to be chilled in a cooler.

If samples cannot be filtered in the field, the samples should not be preserved (no acid preservative added to the plastic bottle). The acid used for preservation can cause metallic ions adhering to the sediment particles to be released and dissolve into the water. Instead, unfiltered samples to be analyzed for dissolved metals should be delivered as soon as possible to the laboratory for laboratory filtering prior to sample analysis.

Suspended Metals

When an analysis for suspended metals is requested, the sample filter is sent to the laboratory to be analyzed. The filter, preferably the membrane type, is placed in a glass jar for delivery to the laboratory. An alternative is to deliver the unfiltered, unpreserved sample to the laboratory quickly and request that the laboratory filter the sample in the process of performing the suspended metals analysis. The sampling plan will specify which option should be used for the site investigation.

Hexavalent Chromium

Samples to be analyzed for hexavalent chromium (Cr^{+6}) are collected separately from other metals. Hexavalent chromium has a very high toxicity compared to trivalent chromium (Cr^{+3}), the other form of chromium found in nature. A hexavalent chromium sample is filtered but collected in a plastic bottle without acid preservative. After collection, it should be chilled to 4°C in a cooler. The holding time is 24 hours: a Cr^{+6} sample must therefore be submitted to the laboratory as soon as possible. The laboratory should always know when to expect the delivery of samples and any special requirements associated with them.

Storing Other Water Samples

The following tables provide the container types, preservation methods and holding times associated with sampling inorganic nonmetallic as well as other contaminants.

—Nutrients (forms of nitrogen and phosphorus): total Kjeldahl nitrate, nitrates, nitrites, ammonia and phosphates.

Parameter	Container	Preservation	Holding Time
Total Kjeldahl nitrate	plastic	nitric acid, cool	28 days
Nitrates	plastic	cool	48 hours
Nitrates +nitrites	plastic	sulfuric acid, cool	28 days
Nitrites	plastic	cool	48 hours
Ammonia	plastic	sulfuric acid, cool	28 days
Orthophosphate	plastic	filter, sulfuric acid, cool	48 hours
Elemental phosphorus	glass (no plastic)	cool	48 hours
Total phosphorus	plastic	sulfuric acid, cool	28 days

Table 6-2.

—Chloride, sulfate, fluoride, hardness and alkalinity. Magnesium and calcium are related to the hardness of the water. Alkalinity is associated with sodium and potassium.

Parameter	Container	Preservation	Holding Time
Chloride	plastic or glass	none	28 days
Sulfate	plastic	cool	28 days
Sulfide	plastic	zinc acetate & sodium hydroxide	7 days
Fluoride	plastic	none	28 days
Hardness	plastic	nitric acid & sulfuric acid	6 months
Alkalinity	plastic	cool	14 days

Table 6-3.

—Cyanide – Should be collected in a plastic bottle with sodium hydroxide and/or ascorbic acid preservatives. The sample's maximum holding time is 14 days.

—Total organic carbon (TOC), total organic halogen (TOX), chemical oxygen demand (COD), and biochemical oxygen demand (BOD).

Parameter	Container	Preservation	Holding Time
TOC	glass	sulfuric acid, cool	28 days
TOX	glass	sulfuric acid, cool	7 days to extraction, 40 days after extraction
COD	plastic	sulfuric acid, cool	28 days
BOD	plastic	cool	48 hours

Table 6-4.

—Bacteriological (total coliform, fecal coliform, fecal streptococcus and iron bacteria).

Samples for bacteriological analysis must be collected in a sterile glass or plastic container. Samples may need to be preserved with sodium thiosulfate. This is especially true if chlorine is present in the water. The holding time for bacteriological samples is relatively short: six hours or less for optimum validity.

— Radiological (alpha, beta, gamma radiation detection; radon; radium; and uranium).

Radiological sample collection and handling is much like that for metals. Radon is an exception since it is a gas dissolved in water. Thus, water samples for radon should be collected in a 40 mL glass vial as for VOCs. The sample is therefore not filtered, and the vial is completely filled with the water until a meniscus is formed; then the lid is screwed on tightly and the container is inspected for bubbles. If an air bubble is observed, the sample needs to be recollected.

Parameter	Container	Preservation	Holding Time
Gross alpha	plastic	filter, nitric acid	6 months
Gross beta	plastic	filter, nitric acid	6 months
Gamma scan	plastic	filter, nitric acid	48 hours
Radon	40 mL glass vial	cool	24 hours
Radium	plastic	filter, nitric acid	6 months
Uranium	plastic	filter, nitric acid	6 months

Table 6-5.

Decontamination

Sample collection requires the use of clean, decontaminated equipment. Technicians should decontaminate equipment after each water sample is collected. This includes instruments used to measure water level, pH, temperature, conductivity or dissolved oxygen.

Samples to document or demonstrate decontamination success (equipment blanks, rinsate blanks and field blanks) may need to be collected. These QA/QC samples will be specified in the sampling plan. The QA/QC samples should be analyzed for the same chemical parameters as the water samples being collected.

Decontamination fluids are investigation-derived wastes and may therefore need to be disposed of as hazardous waste. The sampling plan should specify decontamination procedures as well as disposal instructions.

Checking Your Understanding

1. Describe the sample container and procedure for collecting VOC samples.
2. Explain the role of the trip blank.
3. Explain why samples for metals analysis are collected in plastic containers and samples for organic compounds analysis are collected in glass containers.
4. Identify which samples have the shortest holding times and which have the longest holding times.

Point of View: Cheryl Stith

Cheryl, a Certified Hazardous Materials Manager (CHMM), is environmental health and safety manager for a large national analytical laboratory and an adjunct faculty member at Cosumnes River College in California. She is active at the national level in promoting partnerships in the environmental sciences among educators and industry.

Q. Cheryl, have you had any unusual water sampling experiences?

When you think of water sampling, you might picture a lake or river. But I have collected water samples from manholes. Manholes are "confined spaces," which OSHA regulates because such areas can be so hazardous for workers.

Q. What hazards do manholes present?

There is a whole laundry list of hazards! For example:
— "Slip/trip" hazards. Manholes can be located in awkward places.
— Lifting hazards. A manhole cover is very heavy. So is sampling equipment that is full of water.
— Possible oxygen-deficient atmospheres if you have to enter.
— The unknown! Possibly a toxic contaminant, such as hydrogen sulfide or flammable vapors.
— Biological hazards from domestic waste water. Human waste may be visible as you are sampling. Also, rats, spiders...

Q. What precautions should be taken?

Our sampling staff is trained in OSHA's confined space safety and emergency response requirements, hazard recognition, and the minimization of injury.

Q. Have you heard of a case where someone got in trouble with a hazardous atmosphere in a manhole?

In my hazmat response class, I use a case from the book "Practical Guide to Respiratory Usage in Industry" by Gyan S. Rajhans and David S. L. Blackwell [Butterworth Publishers, 1985]. Briefly, what happened is that a worker entered a manhole near a brewery. He felt weak and tired shortly after entering. He collapsed while trying to climb out. A co-worker saw what happened, grabbed his respirator and went to help. Another worker grabbed another respirator and went in the manhole to attempt rescue. The fire department retrieved bodies.

Q. Do you mean they all died?

They died as a result of oxygen deficiency. They had selected respirators for protecting against ammonia when they needed air-supplied respirators. OSHA's recognition of the history of confined space hazards and accidents led to the Permit-Required Confined Spaces regulations in 29 CFR 1910.146, which took effect in the early 90s. The training and procedures required by these regulations could have prevented these deaths. Usually a safety person is responsible for proper characterization of confined spaces, but workers must be trained to recognize a hazard by themselves! I tell my students all the time to be observant, to ask questions, and to not assume that the person who is supervising you "knows" what he or she is doing. In my opinion, workers must be willing to take some of the responsibility for their own safety.

Usually staff are hired in at the technician level to conduct sampling, but I am glad I, as an EH&S manager, have had to perform the tasks myself, because it gives me firsthand knowledge of the conditions, the hazards, and safety-related issues, such as staff not assessing hazards correctly, and why staff do not wear their PPE, or why they do not follow safety protocol.

Q. We sometimes hear it said by environmental health and safety managers that accidents occur even when safety measures are in place and workers have been trained. Yet some workers don't take potential hazards seriously. They don't think anything can happen to them.

A. The biggest mistake I have observed is failure to recognize confined space hazards. NIOSH and OSHA both make information available on their websites case studies, regulations and training materials. Take advantage of the resources.

6-7 Safety Precautions Recommended in Sampling Water

Concepts

- Sampling technicians and employers must comply with HAZWOPER and any other applicable OSHA standards.
- The Site Safety Plan must be read, understood and followed by the investigation team.

The HAZWOPER standard (29 CFR 1910.120) is relevant to sampling activities, including water sampling. The sampling technician and his or her employer are responsible for complying with the requirements of HAZWOPER and any other applicable OSHA worker protection standards. The written Site Safety Plan that is prepared for the sampling activities must be read, understood and followed by all members of the investigation team.

Among the safety procedures that should be heeded are:

1. Be mindful of overhead utility lines. Some sampling devices, such as the pond sampler or subsurface grab sampler, are on the end of poles that can be extended six to 12 or more feet. Look up and around and avoid utility line locations.
2. Use appropriate devices for real-time monitoring of the air around the work area. Be certain of the proper functioning of monitoring devices through maintenance and calibration before and after each day's sampling activities. Devices with audible and visible alarms are preferred.
3. Be aware of a potentially explosive or dangerous atmosphere, and be ready to leave the area if necessary.
4. Select and use the appropriate level of PPE. Always wear appropriate chemical-resistant gloves while sampling. Water sampling activities may require wearing safety glasses or goggles to protect against inadvertent splashing of the water. Waders and/or other boots may also be appropriate.
5. Chemicals used to preserve water samples are usually strong acids or bases. Use care when handling sample containers and preservative solutions. Read and understand the Material Safety Data Sheets (MSDSs) for the preservative solutions.
6. Some water sampling activities may require wading in flowing water or the use of a boat. Personal flotation devices should be used.
7. When working in public areas, secure the work area with barricades, caution tape, etc. Open wells that are flush with the ground are hazardous as they are easily stepped in. Wear orange vests or other bright clothing. Use the field support vehicle as an additional barrier whenever possible. Fieldwork does not always occur in deserted areas; sometimes the work is at or near a busy commercial facility. Be aware of your surroundings; do not permit passersby to wander into the work area, and be alert to absent-minded motorists who are not accustomed to your presence in an area!
8. Use the buddy system, and schedule check-in times with appropriate personnel communicating by phone or by radio.
9. Do not eat, drink or smoke while performing sampling activities without first going through personal decontamination.
10. Do not use matches or lighters onsite.
11. Follow SOPs for equipment.
12. Use intrinsically safe, explosion-proof electrical devices.
13. Use nonsparking tools.
14. Do not perform hot work in hazardous atmospheres.

Checking Your Understanding

1. Recall the fourteen safety procedures discussed in this section.
2. State two written references regarding site safety that technicians must comply with.

Summary

As we have seen throughout this book, the site sampling plan and site safety plan should always be followed. Consistent sampling protocol will pay off in the form of accurate and representative data. It will also prevent embarrassment and extra expense due to mistakes or poor information. Sampling is one of the most important phases of a site investigation. The information obtained from sampling activities determines whether a site must undergo some form of cleanup activity or falls under additional regulatory requirements.

As part of site characterization activities, the technician must be able to sample groundwater and surface water with appropriate equipment and techniques, while also thoroughly documenting the sampling activities.

Before going onsite, the technician makes advance arrangements, reviews site safety procedures, and prepares sampling equipment, instruments and supplies.

While onsite the technician follows the sampling plan for collecting and storing samples, decontaminates equipment and instruments, and properly handles investigation-derived wastes.

During all sampling-related activities, the protective requirements of OSHA and HAZWOPER apply.

Groundwater and surface water can be contaminated from materials migrating through soil, sediments and rock units. Soil and other geologic materials were covered in Chapter 5. Chemical products or wastes may be stored in containers such as drums and tanks. If these containers leak, water may be affected as well as soil, sediment and rock. Chapter 7 introduces methods and equipment for sampling materials stored in containers.

Application and Critical Thinking Activities

1. Contact your city or county water pollution control plant (the publicly owned treatment works) and request the local standards for the six conventional pollutants as well as current analytical data for those pollutants. Interpret the data in relationship to the standards.
2. How would seasonal changes in the weather affect water levels in a monitoring well?
3. If you have access to a groundwater simulation model*, experiment with it to discover the effects that pumping has upon the movement of contaminants that have leached into the ground and groundwater (the leached contaminants are referred to as "contaminant plumes").

   *A groundwater simulator may be constructed or may be purchased from Western Michigan University. The simulator is an acrylic plastic tank (13" × 20" × 6") filled with sand and gravel layers to simulate soil strata. Plastic tubes are inserted through the layers to simulate wells, ranging from shallow to deep. By injecting dyes and applying water to simulate rain, you can model the formation and paths of contaminant plumes.
4. Check an environmental supply catalog for water-level meters, oil-water meters and various pumps. Compare the features and prices of various models.
5. Arrange a field trip with an environmental firm to observe the process of opening, purging and sampling a monitoring well. During the observations, keep a log of all aspects including equipment, procedures, decontamination, measurements and documentation. Fill out the well sampling data form in this section.
6. What would be your choice of sampling device for a well 80 feet deep with a casing 2 inches i.d.? Why?
7. Discuss the advantages and disadvantages of dedicating a pump to one monitoring well.
8. Suppose you needed to collect a water sample from the surface of a shallow stream that is no more than a foot deep. What sampling device or technique would you use and why?
9. Suppose you needed to collect a water sample from a depth of 4.5 feet in a 5 feet deep manmade holding pond. What sampling device would you use and why?
10. Measure dissolved oxygen (DO) in a body of water, or observe someone taking the measurement with a DO meter. Document your activity or observations in a field log.
11. Conduct the following research:
    —Interview management of the local POTW regarding their concerns about and requirements for industrial discharges to the sanitary sewer and the storm sewer.

—Contact state and local regulatory agencies for their guidelines for granting NPDES permits.
—Contact the state for requirements for certification as a wastewater treatment operator.
—Based upon the information obtained from the research, write a summary of items applicable to wastewater sampling activities.

Obtain one or more Material Safety Data Sheets (MSDS) for preservatives used for water samples. Review and summarize the physical and health hazards and recommended first aid procedures for the preservative solution.

7

Sampling Drums and Containers

Chapter Objectives

Upon completing this chapter, the student will be able to:

1. **Describe** safety precautions, including bonding and grounding, recommended in sampling drums and other containers.
2. **Summarize** procedures for making field observations and taking measurements that must accompany tank, drum and waste pile samples.
3. **Describe** the equipment and calibration techniques used in sampling tanks, drums and waste piles.
4. **Discuss** sampling techniques used to obtain tank, drum and waste pile samples.
5. **Specify** the care and cleaning necessary to maintain the equipment used in sampling tanks, drums and waste piles, including decontamination techniques.
6. **Indicate** the selection process for determining containers, preservatives, and holding times for tank, drum, and waste pile samples.
7. **Discuss** considerations in selection of containers for transfer of waste site materials.
8. **Identify** and distinguish DOT and EPA requirements for containers and drums at hazardous waste sites.
9. **Identify** OSHA and EPA's requirements for handling containers and drums at hazardous waste sites.

Chapter Sections

Figure 7-1: Valley of the Drums, Brooks, Kentucky. About 17,000 drums were removed during EPA-conducted emergency response activities at this uncontrolled hazardous waste site in 1979. The chemical wastes were largely from the paint and coating industries of Louisville.

7-1 Introduction

Concepts

- The EPA's Standard Operating Safety Guides include procedures for sampling drums, tanks and waste piles.
- The technician must comply with the requirements of EPA, OSHA and DOT during drum and container sampling.

So far this textbook has introduced the sampling and monitoring methods for characterizing a site's natural environment – air, soil and water. Chapter 7 introduces the sampling and monitoring of human-generated materials, particularly those potentially dangerous materials contained in **drums**, **tanks** and **waste piles**.

Section 7-2 emphasizes the dangers of sampling drums; hazardous exposures to drum contents at uncontrolled sites can result in worker illness, injury or death. Described in a government publication as "fraught with danger," (see NIOSH/OSHA/USCG/EPA "Occupational Safety and Health Guidance Manual for Hazardous Waste Site Activities," Washington DC, Government Printing Office, 1985) drum sampling must be meticulously planned and carried out. When preparing for drum sampling, technicians should always assume that the most unfavorable conditions exist, and that the danger of explosion and poisoning will be present. The protection of personnel is paramount: wearing protective clothing, working from protective distances and from behind barriers, and using protective procedures.

Section 7-3 introduces tank sampling. Some technicians regard tanks as large drums insofar as the potential hazards are concerned. Tank sampling may involve drum sampling procedures and equipment as well as surface water sampling procedures and equipment. Section 7-3 introduces tank sampling from the point of view of collecting from outside the tank. If it is necessary for technicians to enter a tank or any other confined space, then it is imperative that personnel be trained in – and adhere to – OSHA's Permit-Required Confined Spaces standard (29 CFR 1910.146).

While a waste pile might seem relatively unthreatening compared to drums and tanks, it may harbor unforeseen dangers, both to the natural environment and to workers. Waste piles may contain such problems as batteries leaching acid, fluorescent light ballasts releasing mercury, cans of paint or solvents, lead dust as well as other materials that may cause toxic or explosive situations. Section 7-4 introduces the procedures and equipment for representative sampling of waste piles.Throughout the chapter are also cautionary references to several other special-case manmade containers, such as lab packs, pressurized cylinders or lagoons.

As discussed throughout this text, the technician approaches a sampling assignment by first studying the sampling plan to prepare for the specific operating procedures that will be used onsite. Those procedures are likely to be based on EPA's Environmental Response Team's "Standard Operating Safety Guides." The guides describe standard operating procedures (SOP's) for sampling environmental and waste media. The drum sampling procedure (SOP 2009) provides technical guidance on safe and cost-effective response actions at hazardous waste sites containing drums with unknown contents. Container contents are sampled and characterized for disposal, bulking, recycling, grouping and/or classification purposes. The tank procedure (SOP 2010) provides protocols for sampling tanks and other confined spaces from outside the vessel. The waste pile procedure (SOP 2017) outlines methods and equipment for collecting representative samples from waste piles, sludges, or other solid or liquid waste mixed with soil. Sections 7-2, 7-3 and 7-4 of this chapter are based on the guidance provided by EPA in SOPs 2009, 2010 and 2017.

Besides EPA procedural guidelines, the technician also needs to comply during onsite activities with DOT's container standards, with EPA's hazardous waste storage requirements, and with OSHA and EPA's joint HAZWOPER standard.

7-2 Procedures for Sampling Drums

Concepts

- Technicians should visually inspect drums before any handling.
- All drums should be assumed to be hazardous.
- The drum opening area should be located away from the drum staging area.
- Drums suspected of containing shock sensitive, explosive or radioactive materials require special consideration.
- Each drum is tracked onsite through documentation.
- Manual drum opening is regulated by OSHA and EPA in 29 CFR 1910.120.
- The glass thief and the COLIWASA collect a representative column of material.

Because drum sampling is considered to be potentially dangerous, knowing and following the site health and safety plan is of critical importance. If a preliminary site evaluation does not produce sufficient information to identify the suspected hazards of the site, OSHA and EPA require that an ensemble providing protection equivalent to Level B PPE be worn, and direct reading instruments be used to check for conditions Immediately Dangerous to Life and Health, known as IDLH and mentioned in 29 CFR 1910.120(c)(5).

A site may present neatly arranged drums that pose little problem for technicians moving among them; or it may be a cluttered mess with overturned drums packed together leaving no room for maneuvering. The sampling plan will specify how to accomplish representative sampling for site characterization.

Drums are typically 5-55 gallon metal, fiberboard or plastic containers. Closed-head drums are of unified construction (all one piece) and have two openings: a 2-3 inch bung opening for filling, and a 3/4-inch bung opening to allow the venting of vapors. Open-top drums have removable covers, which are sealed onto the drum by a steel lock-ring. The covers may or may not have bung openings. As a rule of thumb, open-top drums tend to contain solids and sludges, and closed-head drums with bung openings tend to contain liquids.

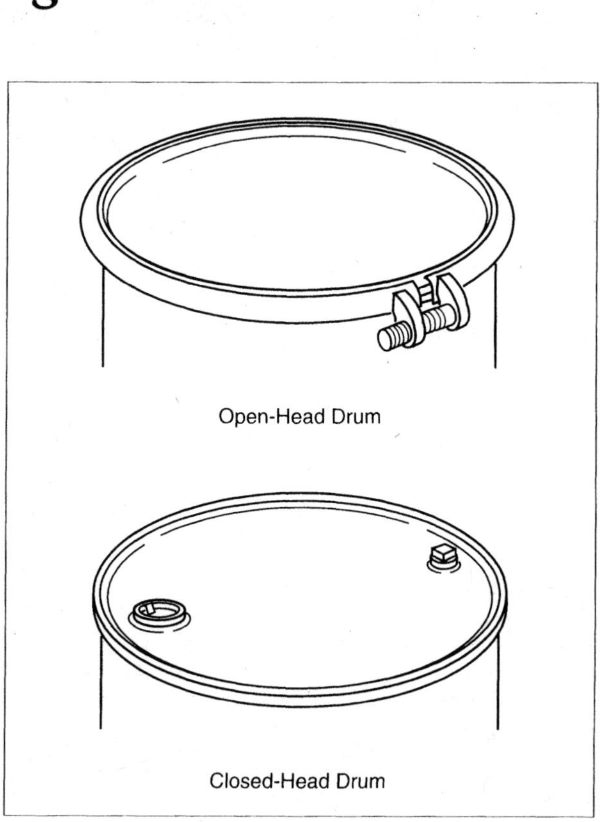

Figure 7-2: Closed-head drum and open-head drum.

Before the technician begins to handle a drum, he or she should visually inspect it for labels or other markings. Even if a label indicates a nonhazardous content, the technician should be cautious; a label on a drum might not accurately describe its contents. If the drum is unlabeled, the technician should assume its contents are hazardous and proceed with the inspection using the utmost safety techniques. The inspection should include monitoring around the drums with a radiation meter, an oxygen and combustible gas indicator and a toxicity monitor.

—The technician should be on the look out for signs of potential trouble;

—Poor drum condition, such as corrosion, rust and leaking contents;

—Symbols, words, or other markings on the drum indicating hazardous content (e.g., explosive, radioactive, toxic, flammable);

—Indications that the drum is under pressure, such

Box 7-1 ■ Lab Packs

A lab pack is a drum that holds individual containers of a variety of chemicals, usually as a result of cleaning out a laboratory. The chemicals may be incompatible or they may be old and unstable. The containers may even be unlabeled. According to NIOSH/OSHA/USCG/EPA's "Occupational Safety and Health Guidance Manual for Hazardous Waste Site Activities," technicians should take the following precautions in handling a lab pack:

— Prior to handling or transporting lab packs, make sure all nonessential personnel have moved a safe distance away.
— Whenever possible, use a grappler unit constructed for explosive containment during initial handling of such drums.
— Maintain continuous communication with the Site Safety Officer and/or the command post until handling operations are complete.
— Once a lab pack has been opened, have a chemist inspect the bottles without opening them, classify them and segregate them according to the hazards of the wastes. The objective of a classification system is to ensure the safe segregation of the contents. Pack the bottles with sufficient cushioning and absorption materials to prevent excessive movement of the bottles and to absorb all free liquids, and ship them to an approved disposal facility.
— If crystalline material is noted at the neck of any bottle, handle it as a shock-sensitive waste, due to the potential presence of picric acid or other similar material, and get expert advice before attempting to handle it.
— Palletize the repacked drums prior to transport. Secure the drums to pallet.

Lab packs may be used to transport any unmarked containers of potentially hazardous materials. For example, materials from a school chemistry lab that has not been properly maintained through the years would be transported in lab packs to a hazardous waste facility.

as bulging, color change, or sounds like hissing, groaning or creaking;

— Indications of shock sensitivity, such as crystalline buildup around openings or unexpected color for a gas or liquid as compared to label information;

— Abnormal readings on detection instruments.

Staging Drums for Sampling

Drums might need to be **staged** for easy access, which means assembled in an orderly, upright manner allowing space for the technician to maneuver during the process. At the staging area, the drums are physically separated into categories

Figure 7-3: The technicians use the HNU photoionization detector (PID) to check for air contaminants around the drums.

Box 7-2 ■ Pressurized Cylinders

The technician should regard the presence of pressurized cylinders at a sampling site as a potentially dangerous situation. If a compressed gas container fails, it may turn into an explosive projectile and the results can be catastrophic. As always, the technicians must follow the procedures from the site-specific sampling plan as well as from the health and safety plan.

As part of the site characterization activities, the technicians will monitor the air around pressurized containers for concentrations of oxygen, flammable vapors and toxic gases. Unless otherwise specified in the site-specific plans, technicians should consider the following concentrations to be cause for action such as withdrawal from the area, contact with the site safety officer, and notification of personnel who have specialized training in handling the materials:

— Oxygen monitor reading that exceeds the customary atmospheric level of 20.8 percent;

— Combustible gas monitor reading that exceeds 10 percent;

— For a pressurized cylinder of a known chemical, a general or chemical-specific toxicity monitor reading that exceeds the pertinent threshold published by OSHA, ACGIH or NIOSH;

— For a pressurized cylinder of an unknown chemical, any general toxicity monitor reading.

The technician should, of course, log all identification numbers on cylinders, record air monitoring data, and describe observations about the cylinders; for example, visible signs of container stress, whistling or groaning sounds, or odors (the technician should not deliberately sniff for odors!).

Employers and workers must comply with OSHA's Compressed Gases general requirements (29 CFR 1910.101), which make reference to procedures in Compressed Gas Association pamphlets and to visual and other inspection procedures as prescribed in DOT's Hazardous Materials Regulations (49 CFR 171-179).

according to their contents: drums with liquid content; drums with solid content; drums with miscellaneous inner containers, such as **carboys** (plastic or glass containers designed to hold caustic materials that would react with metal) and **lab packs** (drums used to hold smaller containers of assorted laboratory chemicals that could react dangerously with each other); gas cylinders; or drums with no contents at all.

Drums suspected of containing explosive or flammable materials should be constantly monitored with a CGI/oxygen meter during staging. The technician should be wary of physical changes that could indicate vapor release or pressure build-up. If the LEL exceeds 10 percent, or if the drum bulges, changes color, leaks gas or liquid, or if it makes whistling or groaning sounds, personnel should retreat and consult the site safety officer.

If a drum is leaking or showing signs of failure, the technician may be able to sample the area on or around the drum. Eventually, hazardous waste operations technicians might place the leaking drum or transfer its contents into an **overpack**, which is a polyethylene or steel container large enough to enclose the drum.

During inspection, technicians classify the drums into categories; color coding may be used to indicate drums that:
— are radioactive

— are leaking or deteriorated

— are bulging

— contain lab packs

— are explosive/shock sensitive.

Opening Drums for Sampling

Where space allows it, the drum opening area should be physically separated from the drum removal and drum staging areas. The technicians move the drums from the staging area to the opening area one at a time using a forklift truck equipped with a device that grabs the drum.

Drums may be opened with 1) a manual bung wrench or a manual **drum deheader** or 2) with a remote device that punctures the head or removes the bung from a distance.

OSHA and EPA hazardous waste site and emergency response regulations (29 CFR 1910.120) permit manual drum opening with bung wrenches or deheaders only if drums are structurally sound and have contents known to be 1) not shock-sensitive, 2) nonreactive, 3) nonexplosive, and 4) nonflammable. For drums containing hazardous contents, the technicians must follow remote opening procedures specified in the site safety plan.

Even if the situation suggests that manual opening is allowed, it is still good practice to first crack open a bung or lid slowly, which provides the opportunity to observe and respond to unexpected reactions of the contents such as fuming or bubbling.

Manual Opening of Drums

Manual Nonsparking Bung Wrench – A universal bung wrench has fittings that remove commonly encountered bungs. Although constructed of nonsparking materials like cast iron, brass, or a bronze-alloy, a nonsparking bung wrench does not completely eliminate the potential for a spark. To operate it, the technician positions the drum upright with the bung up or, for drums with bungs on the side, lays the drum on its side with the bung plug up and then wrenches the bung with a slow, steadily pulling motion. A "cheater bar" may be added to the handle to improve leverage.

Manual Drum Deheader – The drum deheader uses a can opener-like cutting action to remove the lid from a closed-head drum. Drums with removable heads must be opened by other means. The drum deheader is positioned with its cutting edge just inside the top chime or rim and then tightened so that the deheader is held against the side of the drum. Moving the handle of the deheader up and down while sliding the deheader along the chime will cut off the entire top.

Remote Opening of Drums

Remote openers, which are more costly but also more protective of workers, include the backhoe spike, the hydraulic drum opener and the pneumatic drum opener.

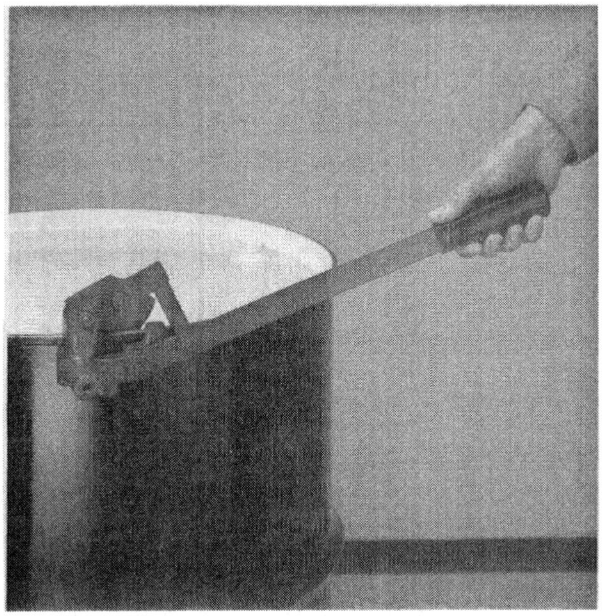

Figure 7-4: Drum opening using a manual drum deheader.

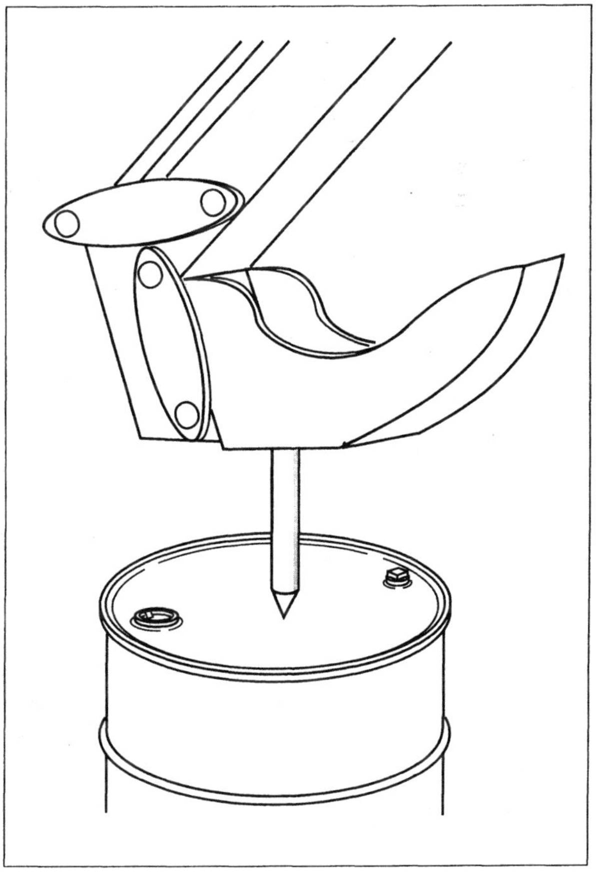

Figure 7-5: A backhoe spike punches the drum head.

Box 7-3 ■ Drum Data Sheet

Drum Data Sheet Form

Source: EPA - SOP #2009

Drum ID#: ____________ Date Sampled: ____________

Estimated Liquid Quantity: ____________ Time: ____________

Grid Location:* ____________

Staging Location: ____________

Sampler's Name: ____________

Drum Condition: ____________

Sampling Device: ____________

Physical Appearance of the Drum/Bulk Contents: ____________

Odor: ____________

Color: ____________

pH: ____________ %Liquid: ____________

Laboratory Analytical Data: ____________ Date of Analysis: ____________

Compatibility: ____________

Hazard: ____________

Waste ID: ____________

Treatment Disposal Recommendations: ____________

Approval

Lab: ____________ Date: ____________

Site Manager: ____________ Date: ____________

* Area of site where drum was originally located.
Based on di Napoli, 1982. Table originally printed in the Proceedings of the National Conference on Management of Uncontrolled Hazards Waste Site, 1982. Available from Hazardous Materials Control Research Institute, 9300 Columbia Blvd., Silver Springs, MD 20910.

Backhoe Spike – A metal spike attached to a **backhoe** bucket is lowered to puncture the drumhead. The spike should be decontaminated after each drum is opened to prevent cross-contamination. A shatter-resistant shield in front of the backhoe operator's cage protects the operator from the splashing of the contents.

Hydraulic Drum Opener – The **hydraulic** opener has a pump which pressurizes oil through a length of hydraulic line. A metal piercing device attached to the end of the hydraulic line punctures the drum. The piercing device may be positioned to puncture either the side or the head of the drum. Some of the metal piercers are hollow tubes that may be left in place as a sampling port.

Pneumatic Bung Opener – A **pneumatic** bung remover delivers compressed air through an air line to a pneumatic drill adapted to turn a bung fitting. The bung fitting is held in place by an adjustable bracketing system. Once the bung has been loosened, the bracketing system is removed before the drum is sampled. The pneumatic bung opener requires the container to be upright and relatively level. Bungs that are rusted shut cannot be removed with this device. Since this opener does not permit slow venting of the drums, the technician must take appropriate precautions.

Venting Swollen Drums

EPA outlines a method for venting swollen drums with a tube-and-spear device. A light aluminum tube (three meters long) is positioned at the vapor space of the drum and held in place by a rigid structure attached over the drum's rim. The spear is inserted in the tube against the drum wall. From behind a protective barricade, the technician delivers a sharp blow to the end of the spear, which drives its sharpened tip into the drum. The vapor vents along grooves created by the spear. Once the pressure is relieved, the technician can remove the bung and sample the drum. After a drum is opened and before sampling, the technician monitors its headspace gases with a combustible gas indicator and a toxicity monitor.

Sampling Drums

Each drum selected for sampling should be monitored before and during the activity, then resealed upon completion of activity. The sampling activities must be documented thoroughly. During sampling, the technician should track the drum by noting in a log the location where it was buried or stored, its movement to the staging and drum-opening areas as well as its ultimate treatment, storage or disposal. Color-coded tags, labels or bands are used to mark similar waste types. Each drum is documented on a drum data sheet, including location, descriptions of the drum's condition and any unusual markings. This data sheet becomes the principal recordkeeping tool for tracking the drum onsite.

Equipment

The glass thief and the COLIWASA are devices that enable the technician to collect a column of material that is representative of a drum's contents.

Glass Thief – The glass thief is a disposable tube, typically 6mm (0.24 inch) to 16mm (0.64 inch) in inside diameter and 1.21 meters (48 inches) long (illustrated in Chapter 3). The procedure for using a glass thief is as follows:

1. Remove the cover from the container to be sampled.
2. Insert the glass thief almost to the bottom of the drum or until a solid layer is encountered. About one foot of the thief should extend above the drum.
3. Allow the waste in the drum to reach its natural level in the thief.
4. Cap the top of the thief with a tapered stopper or thumb, making sure that the liquid does not come into contact with the stopper.
5. Carefully remove the capped thief from the drum; insert the uncapped end into a sample container.
6. Release the stopper and allow the glass thief to drain until the sample container is approximately two-thirds full. Drain any remaining material into the waste disposal drum.
7. Remove the thief from the sample container, break it into pieces, and place the pieces in the waste disposal drum.
8. Reseal the drum and decontaminate the equipment.

In many instances a tank containing waste material will have a sludge layer on the bottom. Slow insertion of the sample tube down into this layer and then a gradual withdrawal will allow the sludge to act as a bottom plug to maintain the fluid in the tube. The plug can be gently removed and placed into the sample container by the use of a stainless steel lab spoon.

Composite Liquid Waste Sampler – The COLIWASA, or Composite Liquid Waste Sampler, is a glass or polypropylene tube with a stopper feature at its lower end that permits representative sampling of multiphase (heterogeneous) wastes from tanks and drums (illustrated in Chapter 3). The stopper is attached to a rod that runs up through the tube. The technician manipulates the top end of the rod to lock and unlock the stopper.

Some COLIWASAs are intended to be disposable. Their construction makes them suitable for sampling liquids. Higher priced models are more likely to be re-used and are appropriate for sampling not only liquids but also slurry-type sludges and semisolids; these are available with 1) a clamp that locks the stopper into the closed position, 2) fittings that couple COLIWASA parts for deeper tank sampling, and 3) purge kits that help extract the sample and clean the COLIWASA. Reusable COLIWASAs are difficult to decontaminate in the field.

Figure 7-6: The COLIWASA is long enough to take a composite sample from the entire depth of a 55-gallon drum.

The procedure for using a COLIWASA is as follows:

1. With the stopper in the open position, slowly lower the COLIWASA into the drum or tank at a rate that permits the inside and outside liquid levels to be about the same. If the level of the liquid inside the COLIWASA is lower than the level outside, then the sampling rate is too fast and results in a nonrepresentative sample.
2. When the stopper hits the bottom of the waste container, push the COLIWASA downward against the stopper to close it. Lock the stopper in the closed position.
3. Slowly withdraw the COLIWASA from the waste container with one hand while wiping it with a disposable cloth with the other hand.
4. Carefully discharge the sample into a suitable sample container by releasing the stopper while it is positioned in the sample container.
5. Reseal the drum and decontaminate the equipment.

Checking Your Understanding

1. Describe the equipment used in sampling drums.
2. Discuss sampling techniques used to obtain drum samples.
3. Summarize procedures for making field observations that accompany drum samples.
4. Explain the appropriate precautions to be taken when using the various drum openers.

7-3 Procedures for Sampling Tanks

Concepts

- Before climbing onto the vessel, the technician should perform a structural survey of the tank as well as an air quality survey.
- Preparation for sampling includes a safety review, grounding of the tank and removal of ignition sources.
- Representative sampling methods can determine where phase changes occur within the material.

Tank sizes range from small storage or process containers less than five feet deep to giant chemical storage units that hold tens of thousands of gallons of product. A tank could be a cargo vessel (on a truck or train), a process tank (a vat) containing chemical inputs or outputs for some manufacturing operation, or a storage structure for chemicals or wastes. Tanks may contain hazardous or nonhazardous chemicals or waste materials in the form of liquids, sludges, still bottoms, or solids of various structures. In EPA's SOP for sampling tanks, the technician is warned that sampling a storage tank requires a great deal of physical coordination, often requiring the sampler to climb to the top of the tank on a narrow vertical or spiral stairway or ladder while wearing protective clothing and carrying sampling equipment. The SOP recommends that before climbing onto the vessel the technician perform a structural survey of the tank to ensure safety and accessibility. Manholes or access portals must be guarded in accordance with 29 CFR 1910.22(c) to prevent personnel from falling into the tank: "Covers and guardrails shall be provided to protect personnel from the hazards of open pits, tanks, vats, ditches, etc."

The technician should also perform an air quality survey prior to sampling to check for toxicity or flammability of the surrounding atmosphere. At no time should sampling continue with an LEL reading greater than 10 percent. If the 10 percent LEL reading is exceeded, the technicians should withdraw and consult with the site safety officer or follow the procedure specified in the site safety plan.

If the tank contains VOCs, the technician must take extreme caution to avoid ignition of the contents by using nonsparking tools and intrinsically safe electronic monitoring instruments. The technician may need to vent excess pressure before opening the tank. Personnel should be located behind a shield to avoid direct contamination from materials forced out by pressure while tank opening is performed.

Entry into a tank is strictly regulated by OSHA's Permit-Required Confined Space entry standard (29 CFR 1910.146). Confined space refers to a space that 1) is large enough for a person to enter it and perform assigned work; 2) has limited or restricted means for entry or exit (for example, tanks, vessels, silos, storage bins, hoppers, vaults and pits); and 3) is not designed for continuous employee occupancy. Hazards include the potential of engulfment, suffocation, explosion or toxic exposures, which can cause death, incapacitation, impairment of ability to self-rescue, injury or acute illness. The confined space entry requirements include, but are not limited to:

—Air monitoring;

—Appropriate equipment such as respiratory protection, protective clothing, safety harnesses, and ropes;

—Forced air ventilation of the space;

—A team approach for entry, including thorough team training;

—Establishment of lifeline communication signals prior to entry;

—Planned rescue procedures; and

—A permit system that limits and controls entry by authorized personnel and that closes off the permit space upon completion of the operation.

Preparing to Sample a Tank

The EPA's SOPs list the following steps to prepare for sampling tanks:

—Inspect the external structural characteristics of each tank and record observations in the site logbook. Potential sampling points should be evaluated for safety, accessibility and sample quality.

—Prior to opening a tank for internal inspection, the tank sampling team should: 1) review safety

Technical Corner ■ Bonding and Grounding

During sampling activities, a technician might need to transfer a flammable liquid from one container to another. A liquid transfer can charge the metal drum or container with static electricity. That electrostatic charge may be released as a spark, which can ignite any flammable vapors in the atmosphere surrounding the liquid transfer. OSHA's 29 CFR 1910.106(e)(6)(ii) requires the **bonding** and **grounding** of equipment used for transfer of Class I flammable liquids.

A common method of protection against the release of the electrostatic charge into a volatile atmosphere involves the use of metal bonding wire with a metal clamp at each end. Before the liquid transfer, the technician 1) bonds the two containers together and 2) grounds the dispensing container by running a bonding wire from it to a metal conductor that contacts the ground (i.e., the earth). All bonding and grounding connections are between areas of clean metal.

Another method of preventing an electrostatic charge is to create the bond by placing the metal nozzle of a metal-lined hose from the dispensing container into direct contact with the metal opening of the receiving container. The effectiveness of this method depends on ensuring that the contact is fully maintained.

Industrial plants or shops that store flammable materials or hazardous wastes will keep those materials in a clearly marked storage shed. Each such storage drum is grounded by attaching a bonding wire from the drum to a metal pole that is in contact with the earth. Note that the technician in the photograph above has also used a bonding wire from the drum to the portable container.

procedures and emergency contingency plans with the safety officer, 2) ensure that the tank is properly grounded, and 3) remove all sources of ignition from the immediate area.

—Follow the site safety plan's methods for safely climbing the tank. Use nonsparking tools to remove manway covers.

—Collect air quality measurements for each potential sample location using a combustible gas indicator/oxygen meter for a lower explosive limit (LEL/0_2) reading and an OVA/HNU for organic vapor concentration. Take readings from the tank headspace, from above the sampling port and in the breathing zone.

—Prior to sampling, use a high volume blower to clear the tank headspace of any toxic or explosive vapor concentration. Do not start sampling until LEL readings are lower than 10 percent.

Sampling the Tank

Once the tank has been determined to be safe for sampling, the technician should follow these steps (the samplers are illustrated in Chapters 3 and 6):

—Determine the depth of the liquid and of the sludge or solid below it.

—From the top of the tank, usually through an opened port or manway, use a subsurface grab sampler or a bomb sampler to collect liquid samples. For representative sampling of liquids deeper than five feet, collect from one foot below the surface, from the liquid's mid-depth, and from one foot above the sludge. For liquids less than five feet deep, collect one representative sample with a COLIWASA or glass thief. If the tank has more than one compartment (such as a vacuum truck tank or a multi-tank process vessel), collect

Box 7-4 ■ Vacuum Trucks

Vacuum trucks are contracted to remove liquid and semiliquid wastes at sites such as waste lagoons, process tanks, hazardous waste storage containers or contained spills. Placing such materials in a vacuum truck is a method of **bulking**, which means mixing similar, compatible materials together for shipment to treatment or disposal facilities. Bulking can take place only after the site materials have been characterized through field or laboratory analysis. If a technician is directed to collect samples from a vacuum truck, NIOSH/OSHA/USCG/EPA's "Occupational Safety and Health Guidance Manual for Hazardous Waste Site Activities" recommends the following precautions:

—Don personal protective equipment before opening the hatch.

—If possible, use mobile steps or suitable scaffolding consistent with 29 CFR Part 1910, Subpart D/Walking, Working Surfaces. Avoid climbing up the truck ladder and walking across the tank catwalk.

—If the truck must be climbed, then raise and lower equipment and samples in carriers to enable workers to use two hands while climbing.

—If possible, sample from the top of the vehicle. If it is necessary to sample from the drain spigot, take steps to prevent spraying of excessive substances. Have all materials on hand in the event of a spill.

The vacuum truck operator removes from the sump an oil-water mixture, which is bulked in the truck with similar materials from other sites.

samples from each compartment as directed by the sampling plan. (Note: Do not collect samples from a lower or bottom valve. If a valve cannot be closed again, the entire tank contents may be released. Furthermore, sampling from a valve does not yield representative samples.)

—Visually compare the three liquid samples (or the one COLIWASA sample) to determine if they indicate different phases or strata. If phase differences appear in the three separate samples, additional samples might be taken halfway between each sampling point to determine more precisely where the changes between strata occur.

—If another sampling port is available, repeat the sampling to verify the phases.

—Replace the port cover.

—Measure the outside diameter of the tank and calculate the waste volume using the depth measurements.

—Record all information on the sample data sheet or site logbook. Label the container with the appropriate sample tag.

—Decontaminate the sampling equipment.

Tank Sampling Equipment

Other equipment used during tank sampling (in which personnel do not enter a confined space) includes the following:

—tape measure

—weighted tape line or equivalent

—camera/film

—stainless steel bucket or bowl

—flashlight (explosion-proof)

—water/oil level indicator

—toxic monitor (such as the Organic Vapor Analyzer or OVA)

—combustible gas indicator/oxygen meter

—high volume blower

Bomb Sampler – The bomb sampler can collect a sample from a specified level in a storage tank. It is a hollow stainless steel cylinder with a weighted plunger that acts as a valve to admit the sample. One line attached to the top of the plunger opens and closes the valve; a second line is attached to the removable top cover which has a locking mechanism to keep the plunger closed after sampling. The procedure for using a bomb sampler is as follows:

1. Attach the sampling line and the plunger line to the bomb sampler.
2. Measure and then mark the sampling line at the desired depth.
3. Gradually lower the bomb sampler by the sampling line until the desired level is reached.
4. When the desired level is reached, pull up on the plunger line and allow the sampler to fill, and then release the plunger line to seal the sampler.
5. Retrieve the bomb sampler by the sample line. Be careful not to pull up on the plunger line.
6. Rinse or wipe off the outside of the bomb sampler.
7. Position the sampler over the sample container and release its contents by pulling upon the plunger line.

Sludge Judge – The sludge judge can collect samples from a large tank (e.g., 15 feet deep). It is made of 3/4-inch clear plastic tubing in five-foot sections marked at one-foot increments. The sections have threaded fittings that allow the sludge judge to be extended. The top section includes a nylon line for raising the sampler. The bottom section has a ball check valve that floats up as the sludge judge is lowered into material and settles back down as the sludge judge is raised. The procedure for using a sludge judge is as follows:

1. Lower the sludge judge to the bottom of the tank.
2. When the bottom has been reached, and the tube has filled to surface level, tug slightly on the rope as you begin to raise the unit. This seats the check valve, trapping the column of material.
3. When the unit has been raised clear of the tank liquid, the amount of sludge in the sample can be read using the one-foot increments marked on the pipe sections.
4. By touching the pin extending from the lower section against the bottom of a sample container, the technician enables material to bypass the ball check valve and to be released from the unit.

Technical Corner ■ Determining Volume of Tanks and Materials

A technician needs to be able to determine the volume of tanks and materials for reporting purposes. The determination may be as simple as reading capacity markings on a tank or referring to a table of tank volumes; or it may be more challenging, requiring the application of mathematical formulas to the measurements.

The most common tanks sampled by technicians are the horizontal or vertical cylinder type – the common shapes for chemical storage tanks – and the three-dimensional rectangle, which is a common shape for process tanks (see figures below).

The formula for the volume of a rectangular tank is V = LWH (length × width × height). For example, a process tank that is 8 feet long by 4 feet wide by 4 feet high is $8 \times 4 \times 4 = 128$ cubic feet.

A formula that can be used to find the volume of a cylindrical tank is $V=0.25\pi D^2H$ (0.25 × 3.14 × diameter squared × height; height is viewed as length in a horizontal tank). For example, a cylindrical storage tank that is 4 feet in diameter by 6 feet high (or long) is $0.25 \times 3.14 \times 4^2 \times 6 = 75.36$ cubic feet.

Volume may be reported as gallons rather than cubic feet. There are approximately 7.5 liquid gallons per cubic foot. In the examples above, 128 cubic feet equals approximately 960 gallons and 75.36 cubic feet equals approximately 565 gallons.

7.5 liquid gallons ≈ 1 cubic foot

The content volume is calculated by subtracting the volume of empty space above the material from the tank capacity. For example, if the empty space in the rectangular tank example measures 8 feet long by 4 feet wide by 1.5 feet high, then the content volume is 128 cubic feet minus 48 cubic feet, which equals 80 cubic feet of material. If the empty space in the vertical cylinder example measures 4 feet in diameter by 2.5 feet high, then the content volume is 75.36 cubic feet minus 31.4 cubic feet, which equals 43.96 cubic feet of material.

The technician is spared the challenge of calculating the content volume of a horizontal cylindrical tank, because tank vendors provide handy tables like the gallonage chart shown below. For example, suppose a technician measures empty space of 30 inches above the content of a four foot diameter by six foot long cylindrical tank. The content is thus 18 inches deep. According to the table, the material volume is 197 gallons.

Another application of volume calculations is the determination of the quantity of soil removed during site activities. A rule of thumb is to measure the hole using the above three-dimensional rectangle formula and then multiply by an expansion factor, such as 1.3. The expansion factor accounts for the tendency of soil to be less compact upon removal from the ground. Soil is normally reported in cubic yards rather than cubic feet. Thus, if measurements are taken in feet, then the resulting volume should be divided by 27 (i.e., 3' × 3' × 3') to convert to cubic yards. The formula may be expressed as 1.5 × (LWH:27) where L, W and H are measured in feet.

INCHES DEPTH	4' x 6' 550 Gals	64" x 6' 1000 Gals	64" x 12' 2000 Gals	64" x 18' 3000 Gals	7' x 14' 4000 Gals
1	2	3	6	9	8
2	8	9	18	27	25
3	15	17	34	51	48

Excerpt from Brown-Minneapolis Tank's gallonage chart for horizontal round storage tanks.

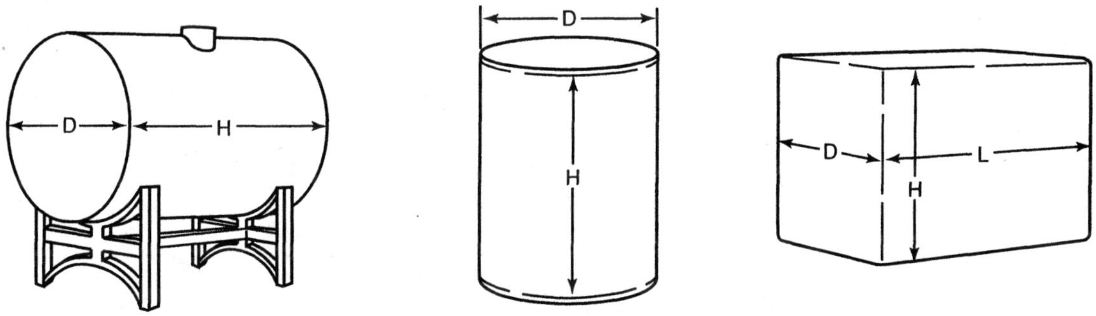

Chemical Tank:
Horizontal Cylinder Shape
$V = 0.25\pi D^2H$

Chemical Tank:
Vertical Cylinder Shape
$V = 0.25\pi D^2H$

Process Tank:
3-Dimensional Rectangle
V = LWH

Box 7-5 ■ Ponds and Lagoons

A past practice of some industries was to construct lagoons, which have the appearance of ponds, to deposit liquid wastes from their processes. If such a lagoon is present on a sampling site, the sampling plan will probably specify surface water sampling equipment and methods. The lagoon's contents, however, may present the potential hazards of a drum or tank; therefore, during monitoring and sampling activities, technicians are warned by NIOSH/OSHA/USCG/EPA's "Occupational Safety and Health Guidance Manual for Hazardous Waste Site Activities" to heed the following:

— Drowning is a very real danger for personnel suited in protective equipment because the weight of protective equipment increases an individual's overall density and severely impairs swimming ability. Where there is danger of drowning, provide necessary safety gear such as lifeboats, tag lines, railings, nets, safety harnesses and flotation gear.

— Stay on shore wherever possible. Avoid going out over the water.

— Be aware that some solid wastes may float and give the appearance of solid cracked mud. Exercise caution when working along shorelines.

Bruin Lagoon, Bruin Borough, Pennsylvania. The wastes are primarily from the white oil manufacturing industry. They contain sulfuric acid, heavy metals and other materials, which have produced contamination of air, surface water and groundwater.

Subsurface Grab Sampler – The subsurface grab sampler allows samples to be collected at a specific depth in a tank or in surface waters. The sampler is constructed of a six-foot aluminum or stainless-steel tube with a polypropylene or Teflon® assembly on its bottom, which in turn attaches to an uncapped one-liter sample container. A cable that runs from the technician's end of the sampler to the head assembly enables the sample container to be opened and closed at a specific depth. Some models offer six-foot extension tubes that enable the sampler to be lengthened up to 30 feet. Warning: The length of the sampler's metal tube can be a hazard if the technician is using it around electrical wires. The technician should be constantly aware of the surroundings and the potential for dangerous contact. The procedure for using a subsurface grab sampler is as follows:

1. Screw the sample bottle onto the sampling head.
2. Lower the sampler to the desired depth.
3. Pull the ring at the top of the tube, which opens the spring-loaded plunger on the head assembly.
4. When the bottle is full, release the ring, lift the sampler and remove the sample bottle.

Checking Your Understanding

1. Describe the equipment used in sampling tanks.
2. Discuss sampling techniques used to obtain tank samples.
3. Summarize procedures for making field observations and taking measurements that must accompany tank samples.
4. List the steps for electrically grounding a tank.

7-4 Procedures for Sampling Waste Piles

Concepts

- A waste pile might be controlled and planned according to 40 CFR 264 Subpart L, or it might be uncontrolled as found on illegal hazardous waste sites.
- A pile generated from an industrial process is not a waste pile until its owner/manager determines it is a waste with no other use.
- Waste pile sampling is similar to soil sampling in equipment and procedures.

A waste pile may be planned and controlled, as described in 40 CFR 264 Subpart L, Waste Piles. These EPA-regulated hazardous waste piles must have a specially designed liner, a **leachate** collection and removal system, and a leachate detection system. On the other hand, a waste pile may be part of an uncontrolled hazardous waste site, created over time by illegal dumping. However, a pile generated from an industrial process is not a waste pile until it is declared a waste by a company manager/owner. For example, a pile of fly ash from a power generation company is not a waste but a by-product potentially suitable for use by another company, such as a cement producer. A manager/owner's responsibility is to keep a process pile homogeneous, to know and document its composition, and to arrange for its eventual use. If the manager determines that the pile will be disposed of as a waste, then he or she must do so according to the regulations.

Point of View ■ David Foss

David Foss, Chief of the Environmental Division, Science and Engineering Directorate, Rock Island Arsenal, is featured in Chapter 2. In the present chapter David reminds us that environmental sampling is not always associated with a formal site investigation process, as evidenced by the following situation.

Eighty bags of sweepings, paint scrapings, paint chips, and shreds and chunks of wood were sitting in a building following a window replacement job. This waste was special. It was not within the Arsenal's routine – that is, it was not a known, tested waste from an Arsenal industrial process. It also was not an ordinary waste such as that generated in office work. The landfill must be consulted about a special waste situation such as these bags of construction debris. In this case the landfill administrator said the material had to be analyzed for lead contamination before the bags could be accepted for disposal.

The question became: What would reasonably constitute a representative sample of the material for TCLP* analysis? Dave asked the contractor to propose a strategy that would, in the contractor's best estimation, result in a representative sample. It was decided that they would remove a trowelful of material from each of several bags along the perimeter of the eighty "containers." The contractor composited the material and placed a portion in a one-quart sample jar, which is an appropriate amount for TCLP analysis. He documented the procedure, the locations of the collected material as well as his rationale. The results of the TCLP analysis were negative for lead contamination or any TCLP parameter; the landfill therefore accepted the waste for disposal as nonhazardous.

This situation illustrates that environmental sampling is not always part of a formal site investigation process. It is not always associated with a written sampling plan. Yet the need for sampling was called for by the RCRA regulations, which govern the disposal of hazardous wastes.

David says that the motto in his department is "When in doubt, test!"

*TCLP is an acronym for Toxic Characteristic Leaching Procedure. The TCLP, which is described in 40 CFR 261.24, is a gas chromatography/mass spectrometry (GC/MS) analysis method, which involves a mathematical computer model that simulates what happens to hazardous waste in a landfill.

A solid waste is considered to be a "toxic characteristic waste" if any of the TCLP-listed chemicals is present at or above its regulatory threshold in a sample of the waste itself or in the leachate of the waste. The GC/MS method as well as the list of TCLP chemicals are described in Chapter 8.

Representative Sampling of a Pile

Waste piles with unknown and/or heterogeneous contents are common in uncontrolled hazardous waste sites. To obtain a representative picture of a heterogeneous waste pile, the sampling planner might specify a stratified random sampling approach. Stratified random sampling, which is based on a three-dimensional grid and random number methods, requires the technician to have access to any point within the waste pile. If a sampling point cannot be accessed safely, the technician must document any deviation from the plan.

A pile may have known or predictable contents as, for example, a soil pile created at a gasoline underground storage tank removal project. An adequate representation of such a pile might be gathered from several simple random samples. Well-documented judgmental sampling is appropriate if technicians detect something – such as a color or odor change in the pile material – that requires more information.

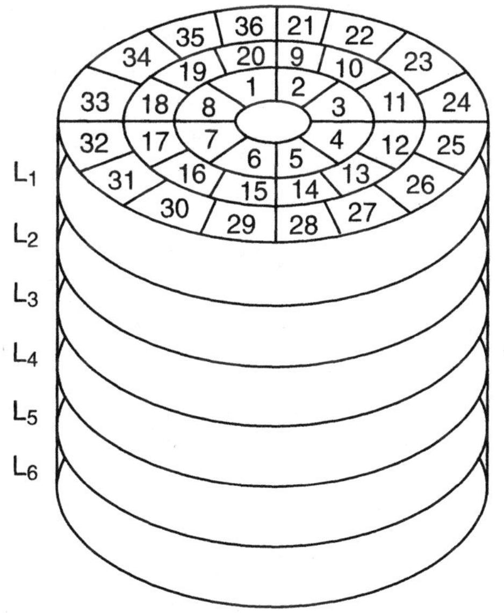

Figure 7-7: To grid a waste pile or a container, visualize the grid in layers (L_1, L_2, etc.) with the top layer divided into numbered units. For a large heterogeneous pile or container, pick the first sampling point randomly and then select additional sampling targets at regular intervals covering all layers. For a large homogeneous pile or contained substance, such as a process waste, randomly select a set of points.

Of course, if sampling is limited to certain portions of the pile, a statistically based sample will be representative only of that portion, unless the waste is known to be homogeneous.

Equipment

The equipment for sampling waste piles is similar to that used for soil; for example, stainless steel spades, shovels, scoops, hand augers and thin-walled tube samplers. Triers or grain samplers may be used for granular substances. Although the procedures for using the equipment are like those for sampling soil as outlined in Chapter 4 (or Chapter 3 for sampling dry contents of bags), the shapes and sizes of some waste piles may present difficulties in access and even safety problems for technicians. If planned sampling points must be changed because of access or safety considerations, then the technician must document the decision-making.

Along with PPE and documentation supplies listed in Chapter 5, the following equipment is standard for waste pile sampling:

—stainless steel, plastic, or other appropriate homogenization bucket or bowl
—canvas or plastic sheet for excess material
—augers (continuous flight or bucket) or thin-wall tube sampler
—scoop, trowel, spatula, spoon, spade or shovel
—extension rods for equipment handles
—trier
—grain sampler

Trier – The trier is formed from a stainless steel tube that is cut away lengthwise along one side. It has a T-handle on one end and is pointed on the other end. The trier is driven into the waste pile and then turned like an auger to extract a core sample. The procedure for using a trier is as follows:

1. Insert the trier into the material to be sampled at a 0° to 45° angle. This orientation minimizes spillage of the sample.
2. Rotate the trier once or twice to cut a core of material.
3. Slowly withdraw the trier, making sure the slot is facing upward.
4. For volatile organic analysis, transfer the sample into an appropriate, labeled sample container with a stainless steel or plastic lab spoon and

Box 7-6 ■ Containers of Fibrous or Other Dusty Materials

Airborne fibers and other dusts can create inhalation hazards as well as explosion hazards. The sampling plan should spell out procedures and equipment to be used in dusty atmospheres. For example, technicians should make sure they are wearing NIOSH-approved dust masks or respirators and use nonsparking equipment when sampling containers that may yield dusts or when working in a dusty atmosphere. The sampling area may qualify as a confined space, in which case the requirements of 29 CFR 1910.146 would apply. If the sampling area is a grain-handling facility, then the requirements of 29 CFR 1910.272 apply. Among the OSHA standards that pertain to fiber or dust hazards are Asbestos in 29 CFR 1910.1001, Lead (29 CFR 1910.1025), Cotton Dust (29 CFR 1910.1043), Inorganic Arsenic (29 CFR 1910.1018) and Cadmium (29 CFR 1910.1027).

secure the cap tightly. Place the remainder of the sample into a stainless steel, plastic or other appropriate homogenization container and mix thoroughly. Then, either place the sample into appropriate, labeled containers and cap tightly or, if composite samples are to be collected, place a sample from another sampling point into the homogenization container and mix both samples together thoroughly. When compositing is complete, place samples into appropriate, labeled containers and secure the caps tightly.

Grain Sampler – The grain sampler consists of two slotted telescoping brass or stainless steel tubes. The outer tube has a pointed tip for penetrating the material being sampled. The sampler is opened by rotating the inner tube so that the slots are aligned. The procedure for using a grain sampler is as follows:

1. With the sampler in the closed position, insert it into the granular or powdered material or waste being sampled from a point near a top edge or corner, through the center, and to a point diagonally opposite the point of entry.
2. Rotate the inner tube into the open position.
3. Wiggle the sampler a few times to allow material to enter the open slots.
4. Close the slots and withdraw the sampler from the material.
5. Place the sampler in a horizontal position with the slots facing upward.
6. Rotate the outer tube and slide it away from the inner tube.
7. Transfer material to sample containers as described for the trier.

Checking Your Understanding

1. Describe safety precautions recommended in sampling waste piles.
2. Describe the equipment used in sampling waste piles.
3. Discuss sampling techniques used to obtain samples from a waste pile.
4. Identify EPA's requirements for containing waste piles.

7-5 DOT and EPA Container and Drum Specifications

Concepts

- If a container transports hazardous waste, it must adhere to DOT regulations.
- If a container is to be used to store hazardous waste, it must adhere to EPA's RCRA regulations.
- 49 CFR 172.101 identifies which containers are required for any particular material.
- 49 CFR 178 Subpart L describes the requirements for each kind of container.
- 40 CFR 264 Subpart I specifies the use and management of hazardous waste storage containers.

This chapter section outlines DOT regulations regarding the containers and drums used in site remedial activities as well as other activities dealing with nonhazardous materials and wastes. The physical and mechanical requirements for containers and drums may differ depending on what the container or drum is used for. For example, if a container is being utilized by a remediation contractor on a site under EPA control, the regulations governing the use of the containers and drums may come from DOT regulations as well as EPA's RCRA or CERCLA regulations. If a technician is involved in sampling of media not under the control of the regulators, it is wise to follow the container regulations in case some time in the future the activity should come under the scrutiny of the regulators.

Generally, one thinks of sampling activities as involving just bottles and small containers; this may not always be the case. Many sampling activities involve transferring a suspect material to a container and then sampling the container. If that container will eventually be involved in transporting the waste material to a disposal site, it must adhere to the DOT regulations. Additionally, if the container's content is to be stored as a hazardous waste, the container must also adhere to RCRA regulations.

Where to find DOT's 49 CFR performance-oriented packaging standards:

§ 178.504 steel drums
§ 178.506 other metal drums
§ 178.507 plywood drums
§ 178.508 fiber drums
§ 178.509 plastic drums and jerrycans
§ 178.510 wooden barrels
§ 178.511 steel jerrycans
§ 178.512 steel or aluminum boxes
§ 178.513 natural wood boxes
§ 178.514 plywood boxes
§ 178.515 reconstituted wood boxes
§ 178.516 fiberboard boxes
§ 178.517 plastic boxes
§ 178.518 woven plastic bags
§ 178.519 plastic film bags
§ 178.520 textile bags
§ 178.521 paper bags
§ 178.522 composite packaging over plastic
§ 178.523 composite packaging over glass

General Container and Drum Specifications

Any container intended for use during remedial activities is acceptable so long as it has been approved by the regulating agency overseeing the project. Therefore, one can expect to find all kinds of containers in use at active remedial sites, ranging from buckets to 55-gallon drums to slide-on or roll-off boxes of the 4-20 yard style.

Container Requirements

DOT's Hazardous Materials Table (49 CFR 172.101) identifies the containers required for each particular material. The performance-oriented packaging standards in 49 CFR 178 Subpart L for nonbulk hazardous materials packagings describe the requirements for each kind of container. For example, steel drum requirements are described in 49 CFR 178.504.

The sampling technician must also be aware of and adhere to EPA's requirements in 40 CFR 264 Subpart I, "Use and Management of Containers" (or 40 CFR 265 Subpart I interim status requirements for facilities that have applied for a hazardous waste permit). These regulations apply to all situations where hazardous waste containers will be used for storage for any period of time. For example, one requirement states that the technician must avoid placing incompatible wastes in the same storage container during sampling activities.

EPA's objective in developing 40 CFR 264 Subpart I is to ensure safety for the environment and everyone handling containers holding hazardous materials. The following list illustrates what the regulation is trying to accomplish:

§ 264.170, Applicability – The subpart applies to any generator or operator of a facility that stores containers of hazardous wastes. One should also be aware of the definition of "empty" container as specified in § 261.7 and § 261.33. A container or drum that was used for another chemical may have 2.5 centimeters of residue left in it, but still be considered empty. To avoid any problems it is recommended that new containers or drums be requested for use in remediation site activities.

§ 264.171, Condition of containers – This requires that the container be in good condition: it must be free from rust, in good structural condition and not leaking. The container must be exchanged if it fails to meet these requirements. This topic is covered by the requirement for weekly inspections, which includes the area supervisor looking for deficiencies. If a waste container accumulation site is to be set up, then an area supervisor should be designated and trained in this requirement. This topic is addressed also in paragraph 264.174.

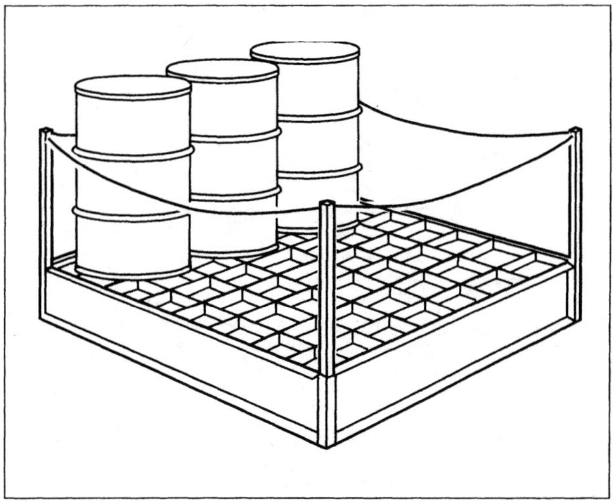

Figure 7-8: The containment catches any spilled material to prevent potential contamination in the adjacent areas.

§ 264.172, Compatibility of waste with containers – The container must be able to hold the waste without being compromised. Therefore, the selection of the container is contingent on the evaluation of the waste to be contained. For example, if in reviewing the technical data for the waste, it is determined that the waste is strongly acidic, then the container must be made of acid-resistant material such as plastic. A container used for alkaline or basic materials would be made of steel. An additional consideration is whether expansion room should be allowed in case the waste gives off a gas or builds up pressure because of temperature variations experienced in the accumulation area or during transit to a disposal site.

§ 264.173, Management of containers – Containers must be kept closed at all times except to remove or add waste. This is the infraction most often cited by agency inspectors: often a funnel or a hose is left in an unsupervised container which is therefore open and unsealed. A container must never be left unsupervised with unsealed bungs or covers. The penalty for violation can be $25,000 per container per day.

§ 264.174, Inspection of containers – The designated employee must inspect the container accumulation site or storage area once per week. The inspection should be recorded in a log and contain the following information: the inspecting supervisor's name, the location, the date of inspection, and the results of the inspection, along with any correction deemed necessary to prevent violation of this section.

§ 264.175, Containment – The container accumulation or storage area should be surrounded by a containment system capable of holding the contents of a leaking container. For liquid wastes, the containment must have sufficient capacity to hold 10 percent of the volume of the container(s) in the containment area, or 10 percent of the largest container in the containment area. The floor should slope toward a collection point, or the containers should be elevated so that they are protected from contact with liquids. If the containment area is outdoors, or not covered with a roof, then it must be of sufficient capacity to hold not only the release volume from the containers but also rainwater. The containment area must have been evaluated to show it is free of cracks, seams or gaps. Solid wastes do not require containment except those that are

specifically designated in § 264.175(d)(1), which include cyanide-bearing wastes and certain other toxic waste streams.

§ 264.176, Special requirements for ignitable or reactive waste – Containers of ignitable or reactive waste must be located at least 15 meters (50 feet) from the facility property line. The actual structure of the accumulation area must adhere to the requirements for nonflammability as noted in 264.17(a); no sparks or open flames may be present. "No Smoking" signs must be posted.

§ 264.177, Special requirements for incompatible wastes – Containers holding wastes incompatible with other materials in piles, open tanks or surface impoundments must be separated by berms, walls, dikes or other devices.

§ 264.178 Closure – The operator of the container accumulation area must remove all hazardous waste and its residues on closure of the area. This means that all containers, liners, bases, soils contaminated with the waste, and its residues must be removed and disposed of properly; otherwise, the accumulation area becomes a waste generator subject to the regulations of 40 CFR 262 through 266.

Checking Your Understanding

1. Identify the relationship of DOT's 49 CFR 172.101 to drums and containers utilized during sampling activities.
2. Identify the relationship of DOT's 49 CFR 178 Subpart L to drums and containers utilized during sampling activities.
3. Identify the relationship of EPA's 40 CFR 264 Subpart I to drums and containers utilized during sampling activities.

7-6 OSHA's Container and Drum Handling Standards

Concepts

- OSHA and EPA regulate the handling of containers and drums in 29 CFR 1910.120(j).
- The HAZWOPER standard in OSHA's 29 CFR 1910.120 discusses the actual handling requirements for containers and drums. OSHA and EPA cooperatively developed the HAZWOPER standard, part of which is reprinted below.

29 CFR 1910.120(j) Handling Drums and Containers

(Source: OSHA, title 29, Code of Federal Regulations, part 1910, Section 120, paragraph j, Washington, D.C., General Printing Office, 1995)

(1) *General.*

(i) Hazardous substances and contaminated soils, liquids and other residues shall be handled, transported, labeled, and disposed of in accordance with this paragraph.

Figure 7-9: Among the items of PPE worn by these technicians are full-facepiece canister-type respirators and hard hats. This indicates Level C, which is appropriate if the atmospheric contaminants and liquid splashes at the site would not adversely affect skin, and if the air contaminants have been identified so that the proper canisters are used.

(ii) Drums and containers used during the cleanup shall meet the appropriate DOT, OSHA and EPA regulations for the wastes they contain.
(iii) When practical, drums and containers shall be inspected and their integrity shall be assured prior to being moved. Drums or containers that cannot be inspected before being moved because of storage conditions (i.e., buried beneath the earth, stacked behind other drums, stacked several tiers high in a pile, etc.) shall be moved to an accessible location and inspected prior to further handling.
(iv) Unlabeled drums and containers shall be considered to contain hazardous substances and handled accordingly until the contents are positively identified and labeled.
(v) Site operations shall be organized to minimize the amount of drum or container movement.
(vi) Prior to movement of drums or containers, all employees exposed to the transfer operation shall be warned of the potential hazards associated with the contents of the drums or containers.
(vii) U.S. Department of Transportation specified salvage drums or containers and suitable quantities of proper absorbent shall be kept available and used in areas where spills, leaks, or ruptures may occur.
(viii) Where major spills may occur, a spill containment program, which is part of the employer's safety and health program required in paragraph (b) of this section, shall be implemented to contain and isolate the entire volume of the hazardous substance being transferred.
(ix) Drums and containers that cannot be moved without rupture, leakage, or spillage shall be emptied into a sound container using a device classified for the material being transferred.
(x) A ground-penetrating system or other type of detection system or device shall be used to estimate the location and depth of buried drums or containers.
(xi) Soil or covering material shall be removed with caution to prevent drum or container rupture.
(xii) Fire extinguishing equipment meeting the requirements of 29 CFR Part 1910, Subpart L, shall be on hand and ready for use to control incipient fires.

(2) *Opening drums and containers.* The following procedures shall be followed in areas where drums or containers are being opened:
(i) Where an airline respirator system is used, connections to the source of air supply shall be protected from contamination and the entire system shall be protected from physical damage.
(ii) Employees not actually involved in opening drums or containers shall be kept a safe distance from the drums or containers being opened.
(iii) If employees must work near or adjacent to drums or containers being opened, a suitable shield that does not interfere with the work operation shall be placed between the employee and the drums or containers being opened to protect the employee in case of accidental explosion.
(iv) Controls for drum or container opening equipment, monitoring equipment, and fire suppression equipment shall be located behind the explosion-resistant barrier.
(v) When there is a reasonable possibility of flammable atmospheres being present, material handling equipment and hand tools shall be of the type to prevent sources of ignition.
(vi) Drums and containers shall be opened in such a manner that excess interior pressure will be safely relieved. If pressure cannot be relieved from a remote location, appropriate shielding shall be placed between the employee and the drums or containers to reduce the risk of employee injury.
(vii) Employees shall not stand upon or work from drums or containers.

(3) *Material handling equipment.* Material handling equipment used to transfer drums and containers shall be selected, positioned and operated to minimize sources of ignition related to the equipment from igniting vapors released from ruptured drums or containers.

(4) *Radioactive wastes.* Drums and containers containing radioactive wastes shall not be handled until such time as their hazard to employees is properly assessed.

(5) *Shock sensitive wastes.* As a minimum, the following special precautions shall be taken when drums and containers containing or suspected of containing shock-sensitive wastes are handled:

(i) All non-essential employees shall be evacuated from the area of transfer.

(ii) Material handling equipment shall be provided with explosive containment devices or protective shields to protect equipment operators from exploding containers.

(iii) An employee alarm system capable of being perceived above surrounding light and noise conditions shall be used to signal the commencement and completion of explosive waste handling activities.

(iv) Continuous communications (i.e., portable radios, hand signals, telephones, as appropriate) shall be maintained between the employee in charge of the immediate handling area and both the site safety and health supervisor and the command post until such time as the handling operation is completed. Communication equipment or methods that could cause shock sensitive materials to explode shall not be used.

(v) Drums and containers under pressure, as evidenced by bulging or swelling, shall not be moved until such time as the cause for excess pressure is determined and appropriate containment procedures have been implemented to protect employees from explosive relief of the drum.

(vi) Drums and containers containing packaged laboratory wastes shall be considered to contain shock-sensitive or explosive materials until they have been characterized.
Caution: Shipping of shock sensitive wastes may be prohibited under U.S. Department of Transportation regulations. Employers and their shippers should refer to 49 CFR 173.21 and 173.50.

(6) *Laboratory waste packs.* In addition to the requirements of paragraph (j)(5) of this section, the following precautions shall be taken, as a minimum, in handling laboratory waste packs (lab packs):

(i) Lab packs shall be opened only when necessary and then only by an individual knowledgeable in the inspection, classification, and segregation of the containers within the pack according to the hazards of the wastes.

(ii) If crystalline material is noted on any container, the contents shall be handled as a shock-sensitive waste until the contents are identified.

(7) *Sampling of drum and container contents.* Sampling of containers and drums shall be done in accordance with a sampling procedure which is part of the site safety and health plan developed for and available to employees and others at the specific worksite.

(8) *Shipping and transport.*

(i) Drums and containers shall be identified and classified prior to packaging for shipment.

(ii) Drum or container staging areas shall be kept to the minimum number necessary to safely identify and classify materials and prepare them for transport.

(iii) Staging areas shall be provided with adequate access and egress routes.

(iv) Bulking of hazardous wastes shall be permitted only after a thorough characterization of the materials has been completed.

(9) *Tank and vault procedures.*

(i) Tanks and vaults containing hazardous substances shall be handled in a manner similar to that for drums and containers, taking into consideration the size of the tank or vault.

(ii) Appropriate tank or vault entry procedures as described in the employer's safety and health plan shall be followed whenever employees must enter a tank or vault.

Checking Your Understanding

1. Identify the standard that contains OSHA and EPA's drum and container handling requirements for protecting workers on a hazardous waste site or emergency response action.

2. Describe OSHA and EPA's drum and container handling requirements for the following:
   a. General.
   b. Opening drums and containers.
   c. Material handling equipment.
   d. Radioactive wastes.
   e. Shock sensitive wastes.
   f. Laboratory waste packs.
   g. Sampling of drum and container contents.
   h. Shipping and transport.

Summary

Drums and containers with unknown contents present the greatest hazard during sampling activities. In fact, EPA advises that any unmarked drum be presumed to contain hazardous materials. Three federal agencies regulate drums and containers:

—DOT is concerned with the selection of proper drums that will be compatible with their contents. DOT is furthermore concerned with the construction and performance of those drums.

—OSHA is concerned with the safety of workers who handle drums and containers at hazardous waste sites and emergency response actions.

—EPA is concerned with the storage procedures for drums containing hazardous waste. Also, EPA cooperates with OSHA in regulating the handling of drums and containers on hazardous waste sites and emergency response actions.

Before and during the sampling of any containers, the technician must observe and evaluate the situation. The constant employment of air monitors is necessary; any indication of a combustible, toxic or radioactive atmosphere before and during container sampling means that technicians must withdraw and consult the site health and safety plan officer.

All debris from sampling must be properly stored and disposed of. Equipment and PPE must undergo appropriate decontamination procedures. Thorough documentation of all phases of the sampling activities is expected.

Application and Critical Thinking Activities

1. Obtain examples of sampling plan procedures for handling drums. Study the requirements regarding drums suspected of containing hazardous contents.
2. Study OSHA's drum handling requirements in 29 CFR 1910.120.
3. A horizontal cylinder-shaped tank is four feet high and six feet long. Measuring through the access portal on top of the tank, a technician determines that the contents begin three feet below. Using the gallonage chart, find the capacity of the tank in gallons and the quantity of the material in gallons.
4. A vertical cylinder-shaped tank is six feet high and four feet in diameter. The tank's contents begin one foot from the top. Calculate the capacity of the tank in gallons. Then calculate the volume of material in the tank.
5. Study the drum and tank sampling equipment shown in an environmental or lab equipment supply catalog. Compare prices and features of a variety of samplers: glass drum thieves, COLIWASAs (including any model with extensions), bomb samplers, and subsurface grab samplers.
6. Given a storage tank and a portable container, demonstrate the grounding procedure for filling the container from the tank.
7. Predict some possible problems or challenges that may be associated with sampling a large heterogeneous waste pile; suggest strategies for addressing those problems.
8. Suppose a drum is needed for handling waste known to be strongly acidic. Consult the Code of Federal Regulations to select an appropriate drum and describe the construction/performance requirements for such a drum.
9. Compare and contrast the focus of each agency – DOT, EPA and OSHA – regarding drums and containers.
10. Select any one of the OSHA Code of Federal Regulations citations referenced in this chapter. Study that part of the CFR and write an explanatory memo that informs class members. Verify your interpretation with the instructor or other qualified professional.

8

Methods of Analysis

Chapter Objectives

Upon completing this chapter, the student will be able to:

1. **Compare** field analysis, mobile laboratories and fixed laboratories in terms of the types of analysis provided as well as the advantages/disadvantages associated with each.
2. **Understand** that different analytical methods produce different levels of data quality.
3. **Know** what qualities to consider when choosing an environmental laboratory.
4. **Describe** procedures used by a laboratory to prepare and analyze environmental samples.
5. **Demonstrate** knowledge of laboratory instrumentation and discuss which instruments are used to analyze for various contaminants in environmental samples.
6. **List** the characteristics used to determine whether or not a sample is considered a hazardous waste.

Chapter Sections

8–1 Analysis Options

8–2 General Laboratory Procedures

8–3 Laboratory Instrumentation

8–4 Methods for Determination of Hazardous Waste

8-1 Analysis Options

Concepts

- Environmental sample analysis can take place in the field, in a mobile laboratory or in a fixed analytical laboratory. These analysis options make use of different equipment and methods and offer different capabilities and limitations.
- The quality of analytical data varies with the method of analysis. Data users must decide what level of data quality is necessary to support project decisions. The EPA has divided analysis options into five levels of quality.
- The choice of an environmental laboratory is an important decision that should be based on evaluation of a laboratory's technology, proficiency, certifications, QA/QC program and customer service.

In previous chapters we learned how to investigate a site – how to implement the elements of a sampling plan and how to collect representative samples. The next step in the investigative process is sample analysis to determine what contaminants, if any, the samples contain. The technician must either perform

Figure 8-1: The extraction method can analyze the solubility of various wastes in solvent.

the analysis while in the field or submit samples to an analytical laboratory. But which type of analysis – field, mobile laboratory, or fixed laboratory – best suits the project needs? Timeliness, cost, environmental regulations and data quality must all be considered in deciding which method of analysis is preferred. When fixed laboratory services are needed, how does the technician go about choosing the best analytical laboratory to undertake the project analysis? What does the laboratory actually do with the samples after they receive them, and what instruments are routinely used to analyze environmental samples? What criteria does the laboratory use to label a sample as "hazardous waste"? Answers to these important questions will be the topics discussed in this chapter.

Field versus Mobile Laboratory versus Fixed Laboratory Analysis

Environmental sample analysis may be classified into three categories according to location:

—**Field Analysis**

—**Mobile Laboratory Analysis**

—**Fixed Laboratory Analysis**

Each of these analysis options makes use of different types of equipment and methods and offers different conveniences, time frames, data quality, capabilities and also limitations. The project planner needs to decide as part of the sampling plan which option to use to best interpret conditions at the site.

Field Analysis

Field analysis consists of measurements made with test kits and hand-held instruments, which are generally easy to use and give direct readout of analytical results.

As we learned in Chapter 4, air monitoring and air sampling activities are conducted in the field by such means as drawing air into syringes, onto filters or into absorbent tubes. Analysis of the air for components including toxic gases and dust may be completed onsite.

Many water monitoring activities may also be conducted with portable equipment. Measurements such as pH, temperature, and conductivity, which usually require analysis immediately following sample collection, can be measured using a combination battery-operated probe. Color comparator kits can be used to determine approximate concentrations of a variety of inorganic compounds containing iron, nitrate, chloride and cyanide ions. These kits usually contain premeasured, prepackaged chemicals and step-by-step instructions. A detectable level of the target analyte will be indicated by a color change. This is then compared to the color of known concentrations of the target analyte to obtain the result.

Single-use **immunoassay** test kits are recently developed field-screening techniques for compounds such as Total Petroleum Hydrocarbons (TPH) and Polychlorinated Biphenyls (PCB). These kits are especially useful in applications where a yes or no answer concerning the presence of contamination is needed, for example to initially define the extent of a contamination plume.

There are many advantages to field analysis, including the fact that the technician can complete the analysis while still onsite. Since results are immediately available, samples can be analyzed in an order of priority; this allows for sampling plan modification as the site investigation progresses. Field analysis techniques have the potential for analyzing a large number of samples in a short amount of time for little expense. Field analysis equipment is less expensive than the one used in fixed laboratory analysis, and much of it is available for rent.

Figure 8-2: The technician performs an immunoassay test to detect the presence of PCBs in soil.

Point of View ■ Mike Steinmaus

Mike Steinmaus is president of Steinmaus Environmental Services in Muscatine, Iowa, an environmental consulting firm that conducts a broad range of environmental studies including impact assessments, site assessments, ecological evaluations, compliance audits, industrial permit applications and training programs. Mike received a B.S. in Resource Development from Iowa State University, and an M.A. in Geography from the University of Iowa. Prior to forming his own business in 1987, Mike acted as Head of the Environmental Analysis Department for Stanley Consultants of Muscatine, Iowa.

Steinmaus Environmental Services is not equipped to perform laboratory analysis. Instead, it relies entirely on environmental laboratories to provide analysis results for the groundwater, wastewater, air, stormwater, soil and solid waste samples that are a part of its projects. Although only ten to twenty percent of the company's projects involve laboratory analysis, sample results of the highest quality providing a solid base for his project decisions are always a priority for Mike.

"I like to use a lab," Mike says, "that I have some type of knowledge of. We get ads in the mail and phone calls all the time from labs offering environmental analysis services. And I probably wouldn't use them unless I had more information about the quality of their work or knew that they could offer something that no one else could provide."

Q. *How many labs does Steinmaus Environmental Services regularly use?*

A. "In the past year we've essentially used one lab; in the past five or six years we've used four different labs. Right now I feel very comfortable with the lab we're using, and that's why we haven't used any other lab."

One consideration in using different labs, Mike adds, is their location in relation to the project site. And he cites a current project for which he has had to search diligently for a lab that could provide a particular air quality analysis.

Q. *What makes you feel comfortable about your choice of environmental laboratory?*

A. "Number one is the quality of the lab. They are a nationally known lab with a good quality control program. That is our primary concern. Number two is that their pricing is reasonable. Number three is their friendliness; their ability to work with us and say, 'Here's what you need.' We can bounce ideas off them and get their opinion on what kinds of things can be done to best meet the needs of our client. Fourth is their turnaround time."

Mike adds that another service that his lab provides at no charge is a carrier for sample container delivery and sample pick-up and delivery to the lab. This saves him time and paperwork and prevents additional charges to his client.

Q. *How did you choose your current lab?*

A. "Initially I obtained literature on the lab. Then I used them once on a trial basis to assess their quality, turnaround time, reporting methods and friendliness. After comparing the lab to others I had used, I decided to become a regular customer."

Q. *What do you do in turn to make your lab's job easier?*

A. "I try to give the lab enough documentation to clarify our objectives. I almost always talk to the lab personnel prior to sending samples to eliminate guesswork. If we have prior contact with the lab, the analysts know the samples are on their way and the lab can prepare for them."

But there are also disadvantages to field analysis. The technician must travel to the site and use the equipment under less-than-ideal analysis conditions. Portable equipment does not have the sensitivity of fixed laboratory equipment, and its accuracy can suffer under the stress of transportation or weather conditions such as humidity or cold. Field activities need to be scheduled efficiently in order to supply the field analyst with a steady supply of samples of the same matrix. Changing sample matrices in the field can cause delays, as recalibration and method changes are usually necessary. One must also consider the potential of field analysis to produce a hazardous waste; if this is the case, the technician will need to deal with proper handling and disposal issues. Finally, most field analysis is not certifiable, and before proceeding with onsite analysis the technician must insure that the results will meet data quality objectives.

Mobile Laboratory Analysis

A mobile analytical laboratory may be contained in a van, trailer or boat. Shipping laboratory equipment to clients' facilities and using their work space as a temporary laboratory is also considered a mobile laboratory. The portability of analytical instruments has improved greatly as technology has advanced, and capabilities of mobile labs may now match those of state-of-the-art fixed laboratories. Mobile laboratories can provide analysis of many sample types for a wide variety of inorganic and organic compounds. Mobile laboratories offer the same quick availability of results as field analysis, and the same opportunity to prioritize sample analysis and modify the sampling plan as work progresses.

Figure 8-3: A technician performs a field screening for volatile organics using a mobile gas chromatograph.

However, mobile laboratories can be costly, as they require not only instrumentation but also qualified analysts to operate the instruments. Many mobile labs charge on a per day rather than a per sample basis, and sometimes have a standby rate for project delays. Logistical concerns, such as an adequate power supply, may complicate their use in more remote locations. Confusion may erupt over who regulates and certifies mobile labs; the answer is dependent on the type of project, its location and the end use of the data generated.

Mobile labs may be wisely utilized in remedial site investigations and underground storage tank site cleanups to assist in determination of the depth and lateral extent of contamination. The presence of a mobile laboratory allows the site manager to better control labor and equipment costs, as there is less down time while everyone awaits the results of samples sent to a fixed laboratory. Mobile labs are also an asset during emergency response situations. Rapid analysis results can help to quickly define the degree of contamination and its path of migration, while also allowing for rapid implementation of safety measures and cleanup procedures.

Fixed Laboratory Analysis

Fixed analytical laboratories have a professional staff and a wide array of specialized instrumentation. These laboratories are equipped to handle analysis of any type of sample for any target compound that a project might require. Fixed laboratories have the most sensitive instruments; they can provide supporting documentation along with the analysis results, and store data for extended periods of time. Information concerning the quality of data generated is available through the lab's quality control program and through certifications held by the laboratory.

A disadvantage to fixed laboratory analysis is the **turnaround time**, or time required for the lab to complete a sample analysis. Turnaround times for routine analysis are generally at least two weeks, with further delays caused by shipping and handling problems. Since samples are sent to the lab by multiple clients, the lab will not be focusing all of its capabilities and efforts on the project samples alone. If priority service is available, it is usually at double or triple the normal analysis cost.

Box 8-1 ■ Guidelines for Selecting an Analytical Laboratory

1. How does the lab approach QA/QC, and is there a written QA/QC plan for you to review? Does the plan include such items as Standard Operating Procedures, preventive maintenance plans, internal and external audits, and a corrective action plan for out-of-control data?

2. Is the lab certified for the parameters to be analyzed? Remember that certification programs do not exist for all types of analysis.

3. What is the lab's capacity - do they have the capability to analyze your samples in a timely manner, even if a laboratory instrument breaks down? Will their standard turnaround time meet your project schedule?

4. Does the lab assure client confidentiality?

5. Are references from other lab clients available for your review?

6. Does the lab have a written policy statement?

7. What are the lab's waste management and health and safety practices? If not handled properly, these could increase your liability.

8. Does the lab have professional liability and/or errors and omissions insurance?

9. Does the lab use subcontractors for some analysis parameters or does it conduct all reported analyses in-house?

10. Will the lab send you the proper sampling containers for your project? Is an analyst available to answer any questions you might have concerning sampling and analysis? Is the lab able and willing to provide assistance with explanation and interpretation of analysis results?

11. Are the lab's analysis prices reasonable?

Five Levels of Data Quality

Ultimately, more important than convenience or cost associated with using a field, mobile or fixed laboratory for sample analysis is the quality of data. Is the project planner interested in simply determining the presence or absence of contamination at the site, or does he or she need to know the exact concentration of the target analyte present? How much supporting documentation is required under the regulations governing the project? How confident will the project planner be that the analysis results represent the true conditions on the site?

Sometimes, the exact method of analysis will be mandated by the agency regulating the environmental project. At other times, the project planner must decide what level of quality is necessary to support project decisions. EPA has set up the various analysis options into five levels. These levels are distinguished by the types of instrumentation used, the amount of documentation required, and the quality of data each provides.

EPA Level I – Field Screening

EPA Level I field screening or analysis with portable instruments such as pH, conductivity, HNU or field test kits provides the lowest data quality of any of the levels, but the most rapid results. The data generated is usually considered qualitative, and may not be compound-specific. The analysis may only provide a yes or no answer concerning the presence of contamination. Documentation of screening analysis consists of the field technician's logbook. Screening is the least costly of the analytical options and is typically used for health and safety monitoring or initial site characterization to locate areas for further sampling and analysis.

EPA Level II – Field Analysis

EPA Level II field analysis uses more sophisticated portable analytical instruments, perhaps set up in a mobile onsite laboratory. Level II provides for rapid results and somewhat better quality data than Level I. There is, however, a wide range of data quality at this level, depending on instrument calibration, sample preparation techniques and the skill of the analyst performing the task. Field analysis can include air, soil and water analysis for a variety of organic and inorganic compounds. This level is typically used for site characterization, engineering design and monitoring of remediation efforts.

EPA Level III – Engineering

In **EPA Level III** engineering, analysis is performed off site in a fixed analytical laboratory using EPA-approved analytical methods. The laboratory may or may not use protocols mandated under the EPA's **Contract Laboratory Program (CLP)**. CLP labs are certified to perform EPA project analysis under rigorous QA/QC and documentation protocols. Although Level III does not include the stringent QA/QC or documentation of a CLP lab, the EPA-approved methods that are used do include QA/QC protocols with accuracy and precision standards that must be met. Level III provides for an intermediate level of data quality with confirmed identity and quantitation of organic and inorganic analytes in air, water, soil and solid waste samples. Engineering data is widely used in environmental projects for risk assessment, health and safety issues, site characterization, establishing cleanup objectives, project design and monitoring during remediation efforts.

EPA Level IV – Confirmation

In **EPA Level IV** confirmation, analysis is conducted under the EPA's Contract Laboratory Program (CLP) and is characterized by rigorous QA/QC protocols and documentation. Level IV provides the highest level of data quality with confirmed identification and quantitation of organic and inorganic analytes in air, soil, water and solid waste samples. Level IV analysis is also the most expensive and requires the longest turnaround time. Confirmation analysis is typically used for risk assessment and evaluation of remedial alternatives and for proof of accuracy of lower-level data. Confirmation analyses may also be required by the regulatory agency as confirmation that cleanup goals have been reached.

EPA Level V – Nonstandard Analysis

Analysis by nonstandard protocols, for example when analysis of an unusual compound is required, is categorized as Level V. These analyses may require that a method be developed or modified, and often require the issuance of regulatory agency approval before use. Turnaround time is generally very long for **EPA Level V** nonstandard analysis and, since the analysis is one-of-a-kind, no comparison with other data is possible.

How to Choose an Environmental Laboratory

When the need for fixed laboratory analysis arises, how does the project planner determine which laboratory is most qualified to provide efficient service and accurate results at a reasonable cost?

The choice of an environmental laboratory should be based on the evaluation of a laboratory's

—Technology

—Proficiency

—Certification

—QA/QC Program

—Customer Service

Technology

Because many environmental laboratories specialize in certain services such as hazardous waste determination, wastewater analysis, or farm chemical analysis, it is important to match a lab's experience with project needs. Many labs offer qualification summaries that can help determine expertise levels and specialties.

When reviewing a laboratory's statement of qualifications, the project planner should note the type of instrumentation the laboratory has available. Not only should the inventory include the correct instruments for the analysis requirements, but there should be backup equipment in case of instrument failure. The education and experience of laboratory employees are another important consideration. Qualification statements also may include a list of past projects in which the lab has participated as well as a reference list of customers.

Proficiency

The quality of data generated by the laboratory is evidenced by precision and accuracy information for internal quality control samples, and by external evaluations such as **performance samples** and laboratory audits. Results of these proficiency measures should be available for customer review. Customers may also conduct their own proficiency examination by sending the laboratory samples of known analyte concentration along with environmental samples.

Certification

Laboratory certification provides assurance that a laboratory is truly capable of performing the analyses for which they hold certification. Unfortunately, certification programs are not available for all types of laboratory analysis.

There are currently no mandatory national programs for environmental laboratory certification. Voluntary national certification can be obtained through such organizations as the American Association for Laboratory Accreditation (A2LA), and the American Industrial Hygiene Association (AIHA).

Most laboratory certification programs are administered by states. States may certify laboratories for a broad category such as wastewater, drinking water or for a specific parameter such as coliform bacteria. To become certified, laboratories must pay certification fees, undergo onsite inspections and pass proficiency examinations consisting of EPA performance samples. EPA performance samples are sent to laboratories twice per year. Laboratories must determine the quantity of a particular compound in the sample and send results to the agency for evaluation. Approximately 80 percent proficiency is required to pass the EPA examination.

Certified laboratories are generally given identification numbers and certificates; they should in turn be able to provide a copy of these as proof of certification. Although there may be no certification program for the specific analysis required, a laboratory's certification for other parameters is still assurance that the laboratory provides accurate and reliable services.

QA/QC Program

The purpose of a laboratory QA/QC program is to ensure the consistent production of reliable data of high accuracy and precision. All laboratories should have a written QA/QC program and should be willing to provide a copy for customer inspection. This program should include:

—The use of standard and approved methods of analysis;

—The use of Standard Operating Procedures (SOP) to train all analysts to perform methods correctly and identically;

—Procedures such as internal quality control samples, performance evaluations, and internal and external audits to check the accuracy and precision of all analyses. These procedures will be discussed in detail in Chapter 9;

—A preventive maintenance plan for laboratory instruments and facilities;

—A corrective action program to deal with data quality problems as they arise.

Customer Service

Issues such as cost of analysis, turnaround time and communication with the lab seem the most important considerations in the choice of a laboratory and – from a business perspective – cost especially is an important consideration. However, nothing should come before the issue of quality of data. The increased cost associated with a reputable lab is usually negligible compared to the liabilities associated with data that is invalid or cannot be supported with documentation. Laboratory price lists will provide a comparison of analysis costs and most labs will prepare a quote of costs associated with analysis for a particular project. Discounts are sometimes given for submission of large numbers of samples.

The pressures associated with environmental projects often increase dramatically while results of sample analysis are awaited. The majority of decisions made by project managers revolve around sample results, and therefore turnaround time is important. Most analytical processes cannot provide instantaneous results, especially since environmental samples are often complex and dirty mixtures. As a business, the laboratory will have to give equal attention to many customers at once. Before choosing a laboratory, it is best to ensure that standard turnaround times offered by the laboratory are compatible with project needs.

A qualified laboratory will be willing to establish interactive and continuous communications with customers before, during and between environmental projects. The laboratory should be willing to provide not only suitable sampling and shipping containers for their customers' use but also assistance with correct sampling technique and interpretation of analytical results. Customers should feel free to call their laboratory with questions, and the laboratory should not hesitate to set up a tour of the lab at a customer's request.

Checking Your Understanding

1. List three advantages and three disadvantages of using field analysis techniques to obtain environmental sample data. Give an example of a field analysis technique useful in analyzing groundwater samples.
2. Why might mobile laboratory analysis be the most expensive analysis option of all?
3. Describe the difference between EPA Level III and Level IV analysis options in terms of methodology and resulting data quality differences.
4. Review the criteria listed in the text for choosing an environmental laboratory and discuss why each of these evaluation points is important in laboratory selection.

8-2 General Laboratory Procedures

Concepts

- Sample preparation techniques used to convert an environmental sample into the correct form for analysis include dilution, extraction, filtration, digestion, drying, particle size reduction and compositing.
- A laboratory's analytical processes include laboratory quality control procedures to alert the analyst to instrument malfunction or methodology problems. Following these QC procedures helps to insure the production of reliable analytical data.
- Analytical methods vary not only in technique, but also in terminologies, sequence of steps, frequency of required quality control measures, and acceptable limits for quality control parameters. Environmental regulations sometimes mandate the use of specific analytical methods. When arranging for sample analysis, it is therefore important to let the laboratory know as much about your project as possible.

What happens to the samples once the laboratory receives them? The laboratory will sign, date and note the time of arrival on the chain-of-custody form that should accompany all environmental samples for laboratory analysis. The samples will then be checked for correctness of bottle type, sampling technique, preservation and holding times. Based on this information, the laboratory will decide if it can accept or if it must reject the samples. Communication with the laboratory prior to sample collection activities should eliminate most of the problems associated with sample rejections. Acceptable samples will be assigned unique laboratory identification numbers for tracking purposes throughout the analysis procedure. The samples will then be placed in the correct storage area, usually a refrigerator, until sample preparation begins.

Sample Preparation

Sample preparation involves getting the sample into the correct form for analysis, based on the method or instrumentation that will be used. Various types of sample preparation techniques are discussed below:

—*Aliquot* – An aliquot is the representative portion of the sample taken for analysis.

—*Dilution* – Dilution means taking a portion of the sample and mixing it with another medium, such as laboratory reagent water, to reduce the concentration of analytes to a level suitable for the laboratory instrument's capabilities. A 1:1 dilution means equal parts of sample and dilutant; a 1:50 dilution means one part sample to fifty parts dilutant.

—*Extraction* – Extraction involves drawing out the target analyte from the sample into another medium more suitable for analysis. A soil sample contaminated with diesel fuel, for example, goes through an extraction process with methylene chloride, which captures the diesel fuel from the soil. Analysis by gas chromatography then proceeds.

—*Filtration* – Many surface water, groundwater and wastewater samples need to be filtered before analysis can proceed. Filtration removes suspended particles that could interfere in the analysis procedure. Filtration is also undertaken when dissolved analyte content (for example dissolved iron content) rather than total analyte content is desired.

—*Digestion* – **Digestion** is the process whereby a chemical agent, often acid, is added to a sample in order to disintegrate and dissolve the sample. Wastewater samples for metals analysis, for example, are placed in a nitric acid solution and then heated in order to dissolve the metals into solution. Soil samples for metals analysis undergo a similar process.

—*Drying* – Some analytical methods are done on a dry-weight basis, that is, with no moisture present. Soil samples, for example, undergo a drying process in a 105°C. oven for one hour or until constant weight is reached before analysis proceeds.

—*Particle Size Reduction* – The process of cutting, grinding or crushing a sample to reduce its particle size is sometimes necessary before analysis can proceed. The TCLP extraction procedure explained later in this chapter, for example, requires a sample particle size capable of passing through a 9.5 mm (0.375 inch) sieve. Therefore, any solid sample for TCLP analysis must be crushed with a mortar and pestle until it meets this size specification.

—*Compositing* – When representative amounts of the same type of sample are combined into one sample for laboratory analysis, the sample has been composited. Compositing can sometimes save a client money. For example, a client may have a particular type of waste stored in six drums. One sample from each drum is sent to the lab for analysis prior to obtaining the waste's disposal permit. Since all of the waste samples are from an identical source, the lab may take equal amounts from each of the six samples and composite them into one sample for laboratory analysis instead of analyzing all six samples separately.

—*Determining Liquid or Solid Sample Type* – Some sludges and oily environmental samples are difficult for the lab to classify into solid or liquid categories. This classification can make a difference in the method used for analysis, and in how analysis results are calculated and reported (solids are reported as mg/kg, liquids as mg/L). A filtering technique with a time limit is used to classify these difficult samples as solids or liquids. If enough of the sample passes through the filter within the prescribed time limit, the sample is classified as a liquid.

Sample Analysis and Laboratory Quality Control

After the sample has been prepared, it is ready for analysis. The flow chart (Table 8-1) shows the

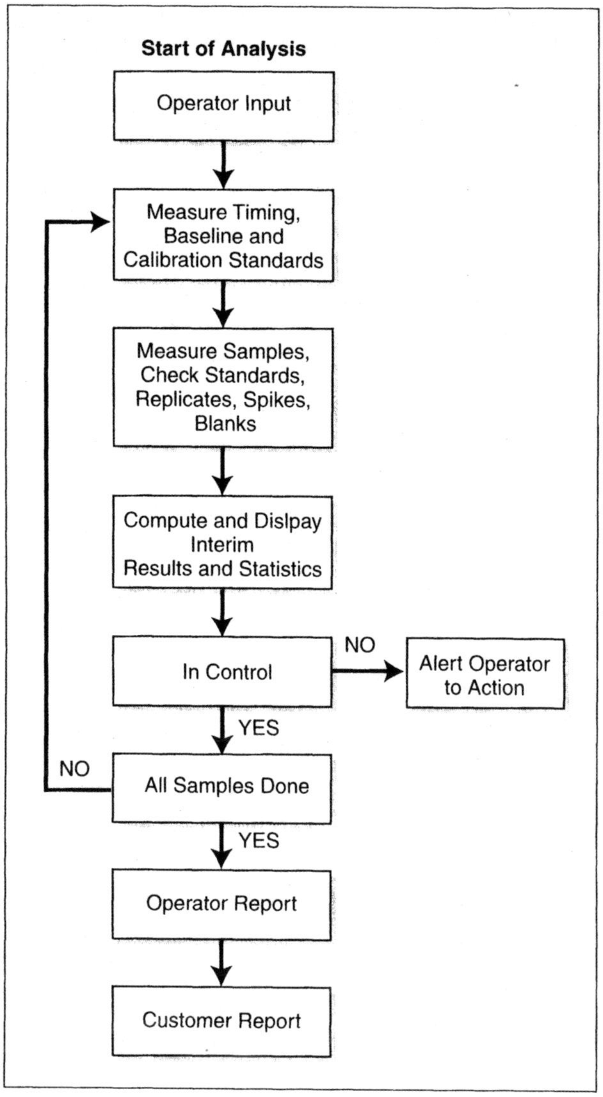

Table 8-1: Flow chart depicting sequence of events during controlled series of laboratory measurements.

sequence of events which take place during a controlled series of laboratory measurements. This flow chart is applicable to any type of laboratory instrumentation.

As the flow chart indicates, the analytical process involves many steps other than actual analysis of the environmental sample. Information such as sample numbers must first be input into the instrument, and the instrument must be set up and calibrated for target analyte determination. Laboratory quality control samples are prepared and analyzed right along with the environmental samples. Results of these laboratory standards, duplicates, spiked samples and blanks alert the analyst to instrument malfunction or methodological problems.

Laboratory personnel are alerted that corrective action may be necessary if:

—Results of quality control samples do not fall within the acceptable range for precision and accuracy;

—Blanks contain target analytes above acceptable limits;

—Undesirable trends are detected in spike sample recoveries;

—Duplicate sample results differ;

—There are unusual changes in the detection limits of an instrument;

—Deficiencies are detected by the lab's QA department during internal or external audits, or from the results of performance samples;

—Inquiries concerning data quality are received from outside the laboratory.

Correction is often handled by the analyst through such actions as checking instrument calibration and sensitivity as well as reviewing sample preparation procedures. Corrective actions taken by the analyst may include:

—Re-analyzing the samples and new quality control samples;

—Re-sampling and analyzing new samples and quality control samples;

—Evaluating and amending the analytical procedures used for the sample analysis;

—Accepting the original data and acknowledging the increased level of uncertainty associated with it.

It is by taking these measures that the analyst can be confident of instrument performance and the generation of quality results. While analysis progresses, the analyst logs necessary information from the instrument display into his or her logbook. From this information the customer report is prepared.

How Methods of Analysis are Determined

The extensive growth of environmental regulations in the recent past has led to a confusing array of analytical methods laboratories must contend with.

While these methods often use the same analytical techniques, they may vary in specifics such as terminology, sequence of analytical steps, frequency of required quality control measures, and acceptable limits for quality control parameters. These differences, which may seem minor to non-laboratory personnel, can in fact have a major impact on laboratory operating and data handling procedures.

When arranging for sample analysis, it is therefore very important to let laboratory personnel know as much about the project as possible:

— Is the sample analysis necessary for a wastewater discharge permit? If so, what city is the regulatory authority? Cities often enact more stringent wastewater discharge limits than those mandated by federal law.

— Is the project a site cleanup regulated by the federal EPA? Are the samples taken from the site initial samples to define the plume of contamination or final samples to prove that cleanup objectives have been met? The latter are likely to require lower instrument detection limits and extensive documentation.

— By what method will the technician be disposing of the solid waste that has been sampled? If it is to be landfilled, it may be necessary to contact the landfill assigned by the project planner to see what analysis it requires before accepting the waste.

Discussing the specifics of a sample before analysis begins is beneficial to both the client and the laboratory. With this information the lab will be able to perform the analysis using methods prescribed by the regulatory authority while providing the necessary level of quality control and documentation.

The federal EPA mandates the use of certain analytical methods for sample analysis under different environmental programs. Most of these method requirements can be found in Part 40 of the Code of Federal Regulations. These methods do not all come from one method source; not only are EPA's own specific methods acceptable, but so are several others. The collective methods from the American Water Works Association (AWWA), American Public Health Association (APHA), and the Water Environ-

Parameter, Units and Method	Reference (Method No. or Page)				
	EPA 1979[3,4]	Std. Methods 16th ED	ASTM	USGS[1]	Other
1. Acidity, as $CaCO_3$ mg/L: Electrometric end point or phenolphthalein end point.	305.1	402(4.a)	1067-82(E)	....	....
2. Alkalinity, as $CaCO_3$ mg/L:					
Electrometric or colorimetric titration to pH 4.5, manual or	310.1	403	D1067-82(B)	I-1030-84	33.014.[2]
automated	310.2	....	....	I-2030-84	....
3. Aluminum–Total,[3] mg/L: Digestion[3] followed by:					
AA direct aspiration	202.1	303C	....	I-3051-85	....
AA furnace	202.2	304	....	....	
Inductively coupled plasma (ICP)...	.....	....	....	....	200.7.[4]
Direct current plasma (DCP) or	.....	....	....	....	Note 33.
Colorimetric (Eriochrome cyanine R)	.....	306A	....	....	....
4. Ammonia (as N), mg/L: Manual distillation (at pH 9.5)[5] followed by:	350.2	417A	....	....	33.057.[2]
Nesslerization	350.2	417B	D1426-79(A)	I-3520-84	33.057.[2]
Titration	350.2	417D	....	....	....
Electrode......................	350.3	417E or F	D1426-79(D)	....	....
Automated phenate or	350.1	417G	D1426-79(C)	I-4523-84	....
Automated electrode............	.....	....	....	....	Note 6.

Table 8-2: Excerpt from 40 CFR 136.3 listing the various methods which the EPA has approved as inorganic test procedures.

ment Federation (WEF) called "Standard Methods for the Examination of Water and Wastewater" are accepted by EPA, as are methods from the American Society for Testing and Materials (ASTM).

Unless specific methods have been mandated by the regulatory authority, the laboratory may utilize several method-reference books as guidelines. A sampling of these guidelines follows:

—Safe Drinking Water Act (SDWA);

—National Pollutant Discharge Elimination System (NPDES) - wastewater:

1) "Methods for Chemical Analysis of Water and Wastes," EPA 600/4-79-020. Method numbers 100 through 400 which include metals and other inorganic parameters such as cyanide, solids, sulfide, and oxygen demand;
2) "Methods for the Determination of Metals in Environmental Samples," EPA 600/4-91-010. This reference includes Method 200.7 for ICP metals analysis (defined in the next section);
3) "Methods for Determination of Organic Compounds in Drinking Water," EPA 600/4-88-039. This reference includes the 500 series methods for volatile and semi-volatile compound analysis;
4) "Standard Methods for the Examination of Water and Wastewater." This reference includes methods for metals determination by both AA (Atomic Absorption is defined in the next section) and ICP. It also includes methods for most other inorganic and organic parameters.

—Resource Conservation and Recovery Act (RCRA) - solid waste:

"Test Methods for Evaluating Solid Wastes Physical/Chemical Methods," EPA SW-846. This four-volume series includes all metal, inorganic and organic parameter methods.

—Underground Storage Tank Programs (UST):

1) "Test Methods for Evaluating Solid Wastes Physical/Chemical Methods," EPA SW-846. The third volume of this reference contains methods for organic analysis, including GC and GC/MS analysis methods;
2) Some states have adopted their own methodologies from EPA methods. The state of Iowa, for example, has developed Method OA-1 for Volatile Gasoline Compounds and Method OA-2 for Diesel and other Organic Compounds. These methods must be referenced on laboratory reports sent to Iowa's underground storage tank program.

—Contract Laboratory Program/USEPA (CLP): This program uses the EPA methods listed above under the NPDES and RCRA programs, but has a separate and more stringent quality control and documentation format that must be followed.

Checking Your Understanding

1. Describe each of the following sample preparation techniques:
   —Dilution
   —Extraction
   —Digestion
   —Compositing
   —Filtration
2. Briefly outline the sequence of events that takes place once an environmental sample is received at the lab. Be sure to include all of the quality control steps involved in a laboratory analysis.
3. How does a laboratory determine the method of analysis to use on an environmental sample? When arranging for laboratory analysis, why is it important to let the laboratory know as much about your environmental project as possible?

8-3 Laboratory Instrumentation

Concepts

- Laboratory instruments routinely used to analyze environmental samples include the Atomic Absorption Spectrophotometer (AA), Inductively Coupled Plasma (ICP), Gas Chromatograph (GC), Gas Chromatograph/Mass Spectrometer (GC/MS), Spectrophotometer, pH Meter and the Specific Ion Analyzer.
- There are four major components in the measurement process of laboratory instruments: the interaction of the sample component with the detector, the detector signal, the signal processor and the readout.
- Wet chemistry procedures such as distillation, titration and gravimetric analysis are also used in the analysis of environmental samples.

Laboratory instruments can be very complex. Students majoring in chemistry often have a full semester class devoted to laboratory instrumentation. In this section we present an overview of some of the major instruments used in environmental sample analysis.

In a broad sense, an instrument for chemical analysis converts a signal that is usually not directly detectable and understandable by the technician into a form that is. Regardless of the complexity of the instrument, there are four major steps in the measurement process:

1. The sample component interacts with the detector producing a change. For example, the ionic activity of a sample interacts with the pH meter's detector creating a change.
2. The detector converts that change into a signal. In the pH meter, for example, the detector is the glass electrode which generates a small electrical signal.
3. The signal processor modifies the detector signal to make it more convenient for the readout

Figure 8-4: AA instrument.

device. The signal processor in the pH meter is the amplifier.

4. The readout can be a chart or a digital display, or even a visual comparison. In a pH meter the amplified signal is converted into a digital or analog display.

Laboratory Instruments

Atomic Absorption Spectrophotometry

Atomic Absorption (AA) Spectrophotometry is used to quantitatively determine the concentration of metal ions in a sample. There are two types of Atomic Absorption instruments – Flame AA and Furnace AA. Table 8-3 presents a comparison of the range of concentrations measured and the lowest concentrations that can be detected by the two types.

AA Flame – The **flame atomic absorption spectrophotometer** analyzes metals in the 0.1 to 100 mg/L (ppm) range. It is currently suitable for analysis of most wastewaters, stream runoff and some soil samples. In the past many EPA-approved methods were based on Flame AA technique, but recent regulatory changes lowering acceptable concentrations of metals in groundwater and drinking water have made the technique obsolete for groundwater and drinking water analysis.

The instrument operates by aspirating a liquid portion of the sample into a flame (air/acetylene or nitrous oxide) where the metal ions are atomized. While the metal ions are in the flame, a selected wavelength of light (specific to each metal) at a known intensity is directed through the flame. The metal ions in the flame absorb some of the radiant energy (i.e., light). By measuring the light intensity before and after the flame, the concentration of the metal in the sample can be calculated. Table 8-3 shows metals typically analyzed by AA Flame.

AA Furnace – The **furnace atomic absorption spectrophotometer** analyzes metals in the 0.1 to 1000 μg/L (ppb) range. It is utilized for groundwater and drinking water analysis. Because of the lower concentrations this instrument can detect, it is an EPA-approved method.

Analysis with this instrument begins with injection of a liquid portion of the environmental sample into a graphite furnace. The furnace atomizes the metals present in the sample in three stages. Furnace temperatures are set specifically for each element being analyzed:

—Stage 1: a light heating is used to drive off the liquids, leaving only metals and organic particles behind;

—Stage 2: a moderate heating is used to char and drive off the organics, leaving only the metals;

—Stage 3: a strong heating is used to atomize the metals.

The atomized metals are then analyzed by measuring the radiant energy absorbance as described for AA Flame analysis.

Inductively Coupled Plasma – Atomic Emission Spectrophotometer (ICP-AES)

The **Inductively Coupled Plasma - Atomic Emission Spectrophotometer (ICP-AES)** is also used to quantitatively determine the concentration of metal ions in environmental samples in a range from 0.01 to 1000 mg/L (ppm). An advantage of ICP analysis is that several metal concentrations can be determined simultaneously from the sample. This technique is suitable for analysis of wastewater, surface water, soil as well as some drinking water parameters. The EPA has approved many ICP methods for drinking water analysis.

Figure 8-5: The ICP instrument analyzes the concentration of metal ions in a sample.

Atomic Absorption Concentration Ranges(1)					
	Direct Aspiration			Furnace Procedure(4,5)	
Metal	Detection Limits mg/L	Sensitivity mg/L	Optimum Concentration Range mg/L	Detection Limit µg/L	Optimum Concentration Range µg/L
Aluminum	0.1	1	5 – 50	3	20 – 200
Antimony	0.2	0.5	1 – 40	3	20 – 300
Arsenic(2)	0.002	–	0.002 – 0.02	1	5 – 100
Barium(p)	0.1	0.4	1 – 20	2	10 – 200
Beryllium	0.005	0.025	0.05 – 2	0.2	1 – 30
Cadmium	0.005	0.025	0.05 – 2	0.1	0.5 – 10
Calcium	0.01	0.08	0.2 – 7	–	– –
Chromium	0.05	0.25	0.5 – 10	1	5 – 100
Cobalt	0.05	0.2	0.5 – 5	1	5 – 100
Copper	0.02	0.1	0.2 – 5	1	5 – 100
Gold	0.1	0.25	0.5 – 20	1	5 – 100
Iridium(p)	3	8	20 – 500	30	100 – 1500
Iron	0.03	0.12	0.3 – 5	1	5 – 100
Lead	0.1	0.5	1 – 20	1	5 – 100
Magnesium	0.001	0.007	0.02 – 0.5	–	– –
Manganese	0.01	0.05	0.1 – 3	0.2	1 – 30
Mercury(3)	0.0002	–	0.0002 – 0.01	–	– –
Molybdenum(p)	0.1	0.4	1 – 40	1	3 – 60
Nickel(p)	0.04	0.15	0.3 – 5	1	5 – 100
Osmium	0.3	1	2 – 100	20	50 – 500
Palladium(p)	0.1	0.25	0.5 – 15	5	20 – 400
Platinum(p)	0.2	2	5 – 75	20	100 – 2000
Potassium	0.01	0.04	0.1 – 2	–	– –
Rhenium(p)	5	15	50 – 1000	200	500 – 5000
Rhodium(p)	0.05	0.3	1 – 30	5	20 – 400
Ruthenium	0.2	0.5	1 – 50	20	100 – 2000
Selenium(2)	0.002	–	0.002 – 0.02	2	5 – 100
Silver	0.01	0.06	0.1 – 4	0.2	1 – 25
Sodium	0.002	0.015	0.03 – 1	–	– –
Thallium	0.1	0.5	1 – 20	1	5 – 100
Tin	0.8	4	10 – 300	5	20 – 300
Titanium(p)	0.4	2	5 – 100	10	50 – 500
Vanadium(p)	0.2	0.8	2 – 100	4	10 – 200
Zinc	0.005	0.02	0.05 – 1	0.05	0.2 – 4

(1) The concentrations shown are not contrived values and should be obtainable with any satisfactory atomic absorption spectrophotometer.
(2) Gaseous hydride method.
(3) Cold vapor technique.
(4) For furnace sensitivity values consult instrument operating manual.
(5) The listed furnace values are those expected when using a 20 µL injection and normal gas flow except in the case of arsenic and selenium where gas is used. The symbol (p) indicates the use of pyrolytic graphite with furnace procedure.

Table 8-3: AA concentration ranges. This chart shows the difference in Detection Limits and measurable concentration between AA Flame and AA Furnace instrumentation. The AA Furnace is able to detect and measure much smaller concentrations of heavy metals in environmental samples. (Source: EPA Methods for Chemical Analysis of Water and Wastes, 600/4-79-020)

In ICP analysis a liquid sample is aspirated into a plasma torch, where the metals in sample are atomized and excited by the extreme heat of the plasma (6000° to 8000°C). Radiant energy is emitted, measured and used to calculate the concentration of metals present in the sample.

Gas Chromatograph – Gas Chromatograph/Mass Spectrometer (GC/MS)

The gas chromatograph (GC) is vital to the environmental laboratory, for it is the instrument used to analyze for organic compounds (volatiles, semi-volatiles, herbicides/pesticides, PCBs) in all types of environmental samples.

Samples to be analyzed for volatile contaminants are introduced into the instrument through a purge and trap unit. With the use of this unit, few volatiles have the opportunity to escape into the atmosphere during the analysis process. The purge unit bubbles a gas through the sample, and volatiles are absorbed into a trap. The trap is then heated, the volatiles are desorbed, and the resultant gaseous vapor is swept through the chromatographic column by a carrier gas.

Samples to be analyzed for semi-volatiles, herbicides/pesticides, and PCBs are subjected to an extraction procedure prior to analysis. A small portion of the extract is then introduced into a heated injector port on the instrument, where sample vaporization takes place. The vapor is then sent to the chromatographic column.

Figure 8-6a: The GC indicates the likely compounds in a sample as peaks on a graph. The computer matches the peaks against those generated by pure forms of numerous compounds. The MS further refines the analysis by determining the molecular weight of each component of the sample.

The chromatographic column is packed with an inert, solid material. Each component of the sample is held within the column's packing for a unique interval, called the **retention time**. As each component leaves the column after its unique retention time, it flows through a detector. Several types of

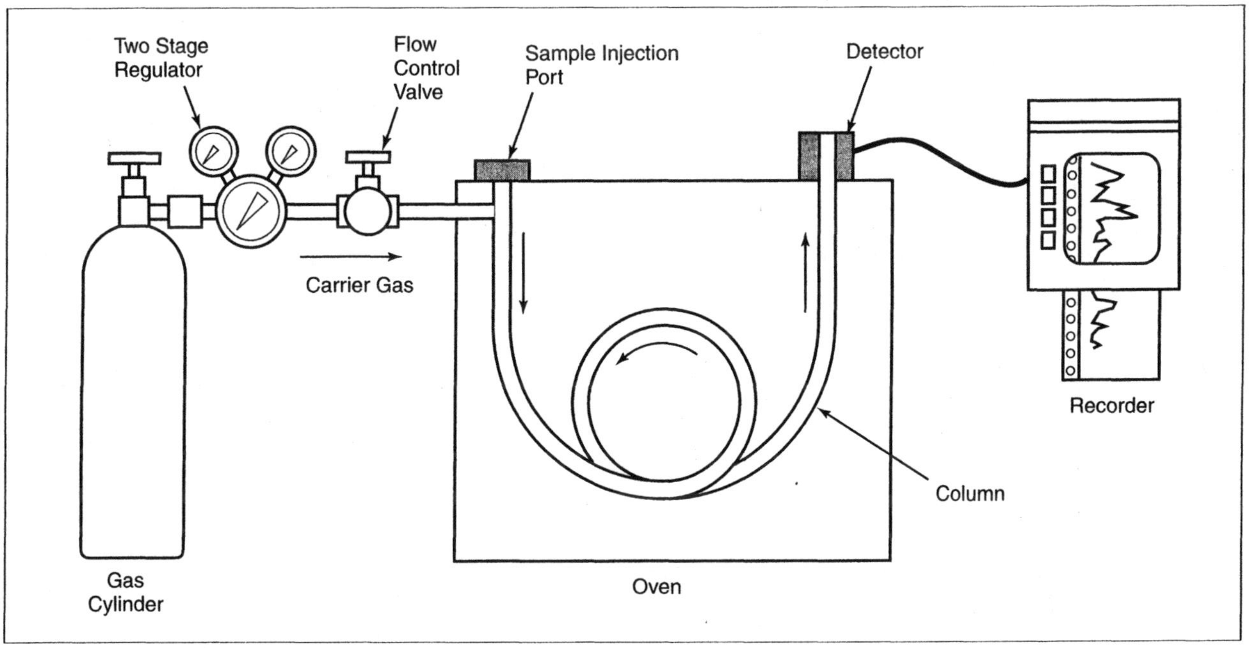

Figure 8-6b: Schematic of GC instrumentation.

Technical Corner ■ The GC/MS Qualitative Analysis

Suppose a groundwater sample is to be analyzed for volatile organic compounds (VOCs) by GC/MS. The analyst properly prepares the sample and injects it into the GC/MS. The GC portion of the instrument produces the following scan:

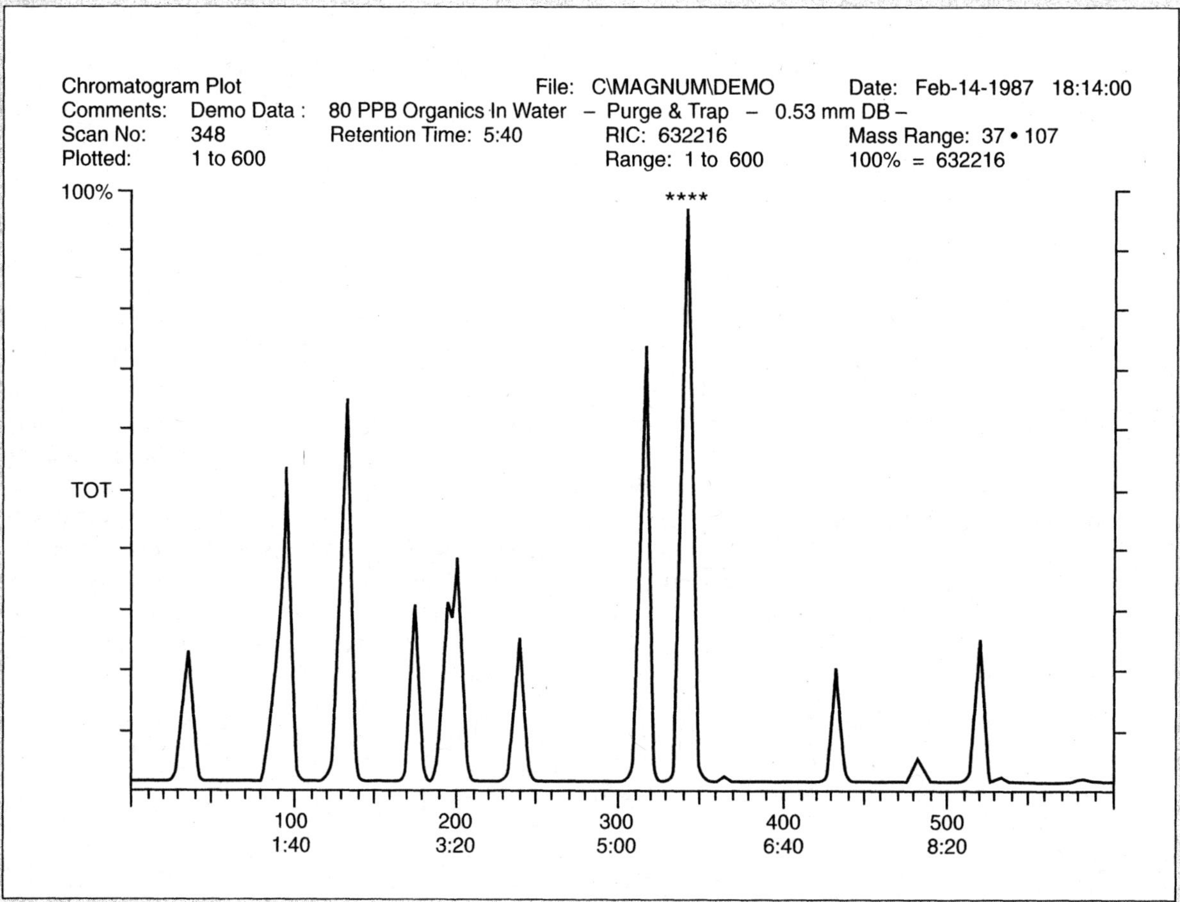

Chromatogram

Each peak that appears on the scan is a VOC found in the sample. The bottom scale on the scan indicates the "retention time," which is the amount of time the compound was retained on the GC column before flowing to the instrument detector. Notice that each volatile compound has a slightly different retention time.

For purposes of this example, we will choose the volatile organic compound with a retention time of 5:40 as the compound of interest. This peak has been marked **** on the **chromatogram**.

Next, the MS portion of the instrument (the detector) produces an **ion scan** of the peak of interest. Ion scans are based on the molecular weight and mass-to-charge ratio of the compound. Each volatile organic compound has a slightly different molecular weight and mass-to-charge ratio. The ion scan for our compound shows a primary ion at 91 (greatest mass-to-charge ratio) and a secondary ion at 106 (lesser ratio). Most compounds have a primary ion as well as one or more secondary ions associated with them. It is sometimes necessary to use these secondary ions to differentiate between similar compounds.

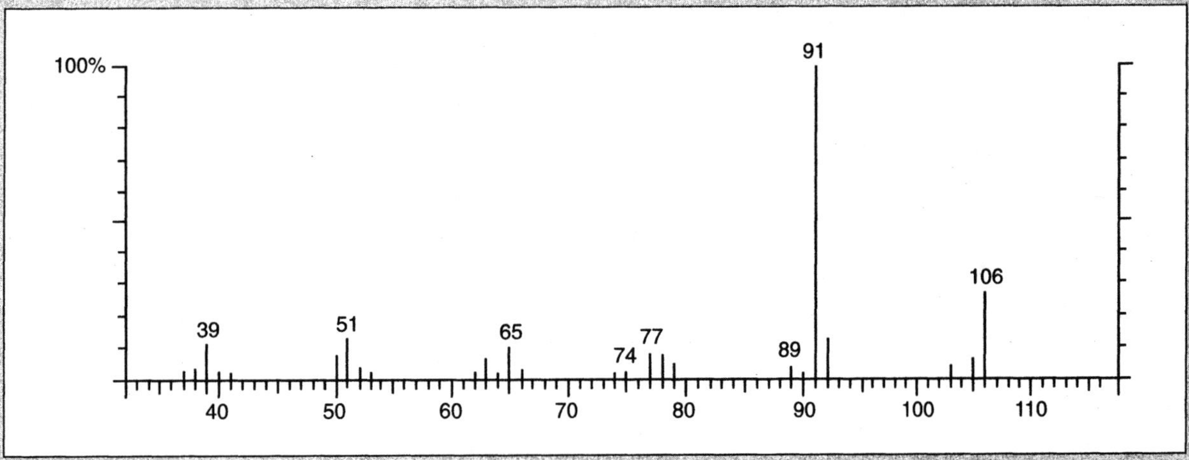

Ion Scan

The analyst calls the ion scan up through the GC/MS computer program. The analyst would create an ion scan for each peak of interest that appears on the chromatogram.

Finally, the analyst asks the GC/MS computer to conduct a "library search" to determine the identity of our compound of interest. The identity is determined by a comparison of the ion scan to the reference ion scans of volatile organic compounds contained in the instrument's library. Based on the "fit" of our compound's ion scan with an ion scan from the library, we identify our compound of interest as ethyl benzene, which has a molecular weight of 106.16 and a carbon weight of 90.5. Notice that these numbers correspond very closely to the ion scan created during this analysis.

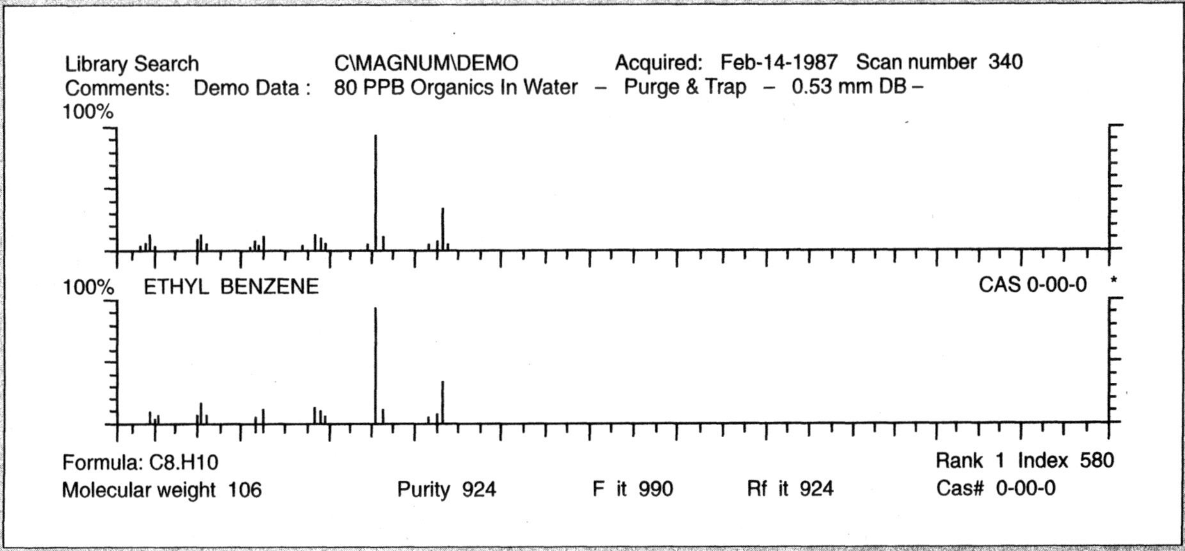

Library Search
Library reference ethyl benzene

In order to turn this analysis into a quantitative one, the analyst would have had to start by running a series of three to five standards containing ethyl benzene at known concentrations through the instrument. This process creates a calibration curve, which can then be used to determine the concentration of ethyl benzene in the sample.

detectors are used (Photoionization Detector or PID, Flame-ionization Detector or FID, Electron Capture Detector or ECD), depending on the analytes of concern. As each sample component is detected, a corresponding peak is recorded on the instrument's computer screen. At the end of the analysis, the sample peaks are matched against peaks of known compounds, and thus the sample components can be identified. This identification can be qualitative and quantitative.

Another laboratory instrument used in conjunction with the GC is the Mass Spectrometer (MS). The combination of the two instruments is called the **Gas Chromatograph/Mass Spectrometer** or **GC/MS**. The Mass Spectrometer accepts the sample as it passes through the chromatographic column and acts as the detector. The source of the mass spectrometer converts the sample components into charged particles. Each sample component has a unique pattern of ion fragmentation that, along with the retention time, can be used to identify the sample components. The extensive ion pattern libraries contained in this instrument's computer make it the ideal choice for contamination identification purposes.

Both GC and GC/MS methods are approved for use by EPA. The GC usually provides results in ppm (parts per million) levels, although some ppb (parts per billion) levels are possible. The GC/MS, which is a more sensitive instrument than the GC, can detect extremely low levels of organic contaminants (ppb levels and below).

Spectrophotometer

Spectrophotometric analysis requires that the target analyte in the sample be colored. If it is not, it must be reacted with a substance to produce a colored solution capable of absorbing light. Light is passed through the colored sample at a specific wavelength, and the amount of light absorbed by the sample is measured. Greater color intensity equals more light absorption, which equals greater analyte concentration. A series of standard solutions containing known concentrations of the target analyte are prepared and analyzed along with the sample. Absorbance readings for the standard solution concentrations are graphed, and the resulting calibration curve is used to determine the sample concentration. The concentration for the sample absorbance reading can be read directly from the calibration curve.

The EPA has approved spectrophotometric analysis for many compounds in wastewater, including cyanide and nitrite.

pH Meter

The **pH meter** consists of a pH electrode, reference electrode and solid state voltmeter. pH is a measurement of the hydrogen ion concentration of a sample, on a scale from 0 to 14. A pH of 7 is considered neutral. Low pH levels (0 up to 7) indicate acidity or an excess of hydrogen ions. The lower the pH, the more acidic the sample. High pH levels (over 7 to 14) indicate alkalinity or an excess of hydroxide ions. The higher the pH, the more alkaline (or caustic) the sample. The pH of a sample is measured as part of almost every environmental analysis method.

Specific Ion Analyzer

The **specific ion analyzer** is used for the rapid determination of many drinking water and wastewater constituents – chloride, fluoride, nitrate, ammonia, dissolved oxygen and cyanide, to name a few. The analyzer consists of an expanded scale pH meter and a specific electrode for each of the constituents to be measured. Electrodes are constructed with a thin membrane interface separating the solution-filled inner electrode from the sample. The electrode's inner solution contains the ion of interest at a constant activity. The sample, on the other hand, has variable ionic activity. A potential difference is created across the membrane by these differing ionic activities. This potential difference is measured in

Item	pH Level
Stomach acid	1.0 - 1.3
Sauerkraut	3.4 - 3.7
Lemon juice	2.2
Vinegar	2.6
Urine	4.7 - 7.9
Human saliva	6.3 - 7.6
Blood	7.2 - 7.3
Pure water	6.9
Milk	6.4
Milk of magnesia	10.5

Table 8-4: The pH of some common substances.

Figure 8-7: Environmental instructors demonstrate the use of titration in a "sheen screen" lab exercise at a professional development workshop.

millivolts and/or concentration (ppm). EPA has approved several methods utilizing specific ion analyzers.

Wet Chemical Analysis

Wet chemical analysis encompasses a variety of analytical methods including distillations, titrations and gravimetric analysis.

Distillation – **Distillation** is a heating and cooling process used to separate or purify the analyte. Distillation is accomplished by first vaporizing a liquid portion of the environmental sample through heat, and then collecting the analyte by differential cooling and condensation. Following distillation, further analysis, usually spectrophotometry or titration, is performed on the sample. Many EPA-approved methods such as determination of cyanide, phenols, and ammonia in wastewater and solid waste, involve the distillation process.

Titration – **Titration** is performed by adding measured amounts of a standard reagent of known concentration to a sample portion until a reaction (usually a color change) occurs. The amount of reagent needed to cause the reaction is then used to calculate the concentration of the analyte. EPA has approved titration methods for the determination of alkalinity and acidity of wastewater.

Gravimetric Analysis – In **gravimetric analysis** the mass of the analyte is determined either by weighing the analyte directly, or by weighing a compound that contains a known portion of the analyte. The gravimetric method for suspended solids determination involves filtering an environmental sample through a pre-weighed filter paper, drying the filter paper to a constant weight, weighing it again, and then calculating the difference in mass before and after filtration. The dissolved solids content of a wastewater sample can be determined by filtering the wastewater sample and then evaporating a known portion of the filtrate to dryness in a pre-weighed container. Again, the difference in mass between the container before and after the process is used to determine the dissolved solids content of the sample. A third gravimetric method involves adding a chemical reagent to the sample to precipitate out (draw out into suspension) the analyte and combine it with the chemical compound. The sample is then filtered through a pre-weighed filter paper, dried to constant weight and weighed again. Here also the calculations to determine the amount of analyte present involve the change in mass. The suspended solids and dissolved solids methods are EPA-approved; the third method is not often used in environmental analysis, although it is extremely reliable.

Checking Your Understanding

1. Distinguish between AA Flame, AA Furnace and ICP instrumentation according to each instrument's capabilities and limitations.
2. Which laboratory instruments are routinely used to measure volatile and semi-volatile organic contamination in environmental samples? Are these measurements qualitative or quantitative?

8-4 Methods for Determination of Hazardous Waste

Concepts

- A waste is considered hazardous if it is listed in 40 CFR 261.31-261.33 or if it exhibits any one of the characteristics of ignitability, corrosivity, reactivity or toxicity.
- The laboratory determination of a Hazardous Waste includes the Toxic Characteristics Leaching Procedure (TCLP) in which a sample is analyzed for eight metals, eight herbicides/pesticides and 25 organic compounds.

RCRA Hazardous Waste: Under the Resource Conservation and Recovery Act (RCRA), a waste is considered hazardous if it is listed in 40 CFR 261.31-261.33 or if it exhibits any one of the following characteristics and is not specifically excluded from regulation as a hazardous waste in 40 CFR 261.4:

—**Ignitability**: a flash point of less than 140°F (60°C).

—**Corrosivity**: a pH of less than 2.0 or greater than 12.5.

—**Reactivity**: unstable, reacts violently with water, sufficiently cyanide- or sulfide-bearing to produce toxic gas, or capable of detonation.

—**Toxicity**: the **Toxic Characteristics Leaching Procedure (TCLP)** extract contains any of the regulated contaminants at or above the regulatory level (see Table 8-5).

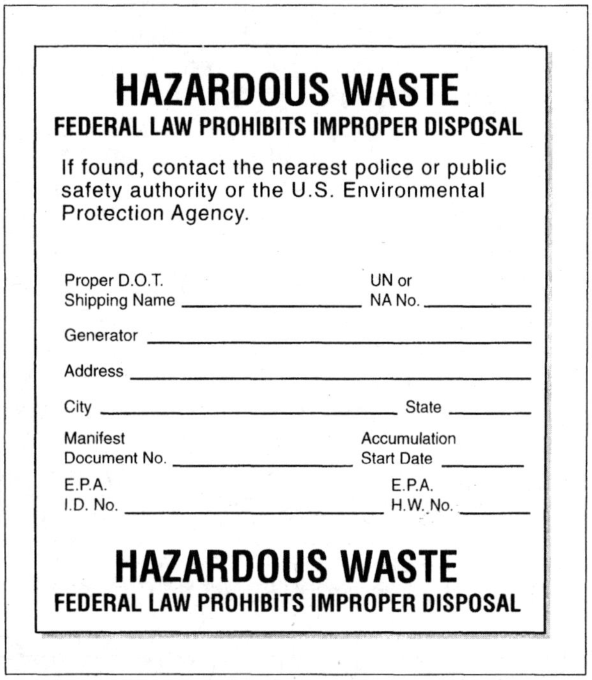
HAZARDOUS WASTE
FEDERAL LAW PROHIBITS IMPROPER DISPOSAL

If found, contact the nearest police or public safety authority or the U.S. Environmental Protection Agency.

Proper D.O.T. Shipping Name ____________ UN or NA No. ________

Generator ____________

Address ____________

City ____________ State ________

Manifest Document No. ____________ Accumulation Start Date ________

E.P.A. I.D. No. ____________ E.P.A. H.W. No. ________

HAZARDOUS WASTE
FEDERAL LAW PROHIBITS IMPROPER DISPOSAL

Figure 8-8: Hazardous waste sticker.

Volatile Compounds	Regulatory Level (mg/L)
Benzene	0.5
Carbon Tetrachloride	0.5
Chlorobenzene	100.0
Chloroform	6.0
1,2-Dichloroethane	0.5
1,1-Dichloroethylene	0.7
Methyl ethyl ketone	200.0
Tetrachloroethylene	0.7
Trichloroethylene	0.5
Vinyl Chloride	0.2
Semivolatile Compounds	
0-Cresol	200.0
m-Cresol	200.0
p-Cresol	200.0
1,4-Dichlorobenzene	7.5
2,4-Dinitrotoluene	0.13
Hexachlorobenzene	0.13
Hexachlorobutadiene	0.5
Hexachloroethane	3.0
Nitrobenzene	2.0
Pentachlorophenol	100.0
Pyridine	5.0
2,4,5-Trichlorophenol	400.0
2,4,6-Trichlorophenol	2.0
Pesticides/Herbicides	
Chlordane	0.03
Endrin	0.02
Heptachlor	0.008
Lindane	0.4
Methoxychlor	10.0
Toxaphene	0.5
2,4-D	10.0
2,4,5-TP	1.0
Metals	
Arsenic	5.0
Barium	100.0
Cadmium	1.0
Chromium	5.0
Lead	5.0
Mercury	0.2
Selenium	1.0
Silver	5.0

Table 8-5: TCLP regulated contaminants.

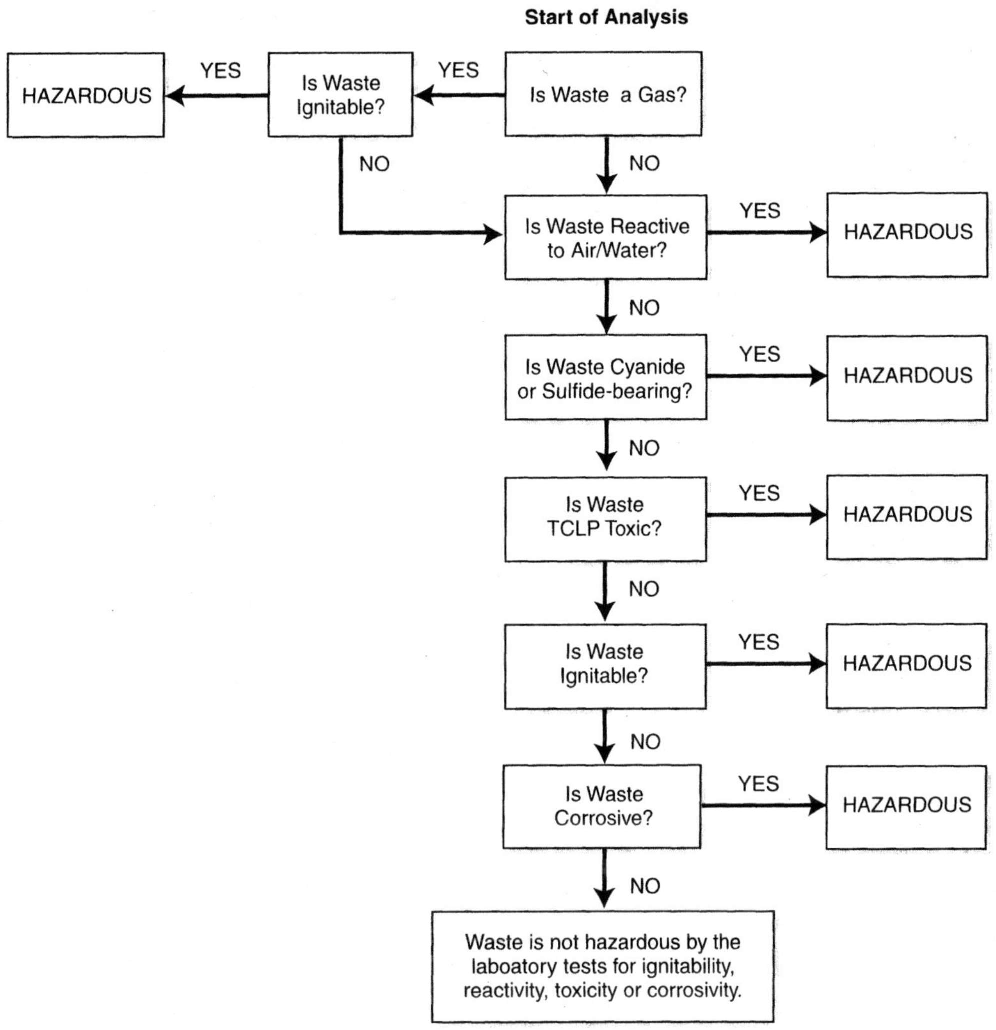

Table 8-6: Flow chart of steps to determine hazardous state of waste.

If the identity of the waste is known, it may be possible to make a hazardous waste determination if the substance is listed in 40 CFR 261.31-261.33. Substances on this list are automatically considered hazardous waste. Most often, however, a hazardous waste determination is made through a series of laboratory analyses, the requirements for which are detailed in 40 CFR 261.20. These analyses include determining the waste's flashpoint, pH, cyanide and sulfide content as well as its toxic components. The flow sheet on Table 8-6 summarizes the steps that a laboratory follows to qualify the waste as nonhazardous or hazardous. An explanation of the chart follows:

1. If the waste is a gas (compressed gas) and can be ignited, then it is a hazardous waste. Examples of ignitable gases include hydrogen and acetylene.

2. If the waste reacts violently with water or produces toxic vapors or fumes that would present a danger to human or animal health or the environment, then it is a hazardous waste. Sodium, for example, can react with water to make hydrogen gas and a caustic solution. Aluminum chloride reacts with water to produce hydrogen chloride gas and heat.

3. If the waste is readily capable of detonation at standard temperature and pressure, then it is a hazardous waste. An example of this is 2-nitrotoluene.

4. The waste is analyzed for cyanide and sulfide to determine reactivity. The laboratory may first analyze the sample for total cyanide and total sulfide. If the results show cyanide or sulfide concentrations above 10.0 ppm, the reactive cyanide and reactive sulfide tests will be run. If these results are above 10.0 ppm, then the material is considered a hazardous waste.

5. The next step in the hazardous waste determination process is the Toxic Characteristics Leaching Procedure (TCLP). This test was implemented by EPA for solid wastes that are intended for land disposal. The TCLP procedure simulates the leaching action that can occur in municipal landfills, potentially causing groundwater contamination. The TCLP procedure includes analysis for eight metals, eight herbicides/pesticides and 25 organic compounds. Often the laboratory will first analyze the waste for total contaminant concentrations, and if any of these total concentrations exceed the TCLP limits, the TCLP extraction procedure will be run. This extraction procedure is an 18-hour process. The liquid extract from the solid waste is then analyzed for the above list of contaminants. If any of the TCLP extraction results exceed the EPA's established TCLP limits, the waste is considered a hazardous waste.

   The portion of the sample taken for the TCLP procedure depends on the solid or liquid state of the waste. A paint filter test must be run on liquid and mixed solid/liquid wastes to determine how much of the waste is free liquid. If the waste is fully a liquid, the liquid portion is considered the TCLP extract. If the waste is mixed liquid/solid, the solid portion is put through the TCLP extraction procedure. Then the liquid portion is added to the extract, and the resultant mixture is analyzed as one sample.

6. The liquid or solid waste is then put through the ignitability determination. Ignitability is determined by testing for a flashpoint. The waste is put in a closed cup container and slowly heated, while a flame is introduced at 2°F intervals. If a flash occurs at or before 140°F, the waste is considered a hazardous waste.

7. The waste also needs to be tested for corrosivity. This is done through a pH analysis. If the waste has a pH less than or equal to 2, or greater than or equal to 12.5, it is considered a hazardous waste.

Checking Your Understanding

1. Explain each of the following characteristics that make a waste hazardous:

   —ignitability

   —corrosivity

   —reactivity

   —toxicity

   What other criteria may be used to determine the hazardous or nonhazardous status of a waste?

2. Discuss the significance of the TCLP test in determining the status of a waste. Briefly outline the TCLP processes.

Summary

Three categories of environmental sample analysis are field, mobile and fixed analysis. Although the method of analysis may be specified by the project's regulatory authority, such as EPA, often the project planner must decide upon the methods according to the level of data quality acceptable for the project. The five levels of data quality developed by the EPA can help the project planner understand the attributes and drawbacks of various analysis options.

When fixed analytical services are required, it is important to choose a laboratory that shows a willingness to communicate with customers, and that will provide good quality data in a timely manner and at a reasonable cost. In choosing a laboratory, project planners should consider elements such as a lab's customer service reputation; its instrumentation and experience; its local, state and federal certifications; the components of its quality assurance/quality control plan; and its proficiency as demonstrated by data precision and accuracy.

For an effective working relationship with an environmental laboratory, and in order to make the most informed decisions concerning project sample analysis, the project planner must have a basic understanding of general laboratory procedures – such as dilution, extraction, filtration, digestion, drying, particle size reduction and compositing–as well as of laboratory instruments that are routinely used to analyze environmental samples – such as the atomic absorption spectrophotometer, inductively coupled plasma, gas chromatograph, gas chromatograph/mass spectrometer, spectrophotometer, pH meter and specific ion analyzer.

A common purpose for environmental analysis is to determine if a waste exhibits any one of the hazardous characteristics of ignitability, corrosivity, reactivity or toxicity.

Application and Critical Thinking Activities

1. Choose two environmental laboratories, call them and ask:
   a. What types of field analysis they offer;
   b. If they have capabilities to perform mobile onsite lab analysis for volatile organic compounds;
   c. The price for analysis of a groundwater sample for lead;
   d. The price for analysis of a soil sample for benzene;
   e. What state certifications they hold;
   f. If they have a written QA/QC program available for your review.
2. Call the agency in your state responsible for certifying laboratories for drinking water sample analysis of coliform bacteria, nitrate and lead. (Your local health department's environmental division should be able to provide you with information about which state agency to call.) Ask for a copy of the requirements to become a certified laboratory for these parameters in your state. Ask the state agency to describe their laboratory inspection procedures. Estimate the total cost to the laboratory for becoming certified.
3. Obtain a copy of the laboratory reference manual entitled "Standard Methods for the Examination of Water and Wastewater". Look up the procedure for determination of oil and grease in wastewater. What sample preparation techniques can you identify as part of this method?
4. Arrange to tour a local analytical laboratory to see firsthand the variety of instruments available for environmental analysis. If possible, observe a laboratory analyst performing sample preparation and/or analysis.
5. Contact a municipal or county landfill in your area. Find out who at the landfill is responsible for determining if a waste is acceptable for disposal at the landfill. What criteria does this individual use to determine acceptability of waste? Does the landfill accept hazardous waste? If so, what laboratory analysis and permits are required before the hazardous waste can be disposed of?
6. Obtain Material Safety Data Sheets (MSDS) for the following substances: benzene, kerosene, methylene chloride, acetone, formaldehyde, perchloroethylene. Which of these substances can be considered hazardous based on the characteristic of ignitability?

9

Data Validation and Interpretation

Chapter Objectives

Upon completing this chapter, the student will be able to:

1. **List** the information – besides the analytical result – that should be included on the laboratory analysis report to make it a legally and technically defensible document.
2. **Describe** the importance of laboratory quality assurance/quality control procedures in the generation of reliable and valid analytical data.
3. **Determine** and compare the accuracy and precision of laboratory data using laboratory quality control information.
4. **Conduct** data validation to determine the soundness of a laboratory analytical result, and its significance to project decisions.
5. **Interpret** the meaning of an analytical result with regard to specific environmental project requirements.

Chapter Sections

9-1 Introduction

Concepts

- Before analytical data can be used to make decisions concerning an environmental project, the quality of the data must be determined. Problems resulting from the use of invalid data can affect project outcome and company reputation.
- Analytical procedures are not perfect, and their results do not always match the true analyte concentrations of the environmental sample.
- Besides the analytical result, the laboratory report contains important information that can be used to determine the quality of the data presented, and make the report legally and technically defensible.

The end product of an environmental project's sampling and analysis effort is a set of analytical data. Based on the data, project managers must make sense of site conditions and plan the next steps to be taken. But before the analytical data can be used to make a decision – be it on the environmental condition of the site, the design of a remediation system for groundwater contamination reduction, or the designation of a waste as hazardous or nonhazardous – the quality of the data must be determined.

Environmental analyses are often complex procedures that measure substances at extremely low concentrations. These measurements are subject to chemical and physical interferences; they are limited by instrument capabilities; and they are dependent on the skill of the analyst.

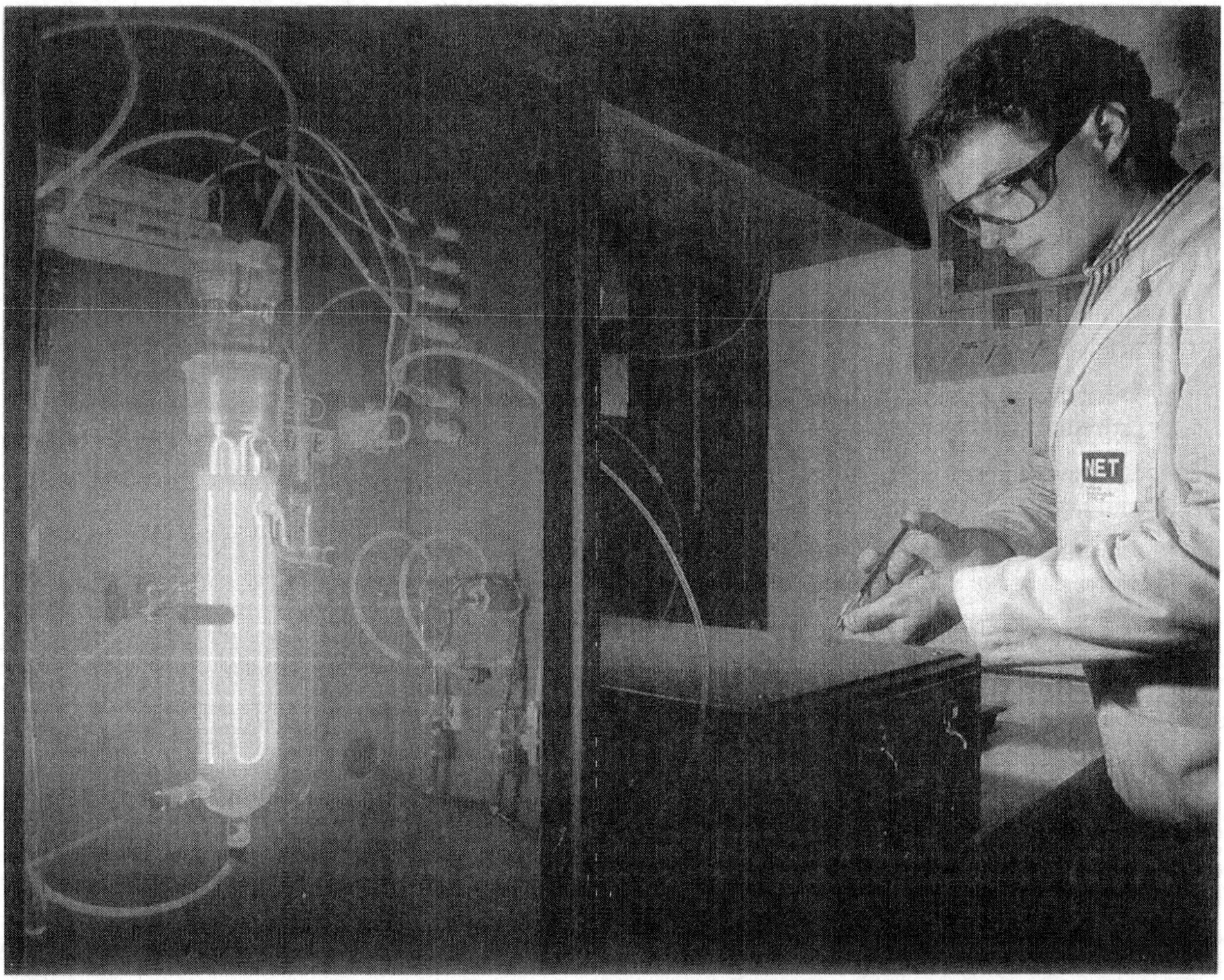

Figure 9-1: The Total Organic Carbon (TOC) Analyzer converts all carbon-based components of a sample to carbon dioxide to determine the sample's "carbon load" (amount of carbon).

How confident can the data user be about the lab results reflecting the true contaminant concentrations in the environmental sample? How precise and how accurate are the analytical values on the laboratory report? Are there any sources of error introduced through the laboratory's analytical procedures and, if yes, what are they? Will the regulatory agency accept the results as indicative of site conditions? Is the waste hazardous or not? These are essential questions regarding data quality. Laboratory clients need to be concerned with data quality, because problems that result from basing project decisions on invalid results can affect a company's finances and reputation, not to mention project design and outcome!

The Role of the Sampling Technician in Achieving High-Quality Data

Chapter 2 introduced the Quality Assurance/ Quality Control program within the written sampling plan. This program defines many of the activities the sampling technician carries out at the site.

Figure 9-2: Environmental Manager Ricky Peer interprets air quality lab reports for a U.S. Army manufacturer and compares the data to Illinois EPA permit limitations.

The analytical laboratory also has a written QA/ QC program, which is designed to ensure that the laboratory produces high quality data. Section 9-2 describes the components of a laboratory QA/QC program, such as internal audits and the use of internal laboratory quality control samples. It also includes several formulas used by the laboratory QA/QC officer to validate the **accuracy** and **precision** of its analytical procedures.

Section 9-3 explains how the data user – in this book, the project planner – can validate analytical results, such as the data's accuracy, precision, completeness, representativeness and comparability. Section 9-3 concludes by helping to answer important questions such as:

—Is the data of sufficient quality and quantity to reduce decision uncertainties?

—Will the data meet regulatory authority requirements?

The underlying theme of Chapters 2 and 9 cannot be overemphasized: both the technician and the environmental and laboratory professionals must strive for accuracy and precision in all tasks and must be able to quantitatively validate the quality of their data at all times.

Defining Analytical Error

In theory, if analytical procedures were perfect, the reported value would represent the "true" concentration present in the sample. In reality, analytical results refer to **true value** as an acceptable range of values rather than an absolute value. Therefore the relationship between reported and accepted concentrations is one of approximation rather than exactness. The difference between the reported concentration and the "true" concentration is called the **analytical error**. Analytical error is the error introduced into the results through the analytical procedure. Analytical measurements have inherent limitations due to the equipment and procedure used, the skill of the person performing the analysis, and conditions under which the analysis is performed. Analytical error can result from factors such as the effect of the sample matrix on the lab instrument, the sample dilution, incorrect instrument calibration, contaminated laboratory equipment, or analyst error such as improper measure-

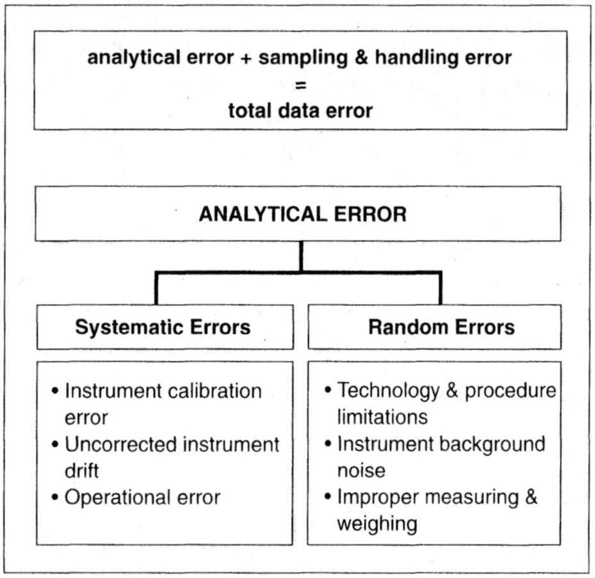

Table 9-1: Analytical error.

ment or methodology. There are two kinds of analytical errors, systematic and random errors. **Systematic errors** are consistent and usually bias the measurement in one direction; they include instrument calibration error, uncorrected instrument drift (gradual change in the output of a circuit) as well as operational errors. **Random errors** have no specific pattern but may occur because of an inherent limitation of the instrument (e.g., instrument "noise" or disturbance) or lack of care in taking the measurements (imprecise weighing or aliquot variability). Other sources of error such as improper sampling procedures, contaminated sample containers, and use of improperly decontaminated sampling equipment are not defined as analytical error, but must also be considered. Without knowledge of the potential magnitude of the analytical error, it is impossible for data users to judge the quality of a reported concentration, or to have enough information to make a confident decision concerning project management.

Defining Data Validation

Data validation is the process of reviewing analytical data in order to:

—Identify errors;

—Determine inconsistencies;

—Compare against quality control samples and background measurements;

—Delete unusable data;

—Explain limitations of remaining project data.

Laboratory personnel undertake data validation before sending the lab report to the client. This validation process evaluates the laboratory's internal quality control procedures, the laboratory instruments' performance, and the resulting precision and accuracy of the data.

The data user or project planner should also conduct validation procedures to determine the accuracy, precision, representativeness, completeness and comparability of the data. The project planner must also interpret the lab report data as they relate to the specific project. **Data interpretation** may involve such tasks as comparing the analytical results with agency-set cleanup objectives, with toxicity thresholds published by regulatory agencies, or with the criteria used to classify hazardous wastes.

Many project planners rely on the analytical report furnished by the laboratory to obtain the information necessary to interpret and validate the data. It is therefore relevant to begin this chapter with a look at the information typically presented in a laboratory analytical report.

Contents of the Laboratory Analysis Report

Although a laboratory may give initial results of environmental sample analysis to a client over the telephone, a written report will always follow this verbal information. This report typically furnishes quality control information with the sample results. Quality control data are used not only to validate the sample results but also to provide information that makes the document legally and technically defensible. Although laboratory reports do not have a standardized format, well-constructed documents will consist of the following elements:

—Laboratory name, address and phone number;

—Client name and address as well as project name if applicable;

—Laboratory sample identification number – a unique number assigned by the laboratory for tracking purposes;

Batch/ sample #	Parameter	Method blank mg/L	Lab control sample % recovery	Duplicates			Matrix spikes	
				Original mg/L	Duplicate mg/L	RPD ⑤	Spike result mg/L	% recovery
AC34 ①	lab control sample Barium	0.01 ②	97% ③	–	–	–	–	–
AC34 9532 ④	Barium, TCLP	–	–	46.14	48.29	4.4%	55.09	89%
AC34	lab control sample Cadmium	0.01	86%	–	–	–	–	–
AC34 9532	Cadmium, TCLP	–	–	1.24	1.32	6.0%	2.26	102%
AC34	lab control sample Silver	0.03	90%	–	–	–	–	–
AC34 9532	Silver, TCLP	–	–	0.26	0.22	15.0%	1.05	79% ⑥

① The QC batch number associated with the Barium analysis of lab sample number 9532.
② The amount of Barium found in the Method Blank associated with this batch of samples.
③ The percent of Barium recovered in the laboratory control sample associated with this batch of samples.
④ The environmental sample that was included in QC batch AC34.
⑤ The Relative Percent Difference (%RPD) calculated from the duplicate analysis of the environmental sample.
⑥ The percent of silver recovered in the sample to which a spike or known amount of silver standard was added.

Table 9-2a: Example of laboratory report – Quality Control data.

Lab Sample No.: 9532 ①

Date Received: 11-19-97

② Date Sampled: 11-18-97

Client Sample ID: FKSS1

Parameter	Sample Preparation				Sample Analysis			Instrument	Result
	Batch ③	Method	Date-time	Analyst	Method	Date-time	Analyst		
Lead	AC34	TCLP	11/20/97 1300	JP	7420 ④	11/21/97 1405	LP	AA V1	10.24
Barium	AC34	TCLP	11/20/97 1300	JP	7080	11/21/97 1405	LP	AA V1	0.53
Cadmium	AC34	TCLP	11/20/97 1300	JP	7130	11/21/97 1405	LP	AA V1	0.68
Chromium	AC34	TCLP	11/20/97 1300	JP	7190	11/21/97 1405	LP	AA V1	2.25
Silver	AC34	TCLP	11/20/97 1300	JP	7760	11/21/97 1405	LP	AA V1	<0.05

① A unique number assigned by the lab for sample tracking.
② Client's sample identification.
③ A number assigned to a group of samples that have been prepared and analyzed as a batch with the appropriate quality control checks.
④ A specific method reference (from EPA's SW-846 document).

Table 9-2b: Laboratory results for one sample in a batch.

—Client's sample identification name or number;

—Sample location;

—Date and time of sample collection; name or initials of collector;

—Date sample was received at the laboratory;

—Batch number – a unique number assigned by the laboratory for internal quality control tracking purposes;

—Date, time and method number for sample preparation;

—Date, time and method number for sample analysis;

—Results of environmental sample and of field quality control sample analysis;

—Laboratory internal quality control results – method blank, laboratory control sample, duplicate, matrix spike, surrogate standard, etc.;

—Method detection limits;

—Data qualifiers and notes concerning analysis;

—Signature and title of analyst;

—Signature and title of person performing final review of report.

The elements of the laboratory report necessary for validating analytical results include: 1) laboratory internal quality control results, 2) field quality control sample results, 3) method detection limits, and 4) data qualifiers. These elements will be discussed in detail in the remaining sections of this chapter.

Checking Your Understanding

1. What are some of the errors that can occur during laboratory sample analysis? Divide them according to whether they are random or systematic.
2. Explain the importance of validating analytical data before they are used to make project decisions. Why should data validation be performed by both the laboratory and the data user?
3. Distinguish between data validation and data interpretation.
4. Evaluate Tables 9-2a and 9-2b and determine if each component of a well-constructed lab report is represented.
5. Identify the elements of a lab analysis report that make it a legally and technically defensible document and state the rationale for each element identified.

9-2 Laboratory Quality Assurance/Quality Control

Concepts

- To ensure the production of reliable and valid analytical data, laboratories follow a written QA/QC plan, which provides rules and guidelines for work conducted at the lab.
- The laboratory QC plan incorporates several internal quality control measurements to assure the generation of reliable data.
- Precision and accuracy data are the types of quality control information field technicians are most likely to encounter when reviewing a laboratory analysis report.
- Each analysis procedure has a detection limit, which is the minimum concentration of an analyte that can be reliably measured by the procedure. The detection limit is never reported as "zero."
- A laboratory will use data qualifiers on the laboratory report to indicate data that do not meet quality control or detection limit criteria. These qualified data may still be useful to project decisions.

In Chapter 2 we learned that the sampling technician uses QA/QC methods in the field to ensure that sampling activities are properly carried out as specified by the written sampling plan. Environmental laboratories also develop and follow a written QA/QC plan, which provides rules and guidelines for work conducted at the lab. Adhering to the QA/QC plan ensures that the lab will generate reliable and valid analytical data.

The Quality Control Program outlines the type and the quantity of internal quality control samples that must be run along with the environmental samples. It also mandates the levels of precision and accuracy that must be achieved to satisfy the lab's QA/QC plan. Laboratories generally base their quality control protocols on EPA methodology and regulations.

The Quality Assurance Program addresses such features as:

—*The laboratory's ethics policy:* for example, statements to the effect that manipulation or falsification will not be tolerated; accidents involving sample analysis will be reported; and client confidentiality with regards to analysis results will be maintained.

—*Standard operating procedures (SOP) for analysis methods:* for example, step-by-step outlines of how to receive samples at the lab; how to preserve and store samples; and how to perform laboratory analysis and determine results of the analysis.

—*The preventive maintenance program for lab instrumentation and facilities:* for example, instrument cleaning and parts replacement schedule or the verification schedule for an instrument's proper operation and calibration.

—*Procedures for reducing raw laboratory data to obtain the reported analytical result:* for example, calculations involved in determining the actual result from the instrument readout; calculations necessary to take into account interferences with the analysis; and calculations to determine the validity of the result based on quality control samples.

—*Levels of internal data review to assure data validity:* for example, how accurate and precise the result must be; and who is responsible for reviewing the data against quality control criteria.

—*Procedures for investigating and correcting out-of-control analytical results:* for example, the steps taken by the analyst when the result does not meet quality control criteria; and the responsibilities of the lab's QA/QC officer in determining lab data quality and correcting data quality problems.

—*Results of audits by external agencies:* for example, documents showing the results of audits by outside agencies for laboratory certification or other purposes.

—*Procedure for lab internal audits:* for example, the mechanisms for monitoring a lab's performance, such as its data precision and accuracy and the overall quality of the lab's data production process.

—*Laboratory proficiency programs*, which use performance evaluation samples described in the next section.

Internal Audits

The laboratory conducts internal audits to determin[e] its overall quality of operation and to monitor th[e]

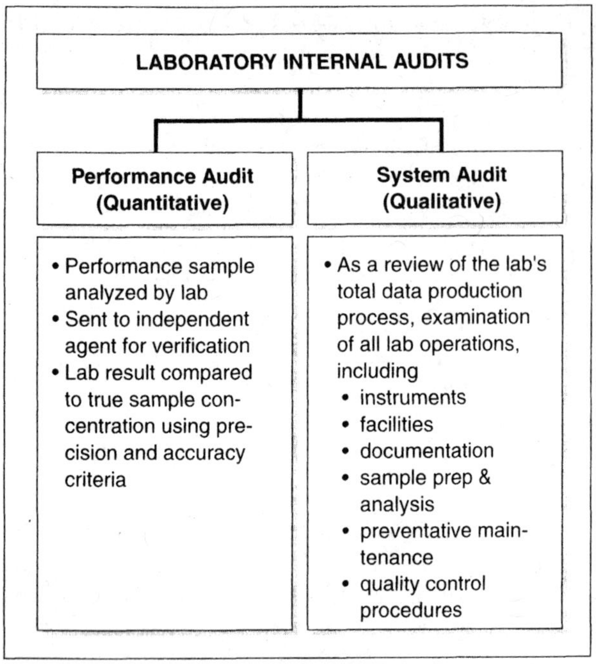

Table 9-3: Laboratory internal audits.

capability and performance of the lab's analytical systems. These audits – performed by the laboratory director or by the lab QA/QC director – are of two types: performance audits and system audits. In a **performance audit**, analytical data are generated using performance evaluation samples. These are unknown samples that are analyzed by the lab and then sent to an independent agent for verification. They provide a quantitative assessment (through precision and accuracy measurements) of the quality of the data generated by the lab.

System audits, on the other hand, provide a qualitative assessment of the lab's total data production process. System audits examine all of the lab's operations, instruments and facilities, including documentation on sample receipt and storage, chain-of-custody, sample preparation and analysis, instrument operating records, quality control sample results and raw data interpretation.

Before any data are placed on the final lab report sent to the client, the laboratory performs in-house data validation under the direction of the lab's QA/QC officer or lab director. This individual is responsible for assessing data quality (generally through statistical determination of accuracy and precision of the analysis), noting any data that do not meet quality control standards or that have been estimated, and investigating and correcting problems that resulted in out-of-control data. The end product of the laboratory's internal data validation process is a quality control report, which is often included as statistical information on the lab report sent to the client.

Accuracy and Precision of Analytical Procedures

The statistical data that field technicians are most likely to encounter when reviewing a laboratory analysis report regard precision and accuracy. Precision and accuracy provide a summary of the analytical error of the laboratory procedure. Lab personnel determine precision and accuracy of analysis results before sending the results to the data user. This explanation of the interpretation of precision and accuracy has been simplified to help the technician understand data that might appear on the lab reports. If the technician's responsibilities include interpretation of quality control data, the study of statistical principles and of environmental reference books such as SW-846 is recommended.

Precision refers to the reproducibility – or the agreement – of individual measurements made under a given set of conditions. In other words, precision examines the spread of analytical results around their

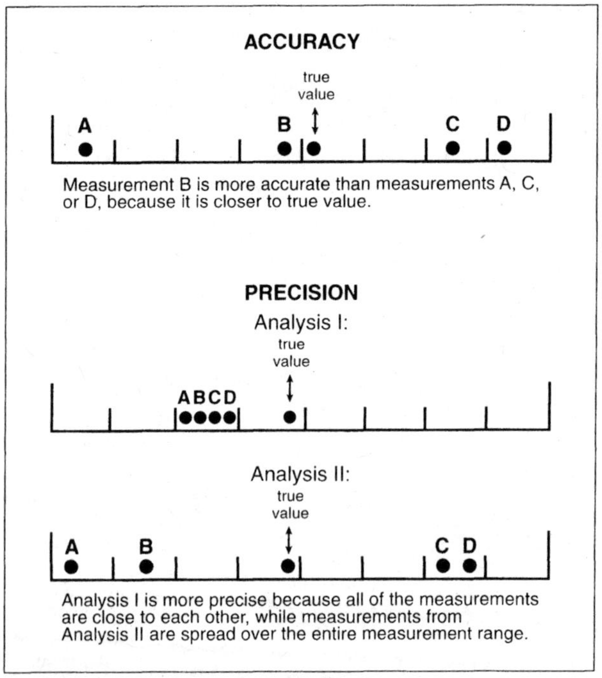

Table 9-4: Illustration of accuracy and precision.

average value. The precision of an analytical procedure is commonly determined through analysis of laboratory duplicate samples.

Laboratories commonly calculate precision as **Relative Percent Difference (%RPD)**. In the formula for %RPD, S = the original sample value and D = the duplicate sample value:

$$\%RPD = \frac{(S - D)}{(S + D)/2} \times 100$$

The lab may include %RPD data as quality control information on the lab analysis report sent to the client. Ideally, the %RPD value should be as close to zero as possible. A low number is indicative of a small difference in analytical results between the original sample and the duplicate sample, which suggests precision of results.

Suppose, for example, 200 ppm is the sample value and 190 ppm is the duplicate sample value obtained from analysis of an organic vapor personal monitoring badge. The application of the formula is as follows:

$$\%RPD = \frac{(200 - 190)}{(200 + 190)/2} \times 100$$

$$= \frac{10}{195} \times 100$$

$$= 5.13\ \%$$

This number, 5.13%, is relatively close to zero; therefore, the data may be considered precise. Acceptable precision levels vary according to method. For example, one laboratory's QAPP states that the %RPD should be less than 20 for water and less than 35 for soils.

Accuracy measures the closeness of an individual measurement to the true value. The lab determines the accuracy of an analytical result through analysis of laboratory control samples and spiked samples. Accuracy can be calculated as the **Percent Recovery (%Rec)** of the laboratory control or spiked sample. The formula for %Rec is:

$$\%Rec = \frac{|A - B|}{C} \times 100$$

where

A = the analyte concentration determined by analysis of the spiked sample

B = the background level determined by analysis of the unspiked sample

C = the amount of the spike added.

The lab may include %Rec data as quality control information on the lab analysis report sent to the client. Ideally, the %Rec value should be as close to 100 as possible. This would indicate that 100 percent of the spike has been recovered.

For example, suppose an environmental sample containing 2.0 ppm iron, which is B in the above formula, is spiked with 3.0 ppm iron standard, which is C in the formula. If the concentration of iron in the spiked sample is 5.0 (A in the formula), then the percent recovery equals 100 percent.

$$\%Rec = \frac{|5 - 2|}{3} \times 100$$

$$= 100\%$$

However, if the concentration of iron in the spiked sample is analyzed at 6.3 (A in the formula), then the percent recovery equals 143 percent.

$$\%Rec = \frac{|6.3 - 2|}{3} \times 100$$

$$= \frac{4.3}{3} \times 100$$

$$= 143.3\ \%$$

According to SW-846 (1.2.2.1.3) spike recoveries should be within ±20 percent (i.e., 80-120 percent). Sometimes, however, a different control sample limit is specified in the laboratory method for a particular compound. For example, one laboratory specifies an acceptable recovery of 71-157 percent for toluene. According to the "Practical Handbook of Groundwater Monitoring" acceptable recoveries may vary depending on the individual compound.

Internal Laboratory Quality Control Samples

The laboratory's QA/QC program requires the processing and analysis of internal quality control samples along with any environmental sample analysis. These internal quality control samples provide the information necessary to determine precision and accuracy of the results. An explanation of various types of laboratory quality control samples follows after the Technical Corner.

Technical Corner ■ Comparing Accuracy and Precision of Four Different Labs

Suppose that a standard sample containing a known amount of analyte is submitted to four different laboratories (Lab A, Lab B, Lab C and Lab D). Each laboratory analyzes ten replicates of the sample using the identical laboratory method. The following results are obtained:

Reported Concentration of Standard (ppm)				
Replicate	Lab A	Lab B	Lab C	Lab D
1	10	13	10	4
2	10	12	14	14
3	10	12	6	15
4	10	11	8	1
5	9	14	12	1
6	8	12	7	6
7	13	11	11	1
8	11	12	5	14
9	12	13	15	11
10	10	12	10	15

The actual concentration of the standard is 10 ppm. The majority of people examining these results would conclude that Lab A provided the "best" results, since Lab A reports either 10 ppm or a value very close to 10 ppm for each replicate.

Accuracy and precision provide a meaningful way of summarizing these replicate data and comparing different analytical procedures and different laboratories.

Accuracy

The reported analyte concentration should equal the actual concentration present in the sample; this means that ideally the analytical method should have no systematic errors. Accuracy measures the systematic error of a method. In the example of the four laboratories, accuracy can be defined as the difference between the average of the 10 reported values and the actual value of 10 ppm. Performing this calculation, the following results are obtained:

Lab	Average of 10 replicates (ppm)	Average error* (ppm)
A	10.5	0.5
B	12.2	2.2
C	9.8	−0.2
D	8.2	−1.8

* Although average error is normally expressed as an absolute value, the results for labs C and D are shown here as negative to be consistent with the calculation.

These results show that on average, Labs A and C yield reported values that are very close to the actual value. Thus, Labs A and C are more accurate than Labs B and D.

Accuracy values can be presented in a variety of ways. The average error shown above is one way of presenting this information; accuracy, however, is more commonly presented as percent bias or percent recovery. Percent bias is a standardized average error; that is, the average error divided by the actual concentration and converted to a percentage. For Lab A in the example, the percent bias is $0.5/10 = 0.05$, or 5 percent. Percent bias is unitless, thus allowing the accuracy of analytical methods to be compared easily.

Percent recovery provides the same information as percent bias. Since accuracy is often determined from spiked samples, laboratories commonly report accuracy in this form. Percent recovery is defined as:

$$\%\mathrm{Rec} = \frac{|A - B|}{C} \times 100$$

Given this definition, it can be shown that %Bias = %Rec − 100. For the example, the observed %Bias and %Rec are as follows:

Lab	Percent Recovery	Percent Bias
A	105	5
B	122	22
C	98	2
D	82	18

Precision

While accuracy measures the average properties of an analytical method, precision examines the spread of the reported values about their mean. The spread of reported values refers to how different the individual reported values are from the average reported value. Precision can thus be seen as a measure of the magnitude of the error.

Precision can be measured in a variety of ways. A simple measure of precision is the "variance." The calculation for variance was also seen in the SW-846 formula in Chapter 2:

$$s^2 = [\{s_1^2 + s_2^2 + s_3^2 \ldots\} - \{(s_1 + s_2 + s_3 \ldots)^2 \div n_s\}] \div (n_s - 1)$$

For the ten replicates sent to four labs in our earlier example, the variances are as follows:

Lab	Variance of the replicates
A	2.3
B	0.8
C	11.1
D	38.4

The results above indicate that Lab B is the most precise (i.e., the least variance from zero variance). This could be determined by examining the ten individual values reported by Lab B, which are all very similar. In contrast, the Lab D reported values are very dissimilar.

Laboratories commonly determine precision from duplicate samples. Thus precision is usually expressed as Relative Percent Difference (%RPD) or Relative Standard Deviation (%RSD).

Summary of Accuracy and Precision

Based on the definitions of precision and accuracy, the performance of each of the four laboratories can be summarized in relative terms:

Lab	Accuracy	Precision
A	High	High
B	Low	Very High
C	High	Low
D	Low	Low

From the summary, it appears that Lab A provides the most reliable values. Notice, however, that although Lab B has low accuracy, its precision is very high. Thus, if the reported values are corrected for the systematic error introduced by the laboratory, Lab B is superior to Lab A. In other words, if 2.2 ppm (which is the average error as calculated previously) is added to each of the values reported by Lab B, then the reported values will both have very high accuracy and precision. This example demonstrates that if the bias (i.e., the average error) of the analytical method is known, it can be easily removed. It is not possible, however, to correct for low precision.

(Source: EPA's Data Quality Objectives for Remedial Response Activities)

Laboratory Control Sample

The **laboratory control sample**, or **laboratory standard**, is prepared by adding a known concentration of the analyte to an aliquot of **laboratory reagent water**. The control sample then goes through each step of the preparation and analysis procedure. Laboratory control samples are used in both inorganic and organic analyses to determine the accuracy of the analytical process in the absence of any matrix-related interferences. Control samples may also be used over time to calculate the precision of ongoing analysis. Results of laboratory control samples are reported as percent recovery (%Rec) for the analyte being determined.

Method Blank

Method blanks provide an indication of contamination that may be introduced into a sample from a variety of sources in the laboratory – unclean glassware, vapors from another area of the laboratory, or residue left inside a laboratory instrument from a previous analysis. The method blank consists of an aliquot of laboratory reagent water processed

Quality Control Parameter	Evaluate Analytical Process	Evaluate Sample Matrix	Determine Analytical Accuracy	Determine Analytical Precision
Laboratory control sample	X		X	X over time
Method blank	X			
Surrogate	X	X	X	
Laboratory duplicate sample		X		X
Matrix spike		X	X	X over time
Matrix spike duplicate		X	X	X

Table 9-5: Assessment functions of QC samples.

through all steps of the preparation and analysis procedure for both inorganic and organic analysis.

Surrogate Standard

A **surrogate standard** is an organic compound that is similar in chemical composition to the analyte being determined, but not normally found in environmental samples. Surrogate standards are added to samples undergoing organic analysis prior to the sample preparation or extraction phase of the analysis as verification of the accuracy of the entire analytical process. The recovery of surrogates provides an indication of the degree of target analyte recovery. Results for surrogate standards are reported as percent recovery (%Rec) of the surrogate compound added to the sample.

Laboratory Duplicate Sample

Samples are analyzed in duplicate in order to evaluate laboratory precision, verify sample homogeneity and detect matrix interferences. Duplicate samples are prepared by processing two distinct sample aliquots from the same environmental sample through the entire analytical procedure. Results of laboratory duplicate samples are reported as relative percent difference (%RPD) or relative standard deviation (%RSD).

Matrix Spike Sample

Matrix spikes are prepared by adding a known concentration of the target analyte to an aliquot of environmental sample, and then processing the

	Levels	Method blanks	Laboratory duplicates*	Matrix spikes*	Surrogates
Organic Parameters	IIIA	1 per matrix batch	1 per 20 or fewer samples	1 per 20 or fewer samples	Every sample
	IIIB	1 per matrix batch Maximum batch size is 20 samples	1 per 10 or fewer samples per matrix	1 per 10 or fewer samples per matrix	Every sample
Inorganic Parameters	IIIA	1 per matrix batch	1 per 20 or fewer samples	1 per 20 or fewer samples	None
	IIIB	1 per matrix batch Maximum batch size is 20 samples	1 per 10 or fewer samples per matrix	1 per 10 or fewer samples per matrix	None

* For organic parameters the analysis of matrix spike/matrix spike duplicates fulfills the requirements for laboratory duplicates and matrix spikes.
Source: Illinois EPA, Bureau of Land Site Remediation Program

Table 9-6: Example of laboratory QC sample frequencies.

Figure 9-3: To generate high quality data, the laboratory technician must follow standard methodology designated by EPA while meeting the lab's quality assurance guidelines.

sample through each step of the analysis procedure. The matrix spike measures accuracy of the analysis procedure, and is reported as %Rec of the spike.

Matrix Spike Duplicate Sample

Precision of the analytical procedure can be measured through analysis of a duplicate portion of the matrix spike sample. The relative percent difference or relative standard deviation of the original and duplicate sample percent recoveries is reported on the laboratory analysis report.

To make the most efficient use of time and instrument usage requirements (setup, calibration, etc.), laboratories often perform environmental analysis in batches consisting of samples belonging to more than one client. Laboratories determine the number and type of quality control samples that must be processed and analyzed along with these environmental samples from their QA/QC plan, or from information given to them by the project manager concerning the project's regulatory authority. Since the QC samples are also run on this batch basis, the statistical information given on the laboratory report is for this set of samples processed and analyzed as a batch. The actual QC measurements (duplicates, spiked samples, etc.) may or may not have been performed on a specific client's sample, but will have been performed on the required number of samples within the batch. However, QC samples are run each time the laboratory instrument is used, even if analysis is performed on one sample only.

Precision and Accuracy Level Level IIIA Objectives				
Analyte	**Method**	**Matrix**	**Precision (RPD)>0[1]**	**Accuracy %REC[2]**
Volatile organic compounds	8260A	Aqueous	<25%	50 – 150
Semi-volatile organic compounds	8270B	Aqueous	<50%	25 – 150
Pesticides & PCBs	8081	Aqueous	<35%	25 – 150
Metals	6010A, 7060A, 7061A, 7062, 7421, 7470A, 7471A, 7841, 9012	Aqueous	<25%	70 – 130
Volatile organic compounds	8260A	Solid	<30%	50 – 200
Semi-volatile organic compounds	8270B	Solid	<60%	25 – 200
Pesticides & PCBs	8081	Solid	<60%	25 – 150
Metals	6010A, 7060A, 7061A, 7062, 7421, 7470A, 7471A, 7841, 9012	Solid	<40%	60 – 140

1 Relative Percent Difference of Duplicate Sample Analysis
2 Percent Recovery of Spike Sample Analysis
Source: Illinois EPA, Bureau of Land Site Remediation Program

Table 9-7: Example of precision and accuracy levels to be achieved under a quality assurance program.

Detection Limits – "No Such Thing as Zero"

Some labs can meaure to the parts per quadrillion level!

0.0000000000001 !

In fact, the subatomic capabilities of science have been opening the boundaries of measurement. However, when you consider the concept of "risk assessment," how refined do environmental measurements need to be?

One concept often misunderstood by the data user is that of detection limits. In former eras of chemistry, the lowest measurements which could be made were determined by the smallest units that could be weighed or the smallest differences in color that could be detected. With the development of electronic technology, it has now become possible to routinely measure analyte concentrations at part per million (0.0001%) and part per billion (0.0000001%) levels. As measurement capabilities become more minute, more and more analytes can be detected in ever smaller amounts in environmental samples. As a result, chemists today are reluctant to use "zero" or "none" when reporting the results of analysis. Instead, if a particular analyte is not observed in an analysis, it is reported as being "not detected" or as "less than the detection limit value." The **detection limit** represents the lowest measurement of analyte concentration the instrument and analytical method can reliably achieve. Results that are reported as less than or equal to this limit should be taken to mean that if the analyte is present, it is present at some concentration less than this limit.

The measurement point at which the detection limit lies depends on instrument sensitivity and sample matrix effects. Measurements made near the detection limit of the instrument are not as reproducible as measurements made at higher analyte concentrations – that is, precision is better at higher analyte concentrations. But samples with very high concentrations of analytes, which require dilution before analysis, have higher detection limits than those within the measurable range of the instrument. To ensure accuracy, the detection limit of the analyte should be well below the analyte's level of concern.

Laboratories determine the detection limits of their instrumentation through a procedure outlined in 40 CFR. This procedure requires analyzing a very low concentration of analyte a total of seven times and averaging the analysis results. To ensure that the instrument is performing at its detection limit capabilities, laboratories routinely analyze instrument calibration standard samples with concentrations at or near the detection limit as part of the internal quality control program. Calibration standard results are not generally included on laboratory analysis reports sent to the client; detection limit information, however, may be included.

The EPA has defined the following terms relating to detection limits:

Instrument Detection Limit – The smallest signal above the instrument's baseline electrical noise that an instrument can detect at a 99% **confidence interval** (i.e., a statistical range with a specified probability that a given parameter lies within the range).

Method Detection Limit (MDL) – The minimum concentration of a substance that can be identified, measured, and reported with 99% confidence that the analyte concentration is greater than zero. This is determined from analysis of a sample of a given matrix containing the analyte. Generally, an MDL is equal to three times the baseline noise.

Limit of Detection – The lowest concentration level that can be determined to be statistically different from a blank; numerically defined as three times the standard deviation of seven replicate measurements.

Level of Quantitation – The level above which quantitative results may be obtained with a specified degree of confidence; numerically defined as ten times the standard deviation of seven replicate measurements. This limit is specific to each individual laboratory.

Practical Quantitation Limit (PQL) – Comparable to the level of quantitation, this is an inter-laboratory concept; numerically estimated at five to ten times the instrument's baseline noise. PQL's are limits that can be routinely achieved on most real-world aqueous samples. They are adversely affected by other interfering compounds in the sample, and by complex sample matrices such as oil/water mixtures. PQL's are derived from laboratory performance of the best laboratories under ideal conditions. They provide performance goals that all labs should strive to meet.

Regulatory programs are not consistent as to which of these types of detection limit is specified in their quality assurance program. For example, analyses under the Water Pollution Control Act are required to meet an MDL, while samples regulated under both the Safe Drinking Water Act and RCRA must meet PQL's, and Superfund projects use sample or contract-required detection limits.

Laboratories that obtain certification under particular regulatory programs must determine and submit the MDL's or PQL's of their laboratory instruments to the certifying agency annually or whenever major instrument components are changed.

Data Qualifiers

Several factors can prevent a lab analyst from achieving the usual detection limits, precision and accuracy:

—High concentrations of contaminants in a sample;

—Lack of sample homogeneity (for example, if a contaminant is not spread consistently throughout the matrix to be sampled);

—Interferences with the sample matrix (such as sludgy mixtures of water, oil and solids interfering with the analysis for a particular chemical contaminant).

Additional factors may also prohibit the analyst from achieving quality control requirements:

—Poor recovery of a surrogate standard;

—Contamination of the blank;

—Instrument calibration problems.

Out-of-control data are investigated and, if possible, corrected as part of the laboratory's internal quality control process. During the laboratory's internal data validation process, the lab director or QA/QC officer reviews analytical results and their associated quality control information using criteria specific to the analysis method performed. Any data that do not meet quality control requirements but that are still to be included on the laboratory report are explained using **data qualifiers**. Data qualifiers explain the uncertainty of the data and alert the data user to how much significance can be placed on the measurements. Data qualifiers may indicate either

Qualifier	Definition	Indicates: Uncertain identity?	Uncertain concentration?
Inorganic chemical data[a]			
B	Reported value is <CRDL, but >IDL.	No	?
U	Compound was analyzed for, but not detected.	Yes	Yes
E	Value is estimated due to matrix interferences.	No	Yes
M	Duplicate injection precision criteria not met.	No	Yes
N	Spiked sample recovery not within control limits.	No	Yes
S	Reported value was determined by the Method of Standard Additions (MSA).	No	No
W	Post-digestion spike for furnace AA analysis is out of control limits, while sample absorbance is <50% of spike absorbance.	No	Yes
•	Duplicate analysis was not within control limits.	No	Yes
+	Correlation coefficient for MSA was <0.995.	No	Yes
Organic chemical data[b]			
U	Compound was analyzed for, but not detected.	Yes	Yes
J	Value is estimated, either for a tentatively identified compound (TIC) or when a compound is present (special identification criteria are met, but the value is <CRQL).	No, for TCL chemicals Yes, for TICs	Yes
C	Pesticide results were confirmed by GC/MS.	No	No
B	Analyte found in associated blank as well as in sample.	No	Yes
E	Concentration exceeds calibration range of GC/MS instrument.	No	Yes
D	Compound identified in an analysis at a secondary dilution factor.	No	No
A	The TI is a suspected aldol-condensation product,	Yes	Yes
X	Additional flags defined separately.	–	–
Inorganic and organic chemical data[c]			
U	Compound was analyzed for, but not detected. The associated numerical value is the SQL.	Yes	Yes
J	The associated numerical value is an estimated quantity.	No	Yes
R	Quality control indicates that the data are unusable (compound may or may not be present). Re-sampling and/or re-analysis is necessary for verification.	Yes	Yes
Z	No analytical result (inorganic data only).	NA	NA
Q	No analytical result (organic data only).	NA	NA
N	Presumptive evidence of presence of material (tentative identification)[d].	Yes	Yes

– Data will vary with laboratory conducting analyses.
a Source: EPA 1988 b
b Source: EPA 1988 c
NA Not applicable
c Source: EPA 1988 d,e
d Organic chemical data only

Table 9-8: Data qualifiers used in EPA's Contract Laboratory Program (CLP).

uncertain identity or uncertain concentration of the analysis result.

Examples of data qualifiers include:

U = compound was analyzed for but not detected above the method detection limit;

N = spiked sample recovery is not within quality control limits;

B = analyte found in associated blank as well as sample.

A more complete list of data qualifiers, taken from the EPA's Contract Laboratory Program (CLP) protocols, is presented in Table 9-8. This list also shows whether the data qualifier indicates uncertain identity or uncertain concentration of the analyte.

The presence of data qualifiers does not always mean that the data is unusable for project decision making, but rather that the data user must consider the specific problem that occurred with the analysis before choosing the degree of significance to place on the result.

Checking Your Understanding

1. Distinguish between the elements of a lab's Quality Control Program and Quality Assurance Program. Explain how these programs are necessary to the generation of reliable and valid data.
2. Distinguish between a performance audit and a systems audit; define qualitative review and quantitative review as the terms apply to audits.
3. Discuss the concept of detection limits, and identify which of these two types of detection limits – MDL (Method Detection Limit) or PQL (Practical Quantitation Limit) – would normally be at the lower concentration.
4. For the following data qualifiers, state whether the uncertainty of the data lies in the analyte identity, the analyte concentration or both. When might data of uncertain concentration still be useful in project decision-making?

   U – compound was analyzed for but not detected.

   N – spiked sample recovery not within control limits.

   • – duplicate sample analysis not within control limits.

   E – concentration exceeds calibration range of GC/MS instrument.

   R – quality control indicates that the data are unusable. Re-sampling and/or re-analysis is necessary for verification.
5. Define accuracy and precision as these terms pertain to laboratory analytical instruments.
6. List the various types of internal quality control samples used by an analytical lab and discuss which data quality is assessed by each.
7. Discuss batch analysis and what it means in relation to quality control samples.

9-3 Validation and Interpretation of Analytical Results by Data Users

Concepts

- Analytical data must be evaluated for precision, accuracy, representativeness, completeness and comparability before it can be used as a base for project decisions.
- The level of data quality needed will vary according to its intended use.
- The data user must consider the quantity of data available and the data's adherence to regulatory requirements.
- Analytical data must be interpreted according to the specific requirements of the environmental project.

Figure 9-4: The sampling technician must perform every field duty with diligence. The expression "Garbage in? Garbage out!" reminds us that the laboratory analysis is a costly waste of time and money if the samples have not been properly collected and stored. An excellent field technician is a great asset to the site characterization team.

Before basing project decisions on analytical data, data users must assess the data by following these three steps:

1. Validate data quality through evaluation of the data's accuracy, precision, completeness, representativeness and comparability.
2. Determine whether the data are of sufficient quality and if enough valid data have been generated to reduce uncertainties involved in interpreting site conditions.
3. Determine whether the data satisfies regulatory authority requirements and interpret the data according to goals and requirements of the specific project, such as cleanup objectives, human toxicity levels or hazardous waste criteria.

Step 1: Validate Data Quality

The first step in the validation and interpretation process involves evaluating the data's accuracy, precision, completeness, representativeness and comparability. The data user begins by comparing individual sample results against a set of performance criteria, such as quality control requirements established by the agency in charge of the project. The quality of individual analytical measurements is defined in terms of their accuracy and precision. The quality of the entire data set is then established through its completeness, representativeness and comparability with other data sets. Once the quality of individual sample results and of the data set as a whole is known, the data can be used to establish a total picture of the site, as, for example, when determining the extent and degree of groundwater contamination from a leaking underground gasoline storage tank.

Accuracy

Accuracy is a quantitative determination. The overall accuracy of a set of data is dependent both on the sampling accuracy and the analytical accuracy. The analytical results depend on everything that goes on: sampling process, potential field contamination, preservation and handling techniques, interferences from the sample matrix as well as sample preparation and laboratory analysis techniques. Therefore, although the laboratory has determined the accuracy of the analytical results prior to sending the analytical report to the data user, it is still necessary for the data user to assess the overall accuracy.

Sampling accuracy is evaluated through the use of field and trip blanks, while analytical accuracy is evaluated through the use of known and unknown QC samples and matrix spikes. Data users should verify that the results of field and trip blanks show no target analyte concentrations. If target analytes are found in field and/or trip blanks, then cross-contamination may have occurred during sampling or sample handling procedures, and sample results may not reflect true site conditions.

The results of matrix spike samples, which the lab prepares and analyzes along with the environmental samples, provide an indication of analytical accuracy and matrix interferences. Data users should verify that the percent recovery of the matrix spike is within acceptable limits (usually 80 to 120 percent) or that the reported measurement carries a data qualifier indicating that this quality control criterion was not met.

Another method for determining analytical accuracy is through the use of performance samples. A **performance sample** is a sample of known concentration (i.e., a standard), which the data user obtains from a third party such as a lab supply company. The data user submits the performance sample to the laboratory along with project environmental samples. The laboratory is not told that the sample set includes a performance evaluation sample. When the laboratory analysis report is received, the data user can compare the lab's reported result for the performance sample with its known concentration. In comparing the reported result with the known concentration, the data user must remember that the known concentration will not be a single "true" value, but rather an acceptable range of values. This is because in analytical chemistry, even standard samples with "known" concentrations have limits of certainty.

The laboratory's accuracy error can be determined using the following formula:

$$\text{Accuracy error} = \frac{|X - TV|}{TV} \times 100\%$$

where

X = laboratory's analysis result for the performance sample and

TV = true value of the performance sample (i.e., the standard).

Precision

The overall precision of a set of data is also a mixture of sampling and analytical factors. Analytical precision is much easier to control and quantify than sampling precision. Analysis results from laboratory duplicate samples provide an indication of analytical precision. The analytical precision of the results has been determined by lab personnel prior to releasing the lab report to the data user. The data user should verify that the relative percent difference (%RPD) or relative standard deviation (%RSD) of laboratory duplicate samples is within acceptable limits (usually within 20%RPD).

Collocated or field replicate sample results provide an indication of the overall precision of the sample collection and data generation process. Data users should calculate the relative percent difference or relative standard deviation of the collocated or field replicate samples in order to verify that the data precision is acceptable.

Precision error (Precision_e) may be calculated using the following formula:

$$\text{Precision error} = \frac{|X_1 - X_2|}{TV} \times 100\%$$

where

X_1 and X_2 = replicate values and

TV = true value.

Suppose, for example, a QC officer knows that the true value of a quality control sample is 200 ppm. The lab analyzes two replicate samples, which results in values of 190 ppm and 188 ppm. When the officer applies the formula for precision error, the error is calculated to be one percent, which may be considered quite precise:

$$\frac{|190 - 188|}{200} \times 100\% = 1.0\% \text{ error}$$

(Note that smaller differences between the replicate values X_1 and X_2 equal a smaller percentage of error and thus indicate better precision.)

The data user may determine sampling precision by subtracting the analytical precision from the overall precision as follows:

Overall Precision – Analytical Precision = Sampling Precision

The data user must keep the level of concern and the end use of the data in mind when reviewing precision and accuracy information. In some cases, even data of poor precision and/or accuracy (measurements that have data qualifiers attached) may be useful in making project decisions. For example, if all of the results are far above the level of concern, the precision and accuracy of the data are not as significant as when all of the results fall close to the level of concern.

Completeness

Completeness is a quantitative determination of the amount of valid data obtained from a measurement system compared to the amount that was 1) expected to be obtained under normal conditions and 2) needed to meet the project's data quality objectives.

The goal of completeness criteria is that a sufficient amount of valid data be generated to allow for prudent project decisions. However, according to the EPA document "Data Quality Objectives for Remedial Response Activities," little information exists on the completeness of individual analytical methods. EPA's Contract Laboratory Program (CLP) data are 80 to 85 percent complete on a nationwide basis. Level III, IV (CLP), and V analytical techniques (see Chapter 8) generate data that are approximately 80 percent complete. Level I and II analytical techniques have lower completeness levels. However, since Level I and II techniques take place onsite and provide immediate results, any out-of-control data can be corrected on

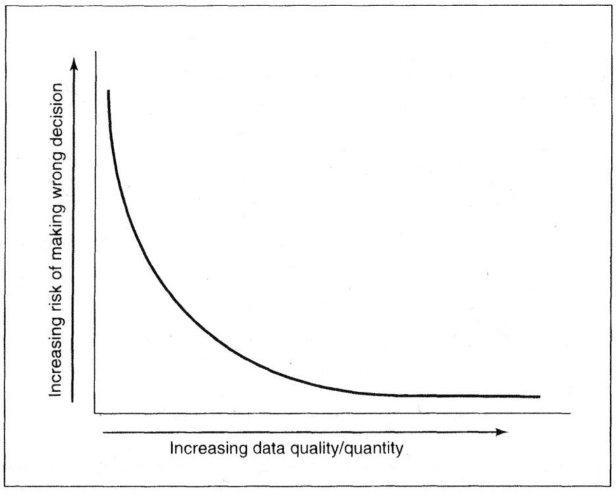

Table 9-9: Relationship between data quality/quantity and the degree of risk.

the spot. This allows for the achievement of a generally high degree of completeness for this type of data. Regulatory programs that list completeness criteria usually require 80 to 90 percent completeness, depending on the analyte.

Information on completeness is not usually included on the laboratory analysis report, but it can be determined by the data user with the following formula:

$$\text{Completeness} = \frac{\text{valid analyses reported}}{\text{total analyses undertaken}} \times 100$$

Completeness involves not only the total number of valid analyses that can be used in the decision-making process but also the qualitative concept of critical samples. **Critical samples** are those for which valid data must be obtained in order to satisfy the objectives of the sampling and analysis activity. An example of a critical sample is an up-gradient well. Without valid results for this sampling point, the degree of contamination already present in the groundwater before it enters the project site cannot be determined, and therefore, the amount of contamination contributed by site constituents remains elusive.

Other critical samples are those involving human health and safety. Without reliable knowledge of the concentration of toxic air contaminants present at the site, field technicians would have to wear full protective gear during all site activity. Data users should assure that valid data has been obtained for sampling points identified as critical in the site's sampling plan.

Representativeness

Representativeness is the degree to which the sample data accurately represent the site conditions. This is a qualitative parameter contingent on both a proper site sampling program and proper laboratory protocols. Representativeness is best achieved by making sure that sampling locations are selected properly and that a sufficient number of samples are collected. Data users should verify that the sampling plan was carried out as designed.

Data users can also assess the degree of representativeness of a set of data by comparing the results of collocated samples (samples collected so that they are equally representative of a given point). Collocated samples can also provide precision data.

Comparability

Comparability addresses the confidence with which one set of data can be compared to another. Comparability is a qualitative parameter that is contingent on the similarity of sampling techniques and analysis methods used for the two sets of data. The goal of comparability is that sample data should be comparable with other measurement data for similar samples and sample conditions. This is achieved through the use of standard techniques for sample collection and analysis.

The difference in age of data sets is one point that must be considered when a comparison is made. Depending on the situation, the elapsed time between two sampling/analysis events may greatly affect the degree of comparability of the results, as could a change in other circumstances surrounding the event. Sampling results collected quarter by quarter or year by year to monitor an industrial wastewater discharge are generally highly comparable. A major change in industrial components, processes, or workload, however, may make the results less comparable.

One type of comparison often made during environmental projects is between background samples and samples collected from the project site. Background samples may be site-specific–from a clean area of the project site or from the surrounding area, or alternatively from local, regional or national averages. Background samples aid in distinguishing site-specific contamination from what is normally associated with the area.

Step 2: Determine if Data Will Reduce Decision Uncertainties

The second step in the validation and interpretation process is to determine whether there are enough high-quality data on which to base project decisions. The required level of quality will vary according to the intended use of the data. Project objectives are defined during development of the project sampling and Quality Assurance plans. The objectives are designed to determine:

—Presence or absence of contamination

—Types of contaminants

—Concentrations of contaminants

—Extent (boundaries) of contaminants

Technical Corner ■ Plotting Data on a Map to Interpret Contamination Levels

(Refer to the map as you study the details, which are adapted from an actual case.)

In 1990 the owner of a convenience store determined that product was leaking from an underground gasoline storage tank. The tank is shown in the upper left corner of the property, just south of Main Street. An existing monitoring well (EMW) is indicated next to the tank pit. That tank and two smaller tanks on the property were removed, and the state environmental agency required the owner to have the site investigated to determine the extent of contamination.

One of the contaminants detected in the water was benzene, which is a component of gasoline. Benzene's regulatory action level (or threshold) is five parts per billion (ppb). The action level (or threshold) is the point at which a chemical becomes subject to regulatory attention. Benzene's 5 ppb threshold is listed in the state underground storage tank (UST) regulations. USTs are covered by RCRA.

The environmental contractor hired to investigate the site mapped the entire area as shown. A historical study revealed two former service stations in the area of the convenience store site. The environmental consulting company collected a sample from the existing monitoring well (EMW) that was already by the leaking UST. The analytical results indicated a benzene level of 9,600 ppb. To begin efforts at defining the extent of contamination, the consultant installed three new monitoring wells (MW) around the contaminated EMW and collected a sample from each. Analysis showed that MW1 had a 2,000 ppb level, MW2 had a 5,700 ppb level, and MW3 had a < 1 ppb level (a level referred to as "nondetect").

Next, MW4, MW5 and MW6 were installed to check the extent of contamination beyond MW2. MW7 and MW8, as well as MW6, checked the extent of contamination beyond MW1. See the Data Table at the end of the Technical Corner for the results for these monitoring wells.

Underground storage tank removal.

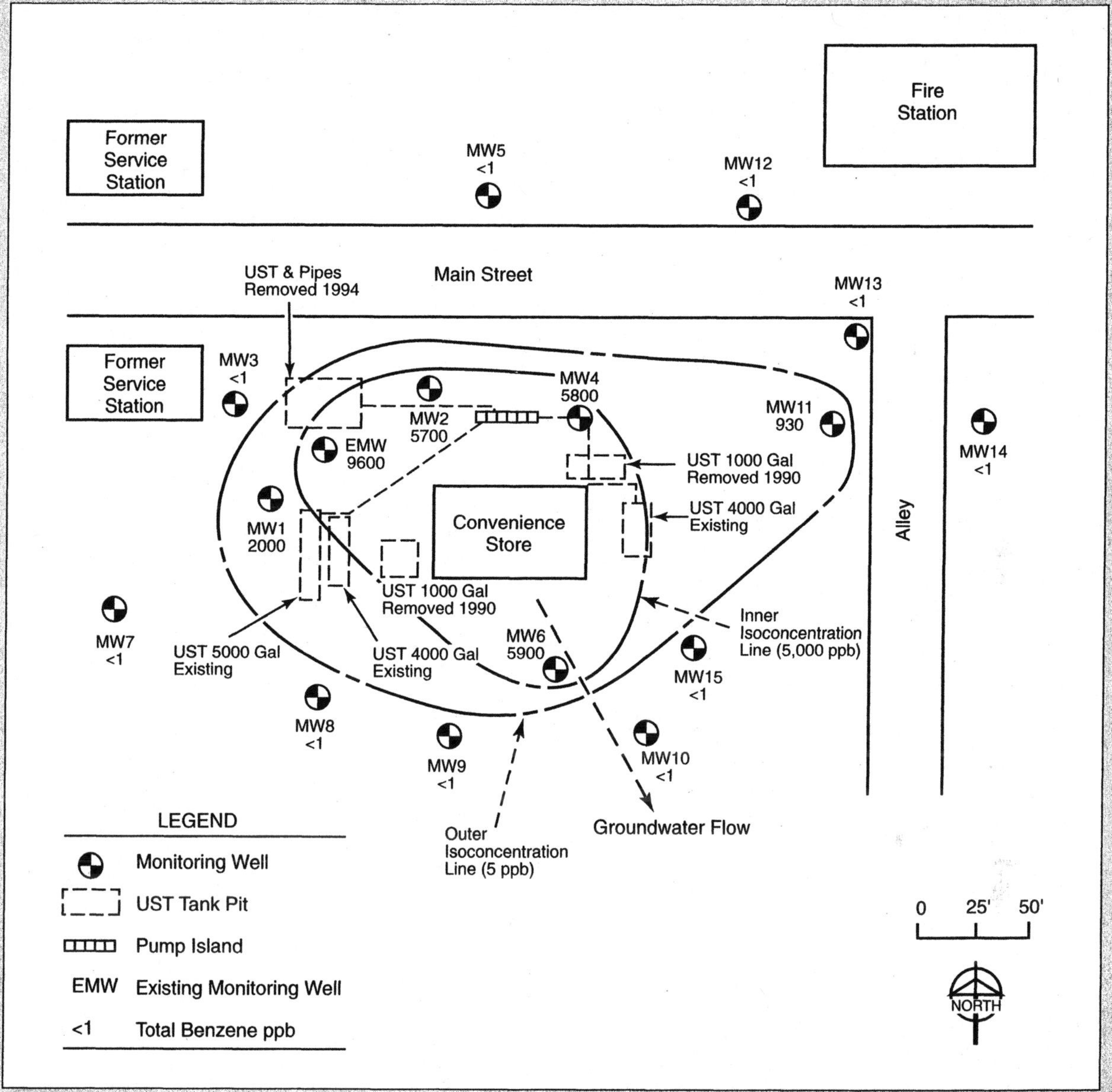

Later, MW9 and MW10 were installed to check the extent of contamination beyond MW6. MW11, MW12, MW13 and MW14 were installed to check the extent of contamination beyond MW4. Additionally, MW15 checked both MW6 and MW4.

When the analytical results for the monitoring wells are received, the consultant marks each ppb level on the site map. He or she then visually interprets the contamination by drawing **isoconcentration lines** to depict relationship among similar levels of contamination. The prefix "iso" in the word isoconcentration means equal or uniform. The intent of the isoconcentration drawing is to indicate areas of uniform, or at least similar, concentration ranges.

On the map, the isoconcentration lines indicate the following ranges of concentration:

— Beyond the outermost line are those monitoring wells showing benzene concentrations of less than the regulatory action level of 5 parts per billion.

— Within the innermost line are those monitoring wells that have the highest concentrations; in this case, the wells testing at more than 5,000 ppb.

Data Table		
Sample ID	MDL*	Benzene, ppb
EMW	1 ppb	9,600
MW1	1 ppb	2,000
MW2	1 ppb	5,700
MW3	1 ppb	< 1
MW4	1 ppb	5,800
MW5	1 ppb	< 1
MW6	1 ppb	5,900
MW7	1 ppb	< 1
MW8	1 ppb	< 1
MW9	1 ppb	< 1
MW10	1 ppb	< 1
MW11	1 ppb	930
MW12	1 ppb	< 1
MW13	1 ppb	< 1
MW14	1 ppb	< 1
MW15	1 ppb	< 1

*Method Detection Limit (MDL) refers to the ability of the laboratory instrumentation to detect benzene at the 1 part per billion level or above.

—Between the outer line and the inner line are wells testing in the range of 5 ppb to 5000 ppb.

Other isoconcentration lines could be drawn between the two lines shown. In this case, however, the environmental professional attempted to show the concentration of benzene with enough detail to illustrate the situation but without cluttering the map with too many lines. The isoconcentration lines depict an area known as a "plume" of contamination. Depending upon the direction of groundwater flow, plumes may vary in shape from somewhat circular to elongated.

The regulatory requirement to define the extent of contamination means that monitoring wells are installed until all wells around the perimeter of the plume are less than 5 ppb (benzene's action level) or "non-detect" (below the method detection limit of 1 ppb).

In this case, the state chose to assign periodic monitoring to the site. The wells classified as nondetect in 1990 are still nondetect seven years later. The benzene levels in the contaminated areas are decreasing gradually over time.

—Health and safety risks from contamination

—Engineering design for contamination remediation

The required level of data quality is also defined in earlier stages of the project as, for example, during development of the project's data quality objectives.

Data users must determine whether the objectives have been met. The quantity and quality of the data are equally important to project decisions. An abundance of data of low quality may not reduce the risk of a wrong decision, just as having very high quality data from only a few sampling points may not be adequate to truly assess site conditions.

Step 3: Compare and Interpret Data to Project Specifics

The final part of the data validation and interpretation process involves comparing the sample results to project-specific factors such as:

—Do the samples show compounds from the EPA's target compound list?

—Are compounds present at levels toxic to humans?

—Are contaminant levels above or below action levels established for the site?

—Do the samples exhibit hazardous waste characteristics?

—Do the data meet regulatory agency requirements; for example, the agency's specified methodology, quality control criteria and detection limits?

Checking Your Understanding

1. Define data validation and list the five points of evaluation that data users should use to validate data quality.
2. Define data interpretation and list five project specifics against which analytical data might be compared.
3. Discuss completeness of a set of analytical data and how it relates to a project's critical samples.
4. List five items that can affect sample data representativeness.

5. When is it possible to confidently compare two sets of data?
6. Discuss the differences in data quality that might be acceptable for the following project objectives (or: Discuss how data quality that is not as complete or precise as desirable might be acceptable for the following project objectives):
   a. Determining presence or absence of contamination at a site.
   b. Assessing the health and safety risks of a site's contamination.

Summary

The data user should follow these steps to validate and interpret analytical data from the laboratory:

1. Determine if sampling, sample preservation and sample handling procedures were performed according to the sampling plan (assesses representativeness).
2. Compare field and trip blank data to detect contamination from sampling and decontamination procedures, or from improper sample handling and storage techniques (assesses sampling accuracy).
3. Compare collocated and field replicate sample data to determine sampling precision.
4. Review the laboratory quality control report, including results of laboratory blanks, matrix spike recoveries, laboratory control samples, and laboratory duplicate samples to ensure that the data is valid from a laboratory quality control perspective (assesses laboratory accuracy and precision).
5. Delete unusable data from data set and attach data qualifiers to remaining data. Explain the limitations of any qualified data for use in project decision making.
6. Review total data set to ensure that a sufficient amount of data has been collected to meet project goals and satisfy completeness criteria. Verify that valid data have been obtained for all the project's critical sampling points.
7. Review analytical methods used to generate the data to ensure that they conform to regulatory authority requirements.
8. Review the detection limits for all sample results to ensure that they are well below levels of concern or cleanup objectives established for the project and that they meet regulatory authority requirements.
9. Review field and laboratory quality control sample types, frequencies and results to assess compliance with regulatory requirements. Ensure that all quality control procedures and results include proper documentation.
10. Identify contaminants of concern among sample results: parameters from EPA's target compound list, compounds associated with industrial processes occurring on the site, compounds for which cleanup objectives have been established, etc.
11. Identify results that are above the level of concern: cleanup objectives, hazardous waste criteria, health and safety toxicity levels, etc.
12. Compare site data to background concentrations and assess the degree of comparability of the two data sets.

Application and Critical Thinking Activities

1. Contact an analytical laboratory and ask for an example of one of their laboratory reports (the lab will delete client identifiers). Compare the report with the list of elements given in the section "Contents of the Laboratory Analysis Report." Note deficiencies and/or additional information that has been provided.
2. Predict some problems that could arise in a project if decisions were based upon invalid data.
3. In the table below are the results from four analytical labs, each of which analyzed the same soil sample for its benzene content. The actual concentration of benzene in the soil is 25 ppb. First, compare the four sets of data: which lab do you think has the most accurate data and which lab has the most precise data? Is it the same lab? Now, using the formulas for average

error and variance found in Section 9-2 Technical Corner (Comparing Accuracy and Precision of Four Different Labs), calculate the accuracy and precision of each lab's data. Was your initial evaluation correct regarding which lab generated the most accurate and most precise data?

Reported Concentration of Benzene in a Soil Sample (ppb)				
Replicate	**Lab A**	**Lab B**	**Lab C**	**Lab D**
1	25	19	25	28
2	29	29	25	27
3	21	30	27	27
4	23	16	25	29
5	27	16	25	29
6	22	21	24	27
7	27	16	28	26
8	20	29	26	27
9	30	26	25	28
10	25	30	27	27

4. Develop a checklist of items you would use if you were conducting an internal system audit of an environmental laboratory.

5. Contact an environmental laboratory and ask for a copy of their QA/QC plan. Using the plan, determine the lab's criteria for the following items:

   a. Acceptable holding time for water samples to be analyzed for heavy metals.

   b. Acceptable control limits for laboratory spiked samples by AA analysis.

   c. The job title of the person at the lab responsible for the internal quality control program.

   d. How long the lab keeps samples on hand after they have been analyzed before disposing of them; how hazardous samples are disposed of.

   e. How often the lab's pH meter must be calibrated:

6. Given the following sample analysis and QC information, determine the accuracy (as %Rec) and precision (as %RPD) of the data:

   a. Sample analysis result for lead 1.60 ppm

   b. Spiked sample analysis result 6.98 ppm

   c. Duplicate sample analysis result 1.05 ppm

   d. Amount of spike added to original sample 5.00 ppm

   e. Method blank 0.01 ppm

7. Practice your data interpretation skills by reviewing the three ABC Laboratory reports in Appendix 7.

   a. What can you conclude about the soil sample from Example 3?

   b. Is John Smith or Sue Johnson's health in danger from workday time spent in the welding booth?

   c. Is the water from K. Smith's kitchen sink safe to drink?

8. Referring to the Technical Corner "Plotting Data on a Map to Interpret Contamination Levels" decide: if you had been in charge of this project, would you have installed the same number of wells and, if no, why?

Appendix 1

Personnel Safety Plan Hazardous Waste Site

Note: This example is a modification of an actual safety plan used on an unnamed hazardous waste site.

Contents

I - III omitted in this sample.

IV Site Organization

The Unnamed Waste Site is located in Unnamed City near Rte. Xxx (See Map, Figure 1). The site is part of a borrow pit operation adjacent to a small stream which flows via a marsh into Country Pond. Within 100 yards of the stream 4300 drums, in various stages of deterioration, were abandoned. These open and sealed drums contain various hazardous substances. Some of these drums, due to their condition and displacement by rainfall, have contaminated the ground in the area where they were abandoned.

During the winter and spring of 1981, EPA undertook operations to move the drums into staging areas preparatory to sampling them. During the course of these operations an HNU was used to obtain organic vapor levels for the site. The ambient air (including background) for the site usually indicated less than 5 ppm organic vapor with frequent increases to 20-30 ppm in the immediate vicinity of the drum movement operations. Some individual drums, checked on a random basis, indicated 2000+ ppm (within 6 inches of the bung).

No areas fit the definition of a confined space.

In order to reduce the potential for contaminant migration and reduce the risk of personnel exposure to hazardous substances, three zones will be established. The three zones are: 1) Exclusion Zone; 2) Contamination Reduction Zone; and 3) The Clean Zone.

V Determination of Zones of Contamination

A. Exclusion Zone (See Site Plan, Figure 2)

The Exclusion Zone is the area southerly of the small stream at point "H." This area encompasses the sand and gravel pit. Within this zone the designated "level of hazard" will be established, necessitating the use of personnel protection equipment.

Due to the condition of the drums and the soil and debris of the former drum storage site, a potential for wind migration of contaminants exists. The Exclusion Zone has therefore been made sufficiently large to encompass foreseeable dispersion based on operations conducted in the spring of 1981. In order to facilitate operations in this zone, three sub-areas will be established:

1. Area C will be the within the Exclusion Zone where only background vapor levels exist. This area serves as a buffer within which wind dispersion of contaminants might occur. In Area C personnel will be required to wear the protective clothing designated for Area B operations and carry an air purifying respirator. A red flag will be flown in a prominent location to serve as a wind reference.
2. Area B is the area within which the wearing of both protective clothing and respiratory protection will be required due to the potential for contamination from the drums during work activities.
3. Area A is the term used to delineate "hot spots" within Area B. A hot spot is a point at which the contaminant levels are, at least periodically, higher than the level of hazard indicated for Area B and require an increase in protection above that provided. These spots will be identified by an orange "bicycle pennant." Personnel working in these areas will wear the respiratory protection and safety clothing necessary for the special degree of hazard. Additionally, personnel will use any necessary monitoring devices and safety tools to complete their specific tasks in a safe manner.

In order to provide an adequate Safety Zone the special level of hazard protection area will extend at least 50 feet from the pennant.

B. Contamination Reduction Zone

The Contamination Reduction Zone serves as a buffer between the Clean Zone and the Exclusion Zone. The zone incorporates the entire borrow pit area and a portion north of the small stream.

Personnel intending to enter the exclusion Area shall don the appropriate protective clothing for the area(s) that they intend to enter as well as obtain the respiratory protective equipment that is necessary prior to crossing the Hotline at point "H" into the Exclusion Area.

Decontamination and Exclusion Zone equipment storage and maintenance will be carried out in this zone. No equipment used in the Exclusion zone will enter the Clean Zone until it has been decontaminated in the Contamination Reduction Zone. To support Exclusion Zone activities three facilities will be situated in the zone:

1. Decontamination Station

The Decontamination Station will be located at the other perimeter of the Contamination Reduction Zone. All personnel who have been within the Exclusion Area shall pass through a decontamination procedure prior to re-entering the Clean Zone (see Decontamination Procedure, Appendix I).

2. Heavy Equipment Park
   A Heavy Equipment Park will be designated near the Decontamination Station for the storage, maintenance, and decontamination of vehicles used in the Exclusion Zone.
3. Weather Shelter
   A Weather Shelter will be located in the Contamination Reduction Zone to provide some protection to personnel when taking breaks, without requiring them to fully decontaminate.

C. Clean Zone

The Clean Zone is the outer area and may be considered clear of contamination. The past history of the site, however, indicates that parts of this zone were used to stockpile drums prior to their being processed through a drum recycling facility owned by the Unnamed Corporation. Ground contamination is therefore present. Work that might be necessary in the former storage areas is to be considered to require protective clothing.

Depending on the location, respiratory protection may not be needed.

Private vehicles not needed for operations are restricted to the parking lot at the Command Post.

The EPA Command Post, or Mobile Laboratory, and an Equipment Trailer will be located within the Clean Zone.

1. The Command Post will serve as the OSC's headquarters and will be equipped with:
   a. First aid supplies
   b. Weather station
   c. Communications
   d. Safety plans
   e. Communication watch.
2. The Mobile Laboratory will be equipped to analyze and categorize the materials at the site. Access will be restricted in order to minimize contamination and interference with analysis.
3. The Equipment Trailer will be used to store safety materials prior to their distribution from the clean side of the Decontamination Station. Other equipment will be stocked here for future use and damaged equipment will be repaired here after decontamination.

VI Determination of the Level of Hazard

The level of hazard will be determined by periodic monitoring of the site for contamination by the Safety Officer.

The investigation, to date, has not indicated the presence of substances that may be absorbed through the skin. The investigation indicated that various substances, primarily industrial solvents, are present in sufficient quantities to cause irritation to the eyes, lungs, and/or intestinal tract, and many are known or suspected carcinogens. Only a small portion of the drums have been sampled; therefore, personnel will be required to wear as a minimum:

A. Gloves
   A 2-glove system will be worn.
   1. Inner Gloves - Nitrile or PVC:
   2. Outer Gloves - Butyl Rubber or Nitrile
   The inner gloves will be worn at all times within the Exclusion Zone.
   The outer gloves will be worn when contact with drums and other contaminated materials is expected.
B. Disposable Splash Suit
   All personnel, as a minimum, within the Exclusion Zone will wear a polylaminated Tyvek disposable coverall.
C. Head Gear
   Hard hats will be worn at the site if overhead work occurs.
D. Eye Protection
   Face shields or goggles will be worn within the Exclusion Zone. "Safety" glasses are not sufficient protection and contact lenses will not be worn at the site.
E. Boots
   Cover boots will be worn in the Exclusion Area. Steel-toed boots will be worn by personnel handling drums, or cover boots over steel-toed work shoes.
F. Respiratory Protection
   The respiratory protective devices used at this site will fall into three categories:
   1. Positive Pressure, Demand, Open Circuit, Self Contained Breathing Apparatus (SCBA or Positive Pressure Demand Airline Respirator)
   2. Air Purifying Respirator
   3. Constant Flow

Only NIOSH/MSHA approved equipment will be used.

The level of respiratory protection to be used will be based upon the use of a photo ionization detector (HNU). In order to provide a maximum of protection, the following procedure will be followed:

1. Positive pressure, demand, open circuit, SCBA will be worn if the level of organic vapors exceeds 25 ppm.
2. An air purifying respirator will be used when the organic vapor level is between 6 ppm and 25 ppm.
3. If the organic vapor level is 5 ppm or less, no device is necessary, but either an air purifying respirator or approved escape device will be carried.

Examples of NIOSH/MSHA approved devices are:

1. SCBA - MSA 401 Pressure Demand TC 13F-30
2. Air Purifying - MSA Ultra Twin Cartridge TC 21C-188 (cartridges and/or canisters must be NIOSH-approved for the respirator)
3. Escape -Robertshaw 5-minute TC 13F-28

# PPM in Breathing Zone	Wear
0-5	None
6-25	Air Purifying
26+	S.C.B.A.

CAUTION: Individual jobs at the site may require personnel to wear an increased level of protection than generally necessary for the site. For example: 1) acid raingear; 2) butyl rubber aprons; 3) SCBA when others on air purifying respirator; and 4) hard hats with face shields.

Note: Operations will be conducted during the summer months when excessive ambient air temperatures may cause personal injury and increased accident probability. Safety procedures to avoid heat stress casualty potential are outlined.

#H-NU calibrated to 9.8 with benzene.

VII Air Monitoring Survey

The Safety Officer will survey the site every second hour, and at such other times as deemed necessary by an alteration of wind speed or direction or the type of work being conducted, using a photo ionization detector.

The Safety Officer shall use a copy of the Site Plan to indicate the location where readings were recorded, velocity of the wind, and the HNU reading. A minimum of 10 locations will be checked during each bihourly survey.

Three activated charcoal filter pumps will be maintained in the Contamination Reduction Zone and sampled twice a day in order to assess the composition of the organic vapors upwind, downwind, and within the Exclusion Zone.

VIII Medical Surveillance

During drum sampling the technicians could be exposed to toxic chemicals, safety hazards, and temperature extremes. The director of the site medical program will maintain a first-aid station for emergency and nonemergency medical treatment. The handling of casualties is described in section IX below.

Medical records will be kept on each technician (see Appendix).

Medical surveillance of each employee will consist of:

A. Pre-employment screening to determine fitness for duty, ability to wear protective equipment, and baseline data. The screening will include an occupational and medical history questionnaire and a physical examination (see Appendix).
B. Periodic medical examinations, frequency and content to vary depending on the nature of the work and exposures (see Appendix).
C. Follow-up examinations, when needed.
D. Examination immediately after any injury or overexposure.
E. Examination after notification by a technician of symptoms possibly related to exposure.
F. Termination examination (see Appendix).

IX Emergency Contingency Plan

On-site emergencies can be expected to result from fire, chemical reaction of drum contents or personnel casualty. If an incident occurs necessitating a response to an emergency, the OSC will sound an air horn. The signal is at least 5 short blasts (each of 1 second duration).

Personnel will assemble at the Decontamination Station to receive SCBA, orders to evacuate, or other assignments.

If the weather deteriorates to the point where the OSC believes work should cease, he will sound 1 prolonged blast (1 of 4-6 second duration) to order the crew to cease operations and assemble at the Decontamination Station.

Fire

If a fire emergency occurs, the crew will assemble at the Decontamination Station, on the 5 blast signal.

The OSC will issue his response orders, having already alerted the Fire Department (phone) and the Police Department (phone) to execute the Town Emergency Plan (See Appendix). Fire fighting materials on-site will include:

1. 20 gallons of a Triple F foam
2. 2 large dry chemical extinguishers (mounted on wheels)
3. 5 20-lb. dry chemical extinguishers
4. 4 MSA 401 SCBAs (for F.D. use only - additional to all others).

<u>Drum Leak</u>
Personnel will assemble at the Decontamination Station, on the 5 blast signal. The QSC will issue his cleanup orders, in order that the problem may be controlled and cleaned up rapidly. In addition to materials on-site for the job, the following materials will be on-site:

Lime - 200 lb.
Speedy Dry - 200 lb.
Overpack Drums - 100
Lab Packs - 25
Reconditioned Hazmat Drums - 100.

<u>Personnel Casualty</u>
Personnel will assemble at the Decontamination Station, on the 5 blast signal, except for one man who will remain with the casualty. The OSC will issue orders for first aid assistance to the casualty. If the casualty has sustained an injury which may involve contact with contaminated material, a sample of the material will be taken for immediate analysis.

<u>Severe Casualties</u>
The OSC will contact the Ambulance (phone) for assistance. If the casualty requires transfer to a hospital, the primary hospital will be the Hospital. The OSC will contact the hospital (phone) and inform them of the incident and the nature of the injury. If Hospital is unable to assist (due to other emergencies), the OSC will contact [alternate] Hospital in Other Town (phone) and alert them.

Decontamination Procedure

The Decontamination Procedure shall be used when contact with contaminants is made or when personnel depart the Standby Zone.

1. Personnel scrub boots at the pans provided outside the DECONSTA.
2. Disposable Tyvek clothing and gloves are disposed of into the dirty trash drum provided outside the DECONSTA.
3. Spent cartridges/canisters are disposed of into the drum provided outside the DECONSTA.
4. The DECONSTA is entered for the decontamination of other equipment in the pans provided in the DECONSTA.
5. Use a new set of inner gloves to clean equipment.
6. Dispose of any generated dirty trash in the drum provided in the contamination portion of the trailer.
7. Depart the DECONSTA via the Clean Room, for the issue of new clothing/material, or depart from the Standby Zone.

Appendix 2

Excerpt from a Sampling and Analysis Plan

Closure Activities of Building 64 Rock Island Arsenal – 1996

Prepared for Rock Island Arsenal
by TCT-St. Louis Consulting Engineers, Scientists, and Analytical Services

TABLE OF CONTENTS

Items marked by an asterisk (*) are omitted in this excerpt from a Sampling and Analysis Plan.

Explanation of acronyms used in the following pages:

MS/MSD: MATRIX SPIKE/MATRIX SPIKE DUPLICATE

COR: CORPS OF ENGINEERS REPRESENTATIVE

LIMS: LABORATORY INFORMATION MANAGEMENT SYSTEM

N.O.S.: NOT OTHERWISE SPECIFIED

SECTION 1

1.0 PROJECT DESCRIPTION

This Final Sampling and Analysis Plan (SAP) for Closure Activities at Building 64 at Rock Island Arsenal in Rock Island, Illinois has been prepared for IEPA's [Illinois Environmental Protection Agency] review and approval. The Scope of Work was issued to TCT-St. Louis under Contract No. DACW62-94-D-0001, Delivery Order 0005. The Scope of Work can be found in Appendix 1-1. The SAP addresses the requirements of Task 6 of the Scope of Work. The SAP has been prepared in order to evaluate and determine whether a release of contaminants has occurred due to prior operations at Building 64 and to assess whether clean closure of Building 64 has occurred.

The SAP was prepared in accordance with Chapter 13 of the Closure Plan for Building 64, April 1992, which can be found in Appendix 1-2. Since the content and requirements of the SAP were not specified by IEPA, TCT-St. Louis contacted Mr. Kevin Lesko, Project Manager for IEPA, to acquire guidance for preparing the SAP. Mr. Lesko recommended that TCT follow the Sampling and Analysis Plan discussed in EPA's SW-846 document. Since this document does not contain detailed information about SAP preparation, IEPA suggested that TCT develop our own SAP utilizing the SW-846 document as guidance.

The Draft Sampling and Analysis Plan was submitted by TCT-St. Louis to Rock Island Arsenal and Nashville District, Corps of Engineers for review in June, 1995. The reviewers had no comments at that time.

This Final Sampling and Analysis Plan has been revised to include the results of a site inspection conducted by TCT-St. Louis and Dennis Stack, PE, on November 6-7, 1995. The interior of Building 64 was inspected after initial cleanup activities were complete in order to identify sampling locations and determine whether a release of contaminants from the Building has occurred.

This Final Plan also incorporates the results of a review conducted by TCT-St. Louis and Dennis Stack, PE of the Crack Identification and Repair Report dated December 21, submitted by August Mack. The purpose of that report was to provide a description and photo documentation of the presence, locations, and repairs of all holes, cracks, and surface deterioration in the floors and secondary containment pits of Building 64.

1.1 Site History and Contaminants

RIA is located on an island in the Mississippi River between Iowa and Illinois. Upon recommendation of the Secretary of War, Rock Island Arsenal was approved by Congress as a western armory on July 11, 1862. The 384 hectares (ha) of Rock Island are divided as follows: Rock Island Arsenal, 367 ha; US Army Corps of Engineers, 4 ha; National Cemetery, 11 ha; and the Confederate Cemetery, 1 ha. The 367 ha under the control of RIA are divided into three management land uses: improved land, 80 ha; semi-improved land, 61 ha; and the remaining land encompasses 226 ha which also includes 68 ha out-leased for a golf course. The Project Location Map is shown on Figure 1-1 and the Rock Island Arsenal Facility is shown on Figure 1-2.

Industrial operations at RIA to the Spanish-American War. Early operations conducted at RIA included forging, smithing, tinning, production of gun carriages and powder cases, the manufacture and repair of small arms, operation of harness shops, and loading of ammunition. Later operations included the manufacture of automotive vehicles, demolitiori bombs and bomb racks, modification of aircraft engines, light tanks, recoil mechanisms, machine guns, rocket launchers, breech mechanisms, and Honest John and Nike Hercules launchers. More recent operations have included production of small arms, artillery components, recoil mechanisms, gun mounts, gun carriages, loaders, grenade launchers, and aircraft weapons subsystems. These operations are geared to prototype assistance, limited production orders, and orders for the production of repair parts with a critical national supply status. Current manufacturing operations include facilities for casting, ferrous and nonferrous metals, sheet metal piercing and forming, forging, welding, plating, painting, operation of proof-firing ranges, paint stripping, machining, cleaning, lubricating, and metal surface treatment.

The metal plating shop in Building 64 had facilities for cadmium, copper, chromium, lead, nickel, and zinc plating, as well as phosphating, black oxide processing, anodizing, and acid pickling. Wastewater resulting from these operations may have contained high metal concentrations, acid or alkaline solutions, and presence of cyanide. The contaminants present due to these processes necessitated segregation of waste streams to the extent practical. The use of cyanide was eliminated from the operations at Building 64 by about 1980. The plating operations in Building 64 were finally phased out in 1991 when the new plating shop (Building 212) was accepted as a replacement facility.

The plating operations associated with Building 64 activities have occurred inside the structure except for an external cyanide destruct system located at the northeast corner of the west wing of the building. The cyanide destruct unit was taken out of service prior to 1974. The various plating tanks and other Areas of Concern (AOC) are located over or adjacent to a network of secondary containment concrete floor pits. The primary objective of the Sampling Plan is to demonstrate clean closure of Building 64.

Under the Closure Plan of RIA Building 64, April 1992 Report, Section 3.2.4, 2nd paragraph, it mentions that a Phase-Out of electroplating operations in Building 64 began in the fall of 1988 and continued until March 15, 1991.

The transfer of production operations to Building 212 occurred on a process-by-process basis. A "Hazardous Waste Log" maintained by RIA staff, which documents waste management activities during closure of Building 64, is presented in the Closure Plan of RIA Building #64 Report, Appendix E.

All the process tanks and/or vats along with the secondary containment pits and their appurtenances were identified within Building 64 as Hazardous Waste Management Areas of Concern (AOCs). The items were identified during site assessment visits and/or records reviews conducted by TCT-St. Louis.

Approximately 106 drums containing chromium solutions were stored in the west wing of Building 64, just west of AOC 128. Seven of these drums had leaked due to freezing and expansion of their contents in the unheated building. Spillage was collected using "oil dry" and were categorized as hazardous waste (D002, D006, and/or D007). Stains on the floor were visible near where the leakage had occurred, along with cracks in the floor and deterioration.

These drums were turned over to the Defense Re-Utilization and Marketing Office (DRMO), manifested, and disposed as hazardous wastes. Sixty-eight (68) drums were picked up on April 4, 1991, and thirty-one (31) were picked up on April 9, 1991 by Metropolitan Environmental and transported to Michigan Disposal Inc., located in Belleville, Michigan.

Contents of tanks other than those containing chromium and cadmium solutions (e.g., copper, nickel, acid and alkaline solutions, rinse tanks, and some other miscellaneous liquid wastes) were not recycled. These materials were selectively combined in tanks (AOCs 55 and 57) in late 1989 and early 1990 for neutralization prior to boiling to reduce their volume. After treatment, these miscellaneous liquid wastes were transferred into AOCs 24-28. In September and October 1990, wastes stored in AOCs 24-28 were transferred to 185 55-gallon drums. These drums were turned over to DRMO, manifested as "Hazardous Waste Liquid" and disposed of as hazardous waste. All 185 drums were shipped to Special Wastes Inc., located in Athens, Tennessee.

A variety of other wastes, including hazardous and non-hazardous wastes, were containerized in September and October, 1990, in 50 55-gallon drums without any treatment. Twenty-five (25) of these drums were categorized as hazardous wastes, three (3) were recycled, and nineteen (19) were considered non-hazardous. Three drums were classified as unknown wastes. The 25 hazardous drums were turned over to DRMO, manifested, and disposed of as hazardous waste to Special Waste Inc., Athens, Tennessee in December 1990 and January 1991.

Chemicals found on the premises within Building 64 were collected from January 17, 1991 through February 7, 1991, and placed into 45 55-gallon drums. The DRMO was responsible for the proper disposal of these wastes.

This facility requires a RCRA closure because it was used for treatment and storage of hazardous wastes without a permit. Cleanup activities are currently being conducted within Building 64. This Sampling and Analysis Plan is designed to verify that Building 64 is clean for RCRA closure.

In order to clarify the site specific terms applicable to this plan, the following definitions will apply:

(A) Closure of a hazardous waste (RCRA) facility refers to the action taken to secure the hazardous waste management units of a factory or plant in a manner which will protect human health and the environment in accordance with the closure requirements of 35 Illinois Administrative Code Parts 724 and/or 725.

(B) Partial Closure means the closure of a discrete portion of a hazardous waste facility, such as a surface impoundment, landfill trench, container storage unit or tank while other units or cells remain operating. In this case the discrete portion of the facility is Building 64.

(C) "Clean" Closure refers to RCRA closure which includes the removal of all wastes, waste residues, leachate, liners and soils (including groundwater) contaminated with waste or leachate that pose a present or potential threat to human health or the environment.

(D) A Hazardous Waste Management Unit (HWMU) is a contiguous area of land on or in which hazardous waste is placed, or the largest area in which there is a significant likelihood of mixing hazardous waste constituents in the same area (51 FR 16424, May 2, 1986). Building 64 is the HWMU for which this Sampling Plan is submitted.

(E) Facility refers to the entire Rock Island Arsenal which includes a large number of buildings and support operations of which Building 64 is an individual HWMV.

(F) Areas of Concern (AOCs) refer to the individual vats, spills, containment pits and other specific items identified during site investigations to be potentially hazardous. These items were inventoried and were presented in Section 3 of the Closure Plan dated April 1992. Figure 4-1 in this document shows AOCs. When taken as a group, the AOCs represent the entire HWMU known as Building 64.

(G) Project File Number (PFN) is a number which has been assigned to each reference document utilized in the preparation of the Closure Plan and this Sampling Plan document. A list of the reference documents for this Plan is presented in Appendix 1-3.

SECTION 2

2.0 PROJECT ORGANIZATION AND RESPONSIBILITIES

...The various quality assurance and management responsibilities of key project personnel are defined below. [Note: Only one job responsibility description is included in this example.]

2.6 Site Manager

The Project Manager will be supported by the RI field team (Site Manager). He is responsible for leading and coordinating the day-to-day activities of the various resource specialists under his supervision. The Site Manager is a highly experienced environmental professional and will report directly to the Project Manager. Depending on the task requirements, one of the following individuals will be the designated Site Manager for a given task: Paul Shetley, Nancy Dickens, Steve Handley, or Mark Sievers. An experienced geologist will be designated Site Manager for all tasks pertaining to borehole excavation and collection of geologic or hydrogeologic data. Specific Site Manager responsibilities include:

—Provision of day-to-day coordination with the Project Manager on technical issues in specific areas of expertise;

—Development and implementation of field-related work plans, assurance of schedule compliance, and adherence to management-developed study requirements;

—Coordination and management of field staff including sampling, drilling, and field laboratory staff;

—Implementation of QC for technical data provided by the field staff including field measurement data:

—Adherence to work schedules provided by the Project Manager;

—Authorship, review, and approval of text and graphics required for field team efforts;

—Coordination and oversight of technical efforts of subcontractors assisting the field team;

—Identification of problems at the field team level, discussion of resolutions with the Site Manager, and provision of communication between team and upper Management;

—Participation in the preparation of the final report;

—Communication with current site owner concerning scheduled activities;

—Identification of sampling and drilling locations as specified in Project Sampling/Analysis Plan;

—Requests to the laboratory to supply sample containers for soil, sediment, and water samples;

—Notify the designated QA laboratory concerning sample container orders for split analysis prior to sampling and shipment of splits;

—Supervision of the collection, labeling, preservation, packing, chain-of-custody, documentation, and proper transportation of soil, sediment, air and water samples, equipment blanks, and duplicates from the site to TCT-St. Louis' laboratory;

—Supervision of all aspects of drilling and sampling activities;

—Maintenance of a daily written log and photo documentation of all sampling and drilling activities:

—Assistance to the Project Safety and Health Coordinator in implementation of the Health and Safety Plan, and;

—Notification of the Project Manager and Corps of Engineers Representative (COR - Tom McGee, USACE, Nashville District, 615-736-7123) of any problems encountered during field activities.

SECTION 3

3.0 SCOPE AND OBJECTIVES

3.1 Scope of Work Activities

Cleanup Sampling Verification Activities are scheduled to begin in the spring of 1996. Activities include:

3.1.1 Source Water Characterization

Two sets of treated potable surface water samples have been collected from Building 50 at Rock Island Arsenal to characterize the water quality and assess its suitability for use during drilling and equipment decontamination. The water supply has been approved for use in other environmental investigations being performed at RIA under contract to Nashville District for submittal to IEPA.

3.1.2 Background Surface and Subsurface Soils

The Sampling and Analysis Plan assumes that no contamination release has occurred from Building 64 into the environment. No background sampling will be performed since background samples are currently being analyzed as part of a USACE-Nashville District remedial investigation already ongoing at RIA. This work is being performed by TCT-St. Louis at the Old Landfill location. Background samples have been analyzed for 23 TAL metals plus cyanide, TCL volatile and semi-volatile organics, nitroaromatics, and pesticides/PCBs. Results of these background sample analyses will be included with Building 64 results when field activities are completed.

3.1.3 Surface and Subsurface Soils

Boreholes will be drilled and soil samples will be collected beneath floors, sumps, trenches, and secondary contaminant pits at Building 64. One borehole will be located just outside of Building 64 in the secondary containment pit of the former cyanide destruction unit. Sample locations were determined after a visual site inspection was conducted upon completion of Cleanup Activities of Building 64 in accordance with Task 5 of the Scope of Work as per Closure Plan Section 13 .3.

3.1.4 Investigation-Derived Waste Characterization

Characterization of investigation-derived wastes, including volatile organic vapor meter sniffing, examination by the Site Geologist, and laboratory analyses, if necessary, of representative samples of soil borings and decon water will be conducted to determine the appropriate method(s) of IDW disposition/disposal.

3.1.5 Laboratory Analyses

Soil samples collected from boreholes beneath Building 64 will be analyzed for cadmium, chromium, copper, nickel, zinc, cyanide, and VOAs (priority pollutants). These contaminants are the primary contaminants present in electroplating wastes generated at RIA.

Analyses will be performed by TCT-St. Louis Laboratory. TCT-St. Louis is validated by the Corps of Engineers, Missouri River Division for all organic and inorganic analyses at hazardous and toxic waste sites under investigation by the Corps of Engineers.

3.2 Objectives

The primary objective of the Cleanup Sampling activity will be to determine whether a release of contaminants has occurred due to prior operations at Building 64 and to assess whether clean closure of Building 64 has occurred.

Data from the Cleanup Sampling activity will be used to characterize the soils on site; and characterize wastes for disposal.

SECTION 4

4.0 FIELD ACTIVITIES

4.1 Water

No groundwater monitoring activities apply to this project. It is assumed that no release of contaminants have occurred; therefore, no monitoring is planned at this time.

Source Water. The source of water used for steam cleaning will be potable water collected from the Building 50 filtration plant at Rock Island Arsenal. The rationale for using this water supply as a source water is to ensure a noncontaminated water will be used during drilling and sampling operations. TCT has previously collected and analyzed two samples of water from Building 50. Analytical test results can be found in Appendix 4-1. The source of water is the Mississippi River, after treatment at Rock Island Arsenal's water treatment plant.

4.2 Subsurface Soil

4.2.1 Soil and Rock Boring Locations

All sampling, sample custody, and transportation procedures are consistent with guidelines published in USEPA "Test Methods for Evaluating Solid Wastes" (SW-846, Third Edition) and "Engineering and Design, Chemical Data Quality Management for Hazardous Waste Remedial Activities," October 1990, CER1110-1-263). A field log will be kept which details the field activities for each day.

Sampling sites have been chosen by TCT and Dennis Stack, P.E., based upon a review of the contractor's Crack Identification and Repair Plan and a Visual Site Inspection of Building 64 performed November 6-7, 1995. The locations for sampling soil were selected based upon site geology, historical maps and drawings of the building, and discussions with IEPA personnel.

Locations of the sampling sites are shown in Figure 4-1. There are 16 sampling sites (15 inside and 1 outside Building 64). The AOCs previously identified within Building 64 are noted in Tables 4-1 through 4-5 and shown in Figure 4-2. The primary objective of the sampling will be to determine whether a release of contaminants has occurred due to prior operations at Building 64 and to assess whether clean closure of Building 64 has occurred.

4.2.2 Rationale for Selection of Sampling Locations

During 1995 Building 64 has undergone environmental cleanup in accordance with Rock Island Arsenal's specifications. TCT-St. Louis has provided Title 1 oversite services for this cleanup. Equipment, tanks, debris, and hazardous and non-hazardous wastes have been removed from Building 64 in accordance with project specifications. The cleanup has progressed to the point where sampling locations can be selected to determine if a release from the building has occurred.

The following criteria were used to select sampling locations to determine if a release from Building 64 has occurred:

1. Samples will be collected at locations where bare dirt was found at the bottom of sumps, pits, or trenches (e.g., Sample Locations Nos. SB-7, SB-8, SB-13, SB-14, SB-15, and SB-16.)

2. Samples will be collected from beneath sumps located in secondary containment pits (eg. Sample Locations Nos. SB-1, SB-5, SB-9, and SB-10.)

3. Samples will be collected from beneath locations where cracks or surface deterioration existed in the floor of secondary containment pits (eg. Sample Locations Nos. SB-2, SB3, SB-11, and SB-12.)

4. A sample will be collected from beneath the floor of the secondary containment pit for in-process waste storage. (Sample Location No. SB-6.)

5. A sample will be collected beneath the floor of the west wing of the building where cracks or surface deterioration was present in the vicinity of a chromium solution storage area. (Sampling Location No. SB-4.)

The 16 sites (SB-1 through SB-16) were selected to assess the presence or absence of contamination beneath the secondary containment pits and/or trenches and cracks in the floor throughout the entire 34,000+ ft^2 enclosure and the adjacent secondary containment pit of the former cyanide destruction unit.

The Sampling Plan assumes that no contamination release has occurred within Building 64. No background sampling will be performed since background samples will be defined from another USACE-Nashville District sampling project already ongoing at RIA related to the Old Landfill. Results of these background samples will be included with the results of Building 64 samples after field activities are completed.

Information concerning each sampling location is summarized in Table 4-6.

4.2.3 Sample Collection and Methodology

Based upon evaluation of soil profile data from one soil boring located directly in front (south) of Building 64, the depth to bedrock is approximately 10 feet. Several of the sump pits are approximately 7-11 feet below the grade of the concrete flooring. If leakage has occurred, it has most likely occurred at the lowest points in Building 64.

At six sampling locations in pits or sumps where bare soil is present (SB-7, 8, 13, 14, 15 and 16), one sample set will be collected from the exposed soil at the surface. At these six locations where bare soil is present, hand or machine augering will be performed to the soil/bedrock interface, or a maximum of ten feet below the exposed surface if bedrock is not encountered, to collect a second set of samples.

At three sampling locations (SB-2, 4, and 6), cracks are present in floors or surfaces at or near the elevation of the floor of the building. At these locations, a coring machine or drill rig will be used to penetrate the concrete surface. Samples will be collected by hand auger or split spoons from the soil immediately beneath the concrete/soil interface.

At two sampling locations (SB-11 and 12), cracks were present in the bottom of a trench approximately 2 feet 4 inches wide and 3 feet 6 inches deep. These cracks have been repaired to allow wet decontamination of the surfaces. At these locations, the concrete trench will be cored to allow collection of one set of soil samples at each location at a depth immediately below the concrete/soil interface

Four sampling locations (SB-1, 5, 9, and 10) are sumps associated with secondary containment pits. These concrete sumps are approximately 2 to 3 feet deep below the floor of the secondary containment pit. At these locations the bottom of the sump will be cored and one set of samples will be collected from the soil immediately below the concrete bottom of the sump.

At one sampling location (SB-3) a deteriorated expansion joint is present in the middle of a secondary containment pit, approximately four feet below the elevation of the Building. At this location, a coring machine or drill rig will be used to penetrate the floor of the containment pit, and one set of samples will be collected from the soil immediately below the concrete floor.

At selected locations sufficient sample will be collected in sufficient quantities to provide for duplicate or MS/MSD samples.

Since trenches and sumps have already been decontaminated, this will reduce the potential for any cross-contamination and allow for the clearing of the interior debris in order to facilitate the effectiveness of the sampling team to approach its desired target locations.

The specific location of the cyanide destruction pit boring will be identified in consultation with the COR in order to avoid an underground sewer pipe which is present. All soil samples will be placed into a clean stainless steel pan, mixed, and placed into appropriate sample containers. The samples for volatile organics will be collected first prior to mixing.

Borings through some of the pits may be accomplished using a concrete boring machine. After concrete is penetrated, the soil samples will be collected using a bucket auger and a stainless steel split spoon in the interval between the bottom of the concrete and bedrock. In the event that the concrete floor of any pit is in contact with bedrock, a 6-inch core of bedrock will be taken by use of the concrete boring machine.

Soil samples will be analyzed for cadmium, chromium, copper, nickel, zinc, cyanide, and VOAs (priority pollutants). See Table 4-7 for the listing of samples, methods, analyses, etc. for the soils. These analytes chosen are the primary contaminants present in electroplating wastes previously generated at RIA. After each boring is complete, it will be sealed from the bottom to the surface with grout.

Investigative-derived waste (IDW) sampling information is described in Section 7 of this Plan. TCT will collect representative samples from the IDW containers to assess waste characteristics for disposal as potentially hazardous waste or for discharge of liquid wastes to publicly-owned treatment works.

4.2.4 Background, QA/QC. and Blank Samples and Frequency

The proposed Sampling Plan assumes that no contamination release has occurred within Building 64. No background sampling will be performed since background samples will be defined from another USACE-Nashville sampling project already ongoing at RIA (e.g., the Old Landfill). In that project approximately 20 surface and subsurface samples were collected and analyzed for a variety of parameters including volatile organics and TAL metals and cyanide. Results of these background samples will be included with the results of Building 64 samples after field activities are completed. In addition, TCT-St. Louis will collect two field duplicate samples, one rinsate sample, five travel blanks, and MS/MSD samples for quality control.

4.2.5 Drilling Methods and Equipment

Each drilling site should be inspected and approved as safe for drilling by the on-site geologist and site safety officer. Prior to placement of the rig on-site, RIA's Public Works Department and the RIA Contracting Officer's Representative will be contacted to insure each drilling location is clear of utility lines. All personnel participating in the on-site sampling program should have documentation of OSHA-required 40-hour training at hazardous waste sites, as well as CPR and first aid training and respirator fit-test documentation. After the rig has been positioned on-site, all safety criteria should be checked. When continuous sampling is performed, the continuous sampler will be removed from the augers and handed to the on-site geologist for sample extraction and classification. If required, another clean continuous sampler should be inserted into the drill stem and the boring should continue. After each sample has been logged, the continuous sampler will be washed with an Alconox soap solution & rinsed with DI water.

Cleaning procedures are as follows:

I. Cleaning Equipment Between Boreholes

Drilling equipment should be cleaned prior to commencement of field exploration activities and between individual explorations. The drilling equipment to be cleaned will include, but not be limited to:

— Casing
— Drill rods
— Drill bits
— Augers
— Tamping hammers
— Measuring tapes
— Soil Sampling devices
— Water lines and hoses

The drill rig and drilling equipment such as pumps, jacks, collars, couplings, and hoses which do not enter the borehole are not required to be cleaned between borings.

II. Cleaning of Sampling Devices During Borehole Advancement

Sampling devices used to advance boreholes and collect soil samples for physical characterization should be cleaned after each soil sample is obtained and before subsequent samples are obtained. Specifically, sampling devices to be cleaned include hand augers, split spoons and thin-walled shelby tubes. A minimum of three split spoon sampling devices shall be available for use during the advancement of each boring. Split spoons should be cleaned after each soil sample is obtained. The samples collected from the borings will be stored and labeled for testing. Classification of all materials and all drilling operations will be recorded and logged.

The soil boring cuttings will be containerized in 55-gallon drums during drilling. Prior to placement in the drums, the cuttings will be placed on plastic sheets laid out across the floor next to the borehole until drilling is complete. Disposal of the drums and their contents of cuttings and plastic will be accomplished in accordance with Section 7.

During drilling activities, once rock beneath the floor of Building 64 is encountered, drilling will cease. An exception would be if concrete flooring of any pit is in contact with bedrock, a 6-inch core of bedrock will be taken by use of the concrete boring machine.

During floor coring and drilling activities, dust will be minimized and controlled and coring and decontamination fluids will be containerized.

4.2.6 Boring Logs

Logs detailing drilling practices shall be maintained in a field notebook and recorded on field boring log forms. The logs shall be prepared by a qualified Geologist or Geotechnical Engineer present during all drilling. Information provided in the logs shall include, but not limited to the following:

— Thickness of each stratum, including thin lenses and layers
— Identification of the material of which each stratum is composed according to the Unified Soil Classification System, or standard rock nomenclature, as applicable
— Depth interval from which each sample was taken
— Depth at which hole diameter (bit sizes) change
— Depth at which groundwater is first encountered, if encountered
— Depth to the static water level and changes in static water level with well depth
— Type of drilling equipment
— Type and size of samplers used
— Depth of grout
— Borehole diameter
— Any sealing off of water-bearing strata (if applicable)
— Drilling dates
— Manufacturer and quantities of all materials used
— Evidence of contamination, i.e., odors, photo ionization detector measurements, staining, etc.

4.2.7 Field Measurement Procedures and Criteria

All sampling and field measurement procedures are consistent with guidelines published in USEPA "Test Methods for Evaluating Solid Wastes (SW-846, Third Edition) and "Requirements for the Preparation of Sampling and Analysis Plans EM200-1-3, 1 September 1994". A field log will be kept which details the field activities for each day.

Field screening of oxygen levels, lower explosive limits, hydrogen sulfide, and organic vapors will be performed during all sampling activities. Data from field monitoring will be used to monitor possible hazardous conditions in the field and to provide general screening of potential contamination for sample collection. Equipment will be inspected on a daily basis and calibrated according to the instrument manufacturer's specifications.

4.2.8 Laboratory Analyses

Soil samples collected from Building 64 will be analyzed for cadmium, chromium, copper, nickel, zinc, cyanide, and VOAs (priority pollutants).

The laboratory selected to conduct the analyses listed above is TCT-St. Louis. They will provide detailed information concerning their Quality Control and data handling procedures, including the number and types of lab duplicates, lab blanks, detection limits, holding times, data reporting formats, etc. in accordance with SW-846. TCT-St. Louis' laboratory has been certified by the Missouri River Division Corps of Engineers to perform analyses for all sites under investigation by the Corps of Engineers.

4.2.9 Sampling Containers and Preservation Techniques

Sampling containers to be used will be in accordance with standard procedures summarized in Appendices 4-2 and 4-3. Once samples have been placed in the appropriate containers, custody sheets will be completed and the containers will be placed directly into a cooler, iced, and shipped to the appropriate lab for analysis. Prior to collection of each sample, sampling equipment (split spoon, stainless steel pan, stainless steel spoon) will be washed with Alconox and deionized (DI) water and rinsed with DI water.

4.2.10 Decontamination Procedures

All equipment will be cleaned before arrival at the work site. Decontamination of the drilling equipment will be accomplished using a portable steam cleaner. All sampling equipment will be decontaminated daily prior to collection of samples. Equipment will be thoroughly washed with Alconox and tap water and rinsed with deionized water (DI). Deionized water will be obtained from TCT-St. Louis and shipped to the field. All water generated from decontamination activities will be placed in 55-gallon drums (DOT 17H) or DOT-approved 34H 15-gallon drums and managed according to Section 7. Decontamination equipment will include plastic sheeting, buckets, brushes, DI sprayer, Alconox, tap water, and DI water. The drill rig and associated support equipment should be field decontaminated following completion of the final boring and prior to demobilization and departure from the vicinity of Building 64.

SECTION 5

5.0 SAMPLE CHAIN-OF-CUSTODY/ DOCUMENTATION

5.1 Field Log Book

The purpose of the field sampling log is to enable the Site Manager to completely reconstruct the sampling events without relying on his memory. Information recorded on other documents, such as boring logs, well construction logs, air monitoring calibration or monitoring forms, or well development forms will not be repeated in log books except in summary form, to avoid transcription errors. Logbooks will be maintained in the possession of the site manager and the Site Safety and Health Coordinator, or in a secure place such as the van glove compartment, during the field work. Following site activities, the logbooks will become a part of the final evidence file.

The following are some suggested topics to be included in the logbook:
— Name and title of author, date, and time of entry.
— Purpose of sample activity
— Name and address of field contact.
— Names and responsibilities of field crew members.
— Names and titles of any site visitors.
— Type of waste, suspected waste concentration if known, and sample matrix.
— Sample collection method.
— Number and volume of sample(s) taken.
— Location, description, and log of photographs (if taken) of the sampling points.
— References for all maps and photographs of the sampling site(s).
— Information concerning sampling changes, scheduling modifications, and change orders.
— Information concerning drilling decisions, not shown on the drill log.
— Information concerning access agreements.

Details of the sampling location (dimensional sketches of sampling locations may be appropriate).
— Date and time of collection.
— Field observations.
— Sample identification number(s).
— Documentation of procedures for preparation of reagents or supplies that become an integral part of the sample (e.g., filters and absorbing reagents).
— Information from containers, labels of reagents used, de-ionized water used for blanks, etc .
— Sampling methodology, including distinction between grab and composite samples.
— Sample preservation.
— Sample distribution and transportation (e.g., name of the laboratory and courier).

All sample documentation such as:
— Bottle lot numbers as received from repository
— Chain-of-custody records numbers
— Decontamination procedures.

All documentation for investigation-derived wastes such as:
— Contents and approximate volume of waste
— Disposal method
— Type and predicted level of contamination

Summary of daily tasks (including costs) and documentation on any cost or scope of work changes required by field conditions.
— Signature and date (entered by personnel responsible for observations).

5.2 Photographs

A photo documentation log will be maintained. Information will be entered in the log for each photo taken. The information will include:
1. Date/time
2. Direction of photo
3. Subject/description
4. Notes
5. Photographer signature
6. Sketch of location
7. Sequential number of photograph and roll number

Where possible, objects of known size will be included in each, if possible, for scale. Unless otherwise specified, all photos will be 4" by 6" color and will be mounted on photo documentation log sheets. The location of each photo will be cross referenced to a view location grid plan.

Subjects to be photographed will include:
1. Each soil sampling location, showing its location in reference to known landmarks
2. Any evidence of contamination found in samples
3. Photos exhibiting personal protective equipment used
4. IDW storage location after completion of waste generation, and/or condition of locations where IDW is replaced into boreholes or allowed to seep into the ground
5. Any damage to property which may be present before, during, or after on-site activities
6. Equipment used on site, e.g., drill rigs
7. Any other noteworthy objects

5.3 Sample Numbering System

The following sample numbering system will be used to identify each sample for chemical analysis. The purpose of this numbering system is to provide a tracking system for retrieval of data on each sample. Sample identification numbers will be used on sample labels, field sheets, sample tracking matrix forms, chain-of-custody records and all other applicable documentation used during the sampling activity.

A listing of all sample identification numbers will be maintained in the field logbook. QC samples will be disguised using the same numbering system selected for the filed samples. An example of the field numbering system is:

Project Code	Year	Sample Type	Site No.	Sample No.
Bldg. 64	96	SB	1	001

The sample identification number for sediment/soil, surface water, and groundwater samples begins with the project code, followed by the year, the sample type, the site number, and the sample number. The project code in this case is Building 64, which stands for Rock Island Arsenal Building 64. The year is included to distinguish this sampling series from past sampling activities. The sample type describes the matrix of the sample, such as SS for a surface soil, SB for subsurface soil, SD for sediment, GW for groundwater, SW for surface water and S for source water. The site number is the number that will identify the particular area on RIA where sampling activities occurred. Finally, the sample number is added to the sample identification number. The sample identification number for subsurface soil samples begins with the project code through the sample number. Following the sample number, the final portion of this sample identification number will be the interval from which the sample was obtained. Therefore, a subsurface soil sample collected from an interval of 3 to 5 feet will be identified in the following manner:

Project Code	Year	Sample Type	Site No.	Sample No.	Interval
Bldg. 64	96	SB	8	002	3-5

5.4 Sample Documentation

5.4.1 Sample Labels

Each sample is identified by affixing a pressure sensitive gummed label and laboratory number on the container(s). The sample collection technique (grab or composite), source of sample, preservative used, the collector(s) signature, date and time of collection, and analyses required will be identified in indelible ink and the label will be covered with clear waterproof tape.
Examples of sample identification labels used by TCT-St. Louis were illustrated in Exhibit 5-2.

The Site Manager is responsible for seeing that samples are properly labeled.

5.4.2 Sample Field Sheets and/or Log Book

A listing of all sample identification numbers will be maintained in the field logbook.

5.4.3 Chain-of-Custody Records

An example of a chain-of-custody transfer form was presented in Exhibit 5-1. The sample containers will be placed in a shipping container (cooler) along with the chain-of-custody record form, pertinent field records, analysis requests, and the name of the shipper. A copy of these forms will be retained by the field crew and transferred to the project files upon completion of the sampling. The shipping bill will be maintained in the project file with the chain-of-custody.

When transferring the custody of the samples, the custodian will sign and record the date and time on the chain-of-custody record. Custody transfers will account for each individual sample, although samples may be transferred as a group. Every person who takes custody will fill in the appropriate section of a chain-of-custody record.

The Site Manager is responsible for seeing that samples are properly packaged and dispatched to the laboratory for analysis. This responsibility includes filling out, dating, and signing the appropriate portion of the chain-of-custody record.

5.4.4 Receipt for Sample Forms

Upon delivery of a sample shipment from the field, the process of sample receipt and identification begins. Prior to opening any shipping container, all custody seals, if present, are to be checked to assure that they are intact. If the field chain-of-custody papers are shipped outside of the shipping containers, custody of the shipment is to be assumed at this time by signature of the TCT-St. Louis sample custodian (or designate). For a shipment that is known to be radioactive, the unopened container is to be scanned for radiation using a beta/gamma and an alpha Geiger-Mueller survey meter. The shipping container is then opened in a well ventilated area, and the condition of the contents surveyed. The temperature conditions inside the shipping container are noted and documented on the custody papers. If the samples require $4° \pm 2°C$ for preservation and ice or the equivalent is not present when the shipping container is opened, the sample custodian will promptly notify the project manager of this shipment irregularity. Field custody papers shipped inside the shipping containers are to be signed and custody assumed at this time. Custody seal numbers (if any) on the shipping containers are to be compared against those recorded on the custody papers; the condition of the seals, the shipping containers, and the shipping contents are to be recorded on the custody papers.

The contents of the shipment are to be removed from the shipping containers and checked against the custody papers and other pertinent shipping documents. Discrepancies in shipment or broken samples are to be reported immediately to the laboratory project manager, who in turn is to contact the client for clarification or advice. The custodian will complete a cooler receipt form for each cooler received. An example of a cooler receipt form was shown in Exhibit 5-3.

5.5 Documentation Procedures

Use of chain-of-custody transfer forms, sample identification labels, and sample receipt forms have been described above.

Stringent sample custody procedures are maintained for all samples received for analysis. The Supervisor of the receiving group is the designated sample custodian. In his/her absence, the laboratory operations director will designate an alternate sample custodian. The second shift supervisor of the inorganic laboratory is the designated sample custodian during the second shift. Due to the evidentiary nature of sample analysis, a documented system for tracking samples from collection through analysis to reporting and disposal is critical.

All samples received at TCT-St. Louis are logged-in by the computerized laboratory information management system (LIMS), which assigns a unique laboratory sample ID number. This sample ID number, which corresponds to the client site or sample ID, is used to track the sample through the laboratory and the data for reporting. Information provided in the LIMS log-in include:
— Client Sample ID
— Client
— Project number
— TCT-St. Louis Laboratory Sample ID number (automatically assigned)
— Date of receipt
— Sample collection date
— Date/Time of log-in
— List of requested analyses
— Required completion date
— Sample matrix (solid, liquid, etc.)

This information is networked throughout the laboratories and is available in various reporting fashions that link analyst and analysis with holding times, and is used to facilitate scheduling and sample loads. Copies of the LIMS log-in report are generated daily for inclusion in the master sample logbook. Sample labels containing the above information are generated by the LIMS computer and placed on each sample. The samples are then placed in appropriate storage.

In addition, the LIMS computer generates the in-house chain-of-custody form and will generate user-defined back-log reports. An example of the in-house TCT-St. Louis chain-of-custody form was presented as Exhibit 5-4. Copies of the work request form are distributed to all appropriate analytical departments as well as to the laboratory director, project manager, and the data management/document control personnel.

The original TCT-St. Louis custody form is kept on file with the sample custodian, and is used at all times when samples are signed in and signed out for analyses. Under this procedure, when an analysis is scheduled, the analyst will request the samples of interest from the sample custodian. The sample custodian will, in turn, remove the samples from storage and sign over custody of the samples to the analyst. Both individuals will place their signatures, along with the date and time of transfer, on the original custody forms maintained by the sample custodian.

At that moment, the analyst accepts custody of the samples. When the analysis is complete, the analyst will return the samples and their custody to the sample custodian with signature, date, and time on the custody form.

Sample digestates and extracts also require a chain-of-custody. At TCT-St. Louis, their custody begins at the moment they are prepared and is documented on the sample extract and sample digestion logs (Exhibits 5-6 and 5-7). A particular analyst or the laboratory section manager maintains custody of the sample extracts and digestants until they are transferred to another analyst or section for analysis. The data generated by the analysis is maintained by the analyst until it is transferred to the laboratory section manager and ultimately to the data custodian, who keeps the original data on file for permanent storage.

5.6 Corrections to Documentation

Before sample work request forms are distributed to the laboratory, the sample custodian delivers the first copy of the form to the project manager who is responsible for the samples which have been received, along with the field/client chain-of-custody forms. The project manager is responsible for checking the log-in information for errors and comparing the chain-of-custody to the work contracted for with the client. The project manager specifically looks for:
— Proper routing list for the work request form
— Ambiguity on the field chain-of-custody form
— Proper preservation of samples for the parameters requested
— Proper turnaround time requested on the Custody Transfer Record
— Correct list of analytes given for each sample received
— Discrepancies between the original agreement with the client and the samples received

Any questions or ambiguities arising from the client chain-of-custody must be resolved by the project manager with the client before the work request form is signed and dated by the project manager. In order to prevent any misunderstandings about the QC or methods required, the project manager may add comments to the allotted space on the form or attach a comment/instruction sheet to the QC specifications form shown in Exhibit 5-5.

Samples are not released to the laboratory or finalized on the LIMS work lists until the project manager has completed the actions described above. It is incumbent on the project manager to designate an alternate to perform this function if the project manager is not available when samples arrive that are associated with an assigned project.

SECTION 6

6.0 SAMPLE PACKAGING AND SHIPPING

6.1 Samples for Chemical Analysis

Unless otherwise specified, samples for chemical analysis will be shipped by TCT-St. Louis personnel to their own laboratory. The shipping address is as follows:

TCT-St. Louis
Attn..: Jim Shetley, Sample Custodian
1908 Innerbelt Business Center Drive
St. Louis, MO 63114-5700
800-377-7344
314-426-0880

6.2 QA Samples

QA samples are not required to be sent to the Corps lab for this project.

6.3 Sample Packaging

Environmental samples will be packaged for shipment as follows:

1. Sample container is adequately identified with sample labels. Sample labels are placed on samples at this time if required.

2. All bottles, except the volatile organic analysis (VOA) vials, are taped shut with electrical tape (or other tape as appropriate).

3. Each sample bottle is placed in a separate plastic bag, which is then sealed. For water samples, each VOA vial is wrapped in a paper towel, and the two vials are placed in one bag. If a trip blank is submitted, it should be wrapped and placed in the bag with the two VOA vials. As much as possible is squeezed from the bag before sealing. Bags may be sealed with evidence tape or custody seals for additional security.

4. A picnic cooler (such as a Coleman or other sturdy cooler) is typically used as a shipping container. In preparation for shipping samples, the drain plug is taped shut from the inside and outside, and a large plastic bag is used as a liner for the cooler. Approximately 3 inches of inert packing material, such as asbestos-free vermiculite, perlite, or Styrofoam beads, is placed in the bottom of the liner. Other commercially available shipping containers may be used. However, the use of such containers (cardboard or fiber boxes complete with separators and preservatives) should be specified in the sampling plan and pre-approved.

5. The bottles are placed upright in the lined picnic cooler in such a way that they do not touch and will not touch during shipment. Cardboard separators may be placed between the bottles at the discretion of the shipper.

6. All samples should be shipped to the laboratory on ice and chilled to 4°C except for the following types of samples, which do not require shipment with ice:
   — Low and medium-concentration water and liquid matrix samples for metals analyses.

   However, because prior knowledge of analyte concentrations is required to apply this exception, it may be prudent to maintain the cooling requirement.

7. Additional inert packing material is placed in the cooler to partially cover the sample bottles (more than halfway). If samples are required to be shipped to the laboratory with ice, ice in double bags must be placed around, among, and on top of the sample bottles. If chemical ice is used, it should be placed in a double plastic bag. The cooler should then be filled with inert packing material and the liner taped shut.

8. The cooler is closed and taped shut with strapping tape (filament-type).

9. At least two signed custody seals are placed on the cooler, one on the front and one on the side. Additional seals may be used if the sampler or shipper thinks more seals are necessary.

10. The cooler is handed over to the overnight carrier. A standard air bill is necessary for shipping environmental samples. The shipper should be aware of carrier weight or other policy limitations.

Hazardous samples are defined as those which are typically highly contaminated, such as oils, sludge, discarded products, and other materials. Contaminant levels in these types of samples are normally greater than 10 ppm. Hazardous samples must be packaged as follows:

1. Sample container is adequately identified with samples labels. Sample tags are placed on samples at this time if required.

2. All bottles, except the VOA vials, are taped closed with electrical tape (or other tape as appropriate). Evidence tape or custody seals may be used for additional security.

3. Each sample bottle is placed in a plastic bag, and the bag is sealed. As much air as possible is squeezed from the bags before sealing. Evidence tape or custody seals may be used to seal the bags for additional security.

4. Each bottle is placed upright in a separate paint can, the paint can is filled with vermiculite, and the lid is

fixed to the can. The lid must be sealed with metal clips or with filament or evidence tape; if clips are used, the manufacturer typically recommends six clips.

5. Arrows are placed on the can to indicate which end is up.

6. The outside of each can must contain the proper Department of Transportation (DOT) shipping name and identification number for the sample. The information may be placed on stickers or printed legibly. A liquid sample of an uncertain nature is shipped as a flammable liquid with the shipping name "FLAMMABLE LIQUID, N.O.S." and the identification number "UN1993". A solid sample of uncertain nature is shipped as a flammable solid with the shipping name "FLAMMABLE SOLID, N.O.S." and the identification number "UN1325." If the nature of the sample is known, 40 CFR 171-177 is consulted to determine the proper labeling and packaging requirements.

7. The cans are placed upright in a cooler that has had its drain plug taped shut inside and out, and has been lined with a garbage bag. Vermiculite is placed on the bottom. Two sizes of paint cans are used: half-gallon and gallon. The half-gallon paint cans can be stored on top of each other; however, the gallon cans are too tall to stack.

8. All hazardous samples should be shipped to the laboratory on ice and chilled to 4°C, except for the following samples which do not require shipment with ice:
   - Medium concentration water and liquid matrix samples for metals analysis.
   - Medium concentration soil and sediment matrix samples for B/N/A, PCBs, and pesticide analyses.

   However, because prior knowledge of the analyte concentration is required to apply this exception, it may be prudent to maintain the cooling requirement.

9. Additional inert packing material is placed in the cooler to partially cover the sample bottles. If samples are required to be shipped to the laboratory with ice, bags of ice must be placed around the cans. The cooler must be filled with packing material and the liner taped shut.

10. The paperwork going to the laboratory is placed inside a plastic bag and taped to the inside of the cooler lid. A copy of the chain-of-custody form should be included in the paperwork sent to the laboratory. The sampler keeps one copy of the chain-of-custody form. The laboratory should be notified if a parallel sample is being sent to another laboratory for dioxin analysis, or if the sample is suspected of containing any substance for which laboratory personnel should take safety precautions.

11. The cooler is closed and sealed with strapping tape. At least two custody seals are placed on the outside of the cooler (one on the front and one on the back). More custody seals may be used at the discretion of the sampler.

12. The following markings are placed on the top of the cooler:
   - Proper shipping name: if unknown, use
   - DOT identification number (49 CFR 172.301).
   - Shipper's or consignee's name and address (49 CFR-172.306).
   - "This End Up" legibly written if shipment contains liquid hazardous materials (49 CFR 172.312).

13. The following labels are required on top of the cooler (49CFR 172.406(e)):
   - Appropriate hazard class label (placed next to the proper shipping name).
   - "Cargo Aircraft Only" (if applicable as identified in 49 CFR 172.101).

14. An arrow symbol(s) indicating "This Way Up" should be placed on the cooler in addition to the markings and labels described above.

15. Restricted-article air bills are used for shipment. The "Shipper Certification for Restricted Articles" section is filled out as follows for flammable solid or a flammable liquid:
   - Number of packages or number of coolers.
   - Proper shipping name: if unknown, use
     - Flammable solid, N.O.S., or
     - Flammable liquid, N.O.S.
   - Classification; if unknown, use
     - Flammable solid, N.O.S., or
     - Flammable liquid, N.O.S.
   - Identification number; if unknown, use
     - UN1325 (for flammable solids), or
     - UN1993 (for flammable liquids).
   - Net quantity per package or amount of substance in each cooler.
   - Radioactive materials section (Leave blank).
   - Passenger or cargo aircraft. (Cross off the non-applicable. Up to 25 lb of flammable solid per cooler can be shipped on a passenger or cargo aircraft. Up to 1 qt of flammable liquid per cooler can be shipped on a passenger aircraft, and up to 10 gal of flammable liquid per cooler can be shipped on a cargo aircraft.)
   - Name and title of shipper (printed).
   - An emergency telephone number at which the shipper can be reached within the following 24 to 48 hrs.
   - Shipper's signature.

6.4 Sample Shipment

All samples will be shipped by overnight courier delivery service. In some cases (e.g., at the conclusion of a sampling event), it may be more convenient for samples to be transported directly from the site to the TCT laboratory by TCT personnel in their van. Samples will be delivered to the laboratory within 24 hours of collection.

6.5 Chain-of-Custody

All sample sets will be accompanied by a chain-of-custody record. When transferring possession of samples, the individual receiving the sample should sign, date, and note the time that he/she received the samples on the chain-of-custody record. This chain-of-custody record documents transfer of custody of samples from the field investigator to another person, other laboratories, or other organizational units. Samples will be properly packaged for shipment and delivered or shipped to the designated laboratory for analyses. Shipping containers will be secured by using nylon strapping tape and custody seals. The custody seals will be placed on the container so that it cannot be opened without breaking the seals. The seal will be signed and dated by the field investigator. All samples will be accompanied by the chain-of-custody record. The U.S. Army Corps of Engineers (USACE) tracking number that is used in conjunction with the government QA sample shipment will be written on the QA sample's chain-of-custody record. The original and one copy of the record will be placed in a plastic bag taped to the inside lid of the secured shipping container. One copy of the record will be retained by the field investigator or project leader. The original record will be transmitted to the field investigator or project leader after samples are accepted by the laboratory. This copy will become a part of the project file. Receipts from post offices and air bills should be retained as part of the documentation of the chain-of-custody. The air bill number or registered mail serial number should be recorded in the remarks section of the chain-of-custody record.

6.6 Sample Receipt

The receiving laboratories will document the condition of field samples upon receipt to the laboratory. This enables verification of numerous items including correct sample volumes, preservation applied, cooler temperature, etc; chain of custody completeness and accuracy; and overall packaging techniques. A cooler receipt form is presented in Exhibit 5-3.

SECTION 7

7.0 INVESTIGATION-DERIVED WASTE

Investigation-derived wastes may include drill cuttings, steam cleaning condensate, decontamination fluids, and personal protective equipment. These wastes will be generated at all drilling locations at Building 64.

7.1 Drill Cuttings

All drill cuttings will be continuously logged and inspected by the Site Manager/Qualified Field Geologist. An HNU organic vapor meter will be used to evaluate potential presence of organic contamination. The test will be performed by placing a representative sample of the drill cuttings into a plastic bag and then placing the HNU probe into the headspace of the bag. Drill cuttings will be containerized in suitable DOT-approved 17H 55-gallon steel drums, and filled to no more than 10 inches from the top to prevent rupture during freezing weather. Wastes from different holes will not be segregated. The drums will be moved to a suitable storage area to be provided by RIA. The borehole will be filled with bentonite in accordance with procedures described in Corps Guidance Manual "Monitor Well Design, Installation and Documentation at Hazardous and/or Toxic Waste Sites" EM1110-1-4000, 31 August 1994.

7.2 Steam Cleaning and Equipment Decontamination Water

Steam condensate and equipment decontamination water will be collected at the decontamination station and inspected by the Site Manager. An HNU organic vapor meter will be used to evaluate potential presence of organic contamination. The test will be performed by placing a representative sample of the water into a jar and then placing the HNU probe into the headspace of the jar. The water will be containerized in suitable DOT 17H 55-gallon drums or DOT approved 34H 15-gallon polydrums, allowing sufficient freeboard in case of freezing weather, and moved to a suitable storage area to be provided by the RIA facility.

7.3 Personal Protective Equipment

Personal protective equipment investigation derived waste includes Tyvek and/or Saranax coveralls, disposable gloves, boot covers, and respirator cartridges. Paper towels, plastic sheets, garbage bags, paper cement bags, and other disposable tools or supplies may also fall into this category. In no case will trash or other wastes be allowed to remain at the work site overnight.

Wastes will be segregated into potentially contaminated and non-contaminated streams. Potentially contaminated wastes, such as Tyveks soiled with solid or liquid wastes which were considered contaminated due to presence of elevated organic vapors, odors, stains, free product, or unusual discoloration, will be placed in trash bags at the work site and stored in 55-gallon sealed 17H DOT-approved steel drums during the duration of the project in a suitable storage area to be provided by the RIA facility. At the conclusion of the project these wastes will be transported to TCT's office for proper disposal. If, in the judgment of the Site Manager, the personal protective equipment or other wastes are not contaminated, they will be disposed of on a daily basis in a suitable dumpster to be provided by the RIA facility.

7.4 Waste Sampling and Analysis

Wastes will be stored at a location identified by RIA. Each container will be labeled as to its contents, source(s), and RIA identification number. After field investigations are completed, a representative composite sample will be collected from representative containers for each media, e.g., solids and liquids. Samples will be analyzed as follows: TCLP metal - Methods 1311, 6010A, 7470; TCLP VOA - Methods 1311 8260A; ignitability - Method 1010; reactivity Method 9030, 9010; corrosivity - Methods 9045, 1110; paint filter test - Method 9095 (see Table 7-1). These samples are being collected to assess waste characteristics for disposal as potentially hazardous or special waste.

Investigation-derived liquid wastes will also be analyzed for the parameters listed in Table 7-2.

A composite soil sample will be collected with a stainless steel spoon and bucket. A minimum of three scoopfuls of soil from each drum being sampled will be placed into a stainless steel bucket, thoroughly mixed, and then placed into the appropriate sample containers. Samples for analysis for VOAs will be collected first prior to mixing.

Water samples will be collected by extracting equal aliquot from each drum of water being composited using a glass drum thief.

All analytical results will be furnished to U.S. Army Corps of Engineers to serve as a basis for proper disposal.

7.5 Waste Disposal

All generated waste from this project will be containerized pending results of analyses. Waste containers will be stored at a secure location in Building 64 or elsewhere to be supplied by RIA. Analytical results will be reviewed to assess the optimal method of disposal. TCT-St. Louis will advise the COR concerning the optimal method of disposal at that time. RIA will supply the containers as well as be responsible for waste disposal.

Appendix 3

Table A: Tabulated Values of "t" for Evaluating Solid Wastes*

Degrees of freedom (n–1)[a]	Tabulated "t" value[b]
1	3.078
2	1.886
3	1.638
4	1.533
5	1.476
6	1.440
7	1.415
8	1.397
9	1.393
10	1.372
11	1.363
12	1.356
13	1.350
14	1.345
15	1.341
16	1.337
17	1.333
18	1.330
19	1.328
20	1.325

Degrees of freedom (n–1)[a]	Tabulated "t" value[b]
21	1.323
22	1.321
23	1.319
24	1.318
25	1.316
26	1.315
27	1.314
28	1.313
29	1.311
30	1.310
40	1.303
60	1.296
120	1.289

*More detailed "t" tables are available in statistics books.

Statistical Notes:

[a] Degrees of freedom (DF) are equal to the number of samples (n) collected from a solid waste less one.

[b] Tabulated "t" values are for a two-tailed confidence interval and a probability of 0.20 (the same values are applicable to a one-tailed confidence interval and a probability of 0.10).

Appendix 4

Table B: TCLP Regulatory Thresholds (RT) 40 CFR 261.24

Constituent (mg/L)	RT (mg/L)
Arsenic	5.0
Barium	100.0
Benzene	0.5
Cadmium	1.0
Carbon tetrachloride	0.5
Chlordane	0.03
Chlorobenzene	100.0
Chloroform	6.0
Chromium	5.0
o-Cresol	200.0
m-Cresol	200.0
p-Cresol	200.0
Cresol	200.0
2,4-D	10.0
1,4-Dichlorobenzene	7.5
1,2-Dichloroethane	0.5
1,1-Dichloroethylene	0.7
2,4-Dinitro toluene	0.13
Endrin	0.02
Heptachlor (and its hydroxide)	0.008

Constituent (mg/L)	RT (mg/L)
Hexachlorobenzene	0.13
Hexachloro-1,3-butadiene	0.5
Hexachloroethane	3.0
Lead	5.0
Lindane	0.4
Mercury	0.2
Methoxychlor	10.0
Methyl Ethyl Ketone	200.0
Nitrobenzene	2.0
Pentachlorophenol	100.0
Pyridine	5.0
Selenium	1.0
Silver	5.0
Tetrachloroethylene	0.7
Toxaphene	0.5
Trichloroethylene	0.5
2,4,5-Trichlorophenol	400.0
2,4,6-Trichlorophenol	2.0
2,4,5-TP (Silvex)	1.0
Vinyl Chloride	0.2

Appendix 5

Sampling Plan-Table of Contents

1. Title page
2. Submittal letter (*to Agency*)
3. Concurrence page (*for signatures of affected offices at the remediation site*)
4. Certification page (*for the professional engineer's signature and seal*)
5. Project Description (*for site history and identification of contaminants*)
6. Project Organization and Responsibilities (*for identification of personnel involved*)
7. Scope and Objectives (*for description of the work required to conduct the clean-up*)
8. Field Activities (*for description of groundwater and soil sampling requirements*)
9. Sample Chain-of-Custody/Documentation (*for description of sample handling from field log books to the numbering system to correction of documentation*)
10. Sample Packaging and Shipping (*for description of the origin of samples, packing, and shipment*)
11. Investigation-derived Waste (*for description of the handling of borings and groundwater and waste disposal methods*)
12. Field Report Content (*for description of the field report*)
13. Project Schedule (*for description of the project schedule steps*)
14. List of Tables
15. List of Figures
16. List of Exhibits (*for identification of forms used in the project*)
17. List of Appendices (*for description of documentation included for clarification of the work to be conducted and for the contract written to conduct the project*)

Appendix 6

Sampling Usage Guides

Container abbreviations:

HDPE	High Density Polyethylene Bottles
BR	Boston Round
cwm	Clear Wide Mouth
AJ	Amber Jug
V	Vial

Note: In Sample Volume columns, the first column is for solid waste methods; the second, for water/wastewater methods.

Solid Waste Method	Description	Water/ Wastewater Method	Sample Volume Required		Container Recommended by EPA		To Preserve	Holding Time
SW-846	Parameters	EPA-600/4-79-020	Solid	Liquid	Solid	Liquid		
	Acidity	305.1	.	100 mls		250 ml HDPE	Cool 4°C	14 days
	Alkalinity	310.1	.	100 mls		250 ml HDPE	Cool 4°C	14 days
	Ammonia	350.2	.	500 mls		1 L HDPE	4°C; H_2SO_4 <2	28 days
	BOD	405.1	.	1 liter		2 L HDPE	Cool 4°C	48 hours
	Bromide	320.1	.	100 mls		250 ml HDPE	None	28 days
	BOD (carbonaceous)	405.1	.	1 liter		2 L HDPE	Cool 4°C	48 hours
	COD	410.1	.	50 mls		125 ml HDPE	4°C; H_2SO_4 <2	28 days
9250	Chloride	325.3	.	50 mls	4 oz cwm	125 ml HDPE	None	28 days
	Chlorine (total residual)	330.1	.	200 mls		500 ml HDPE	None	Immediate analysis
	Color	110.1	.	50 mls		125 ml HDPE	Cool 4°C	48 hours

Solid Waste Method	Description	Water/ Wastewater Method	Sample Volume Required		Container Recommended by EPA		To Preserve	Holding Time
SW-846	Parameters	EPA-600/4-79-020	Solid	Liquid	Solid	Liquid		
9010	Cyanide (total and amenable)	335.2, 335.1	10 g	500 mls	1 L BR	1 L HDPE	NaOH to pH 12 .6g ascorbic acid	14 days
	Fluoride	340.2	.	300 mls		500 ml HDPE	None	28 days
	Hardness	130.1	.	100 mls		250 ml HDPE	HNO_3 <2 or H_2SO_4 <2	6 months
9040, 9045	Hydrogen ion (pH)	150.1	100 g	25 mls	4 oz cwm	60 ml HDPE	None	Immediate analysis
	Kjeldahl and organic nitrogen	351.3	.	500 mls		1 L HDPE	4°C; H_2SO_4 <2	28 days
7196	Chromium VI	218.4	10 g	100 mls	8 oz cwm	250 ml HDPE	Cool 4°C	24 hours
7470	Mercury	245.1	10 g	100 mls	8 oz cwm	250 ml HDPE	HNO_3 <2	28 days
7000 series or 6010	Metals (except Chromium VI & Hg)	200 series or 200.7	10g	200mls	16 oz BR	500 ml HDPE	HNO_3 <2	6 months
9200	Nitrate	352.1 or 300	-	100 mls	8 oz cwm	250 ml HDPE	Cool 4°C	48 hours
9200	Nitrate-nitrite	353.3 or 300	-	100 mls	8 oz cwm	250 ml HDPE	4°C; H_2SO_4 <2	28 days
	Nitrite	354.1 or 300	-	50 mls		125 ml HDPE	Cool 4°C	48 hours
9070	Oil and grease	413.1	10 g	1 liter	2 L cwm	no plastic	4°C; H_2SO_4 <2	28 days
9060	Organic carbon	415.1	10 g	25 mls	4 oz cwm	125 ml HDPE	4°C; HCl or H_2SO_4 <2	28 days
	Orthophosphate	365.2	-	50 mls		125 ml HDPE	Filt. immed.; 4°C	48 hours
	Oxygen, dissolved probe	360.1	-	300 mls		No plastic	None	Immediate
	Winkler, D.O.	405.1	-	300 mls		300 ml BOD bottle	Fix on site Store in dark	8 hours
9066	Phenols	420.1	-	500 mls	1 L BR	No plastic	4°C; H_2SO_4 <2	28 days
	Phosphorus (elemental)	365.2	-	50 mls		No plastic	Cool 4°C	48 hours
	Phosphorus (total)	365.2	-	50 mls		125 ml HDPE	4°C; H_2SO_4 <2	28 days
	Residue (total)	160.3	-	100 mls		250 ml HDPE	Cool 4°C	7 days
	Residue (filterable)	160.1	-	100 mls		250 ml HDPE	Cool 4°C	7 days
	Residue (nonfilterable)	160.2	-	100 mls		250 ml HDPE	Cool 4°C	7 days
	Residue (settlable)	160.5	-	1 liter		Imhoff cone	Cool 4°C	48 hours
	Residue (volatile)	160.4	-	100 mls		250 ml HDPE	Cool 4°C	7 days
	Silica	370.1	-	50 mls		125 ml HDPE	Cool 4°C	28 days
9050	Specific Conductance	120.1	100 g	100 mls	8 oz cwm	250 ml HDPE	Cool 4°C	28 days
9035-38	Sulfate	375.3	-	50 mls	4 oz BR	125 ml HDPE	Cool 4°C	28 days
9030	Sulfide	376.2	100 g	500 mls	1 L BR	1 L HDPE	ZnAc plus NaOH 9	7 days
	Sulfite	377.1	-	50 mls		125 ml HDPE	None	Immediate analysis
	Surfactants	425.1	-	250 mls		500 ml HDPE	Cool 4°C	48 hours
	Temperature	170.1	-	1 L		2 L HDPE	None	Immediate analysis
	Turbidity	180.1	-	100 mls		250 ml HDPE	Cool 4°C	48 hours

Solid Waste Method SW-846	Description Parameters	Water/ Wastewater Method EPA-600/4-79-020	Sample Volume Required Solid	Liquid	Container Recommended by EPA Solid	Liquid	To Preserve	Holding Time
8010, 8240	Purgeable halocarbons	601, 624	5 g	40 mls	4 oz cwm	40 ml Glass V	.008%$NA_2S_2O_3$	14 days
8020	Purgeable aromatic hydrocarbons	602	5 g	40 mls	4 oz cwm	40 ml Glass V	.008%$NA_2S_2O_3$; HCl <2	14 days
8030	Acrolein & Acrylonitrile	603	5 g	40 mls	4 oz cwm	40 ml Glass V	.008%$NA_2S_2O_3$; pH 4.5	14 days
8040	Phenols	604	10 g	1 L	8 oz cwm	2-1/2 L AJ	.008%$NA_2S_2O_3$ 4°C	7 days until ex; 40 days after ex.
8270	Benzidines	605	10 g	1 L	8 oz cwm	2-1/2 L AJ	.008%$NA_2S_2O_3$; 4°C	7 days until ex; 40 days after ex.
8060	Phtalate Esters	606	10 g	1 L	8 oz cwm	2-1/2 L AJ	Cool 4°C	7 days until ex; 40 days after ex.
8070	Nitrosamines	607	10 g	1 L	8 oz cwm	2-1/2 L AJ	.008%$NA_2S_2O_3$ Store in dark	7 days until ex; 40 days after ex.
8080	PCB's	608	10 g	1 L	8 oz cwm	2-1/2 L AJ	Cool 4°C	7 days until ex; 40 days after ex..
8090	Nitroaromatics & Isophorone	609	10 g	1 L	8 oz cwm	2-1/2 L AJ	.008%$NA_2S_2O_3$ Store in dark	7 days until ex; 40 days after ex.
8100	Polynuclear aromatic hydrocarbons	610	10 g	1 L	8 oz cwm	2-1/2 L AJ	.008%$NA_2S_2O_3$ Store in dark	7 days until ex; 40 days after ex.
8270	Haloethers	611	10 g	1 L	8 oz cwm	2-1/2 L AJ	.008%$NA_2S_2O_3$ 4°C	7 days until ex; 40 days after ex.
8120	Chlorinated hydrocarbons	612	10 g	1 L	8 oz cwm	2-1/2 L AJ	Cool 4°C	7 days until ex; 40 days after ex.
8150	Chlorinated herbicides		10 g	1 L	8 oz cwm	2-1/2 L AJ	Cool 4°C	7 days until ex; 40 days after ex.
8280	Dioxins & Furans (T.C.D.D.)	613	10 g	1 L	8 oz cwm	2-1/2 L AJ	.008%$NA_2S_2O_3$	7 days until ex; 40 days after ex.
9020	TOX Total organic halogens		10 g	250 mls	4 oz cwm	16 L BR	4°C; H_2SO_4 <2	7 days until ex; 40 days after ex.
8080, 8140	Pesticides, chlorinated	608	10 g	1 L	8 oz cwm	2-1/2 L AJ	Cool 4°C; pH 5.9	7 days until ex; 40 days after ex.
9310, 9315	Radiological test Gross alpha		-	1 L	2-1/2 L jug	2 L HDPE	HNO_3 <2	6 months
9310	Beta		-	1 L	2-1/2 L jug	2 L HDPE	HNO_3 <2	6 months
9320	Radium (total)		-	1 L	2-1/2 L jug	2 L HDPE	HNO_3 <2	6 months
9131, 9132	Coliform, fecal & total	909.A, 909.C STD	-	100 mls	4 oz cwm	250 ml HDPE	.008%$NA_2S_2O_3$; 4°C	6 hours
	Fecal streptococci	910A, 910B	-	100 mls		250 ml HDPE	.008%$NA_2S_2O_3$; 4°C	6 hours
8015, 8240	Nonhalogenated volatile organics		5 g	40 mls	4 oz cwm	40 ml Glass V	Cool 4°C	14 days
8240	Volatile organics	624, 1624	5 g	40 mls	4 oz cwm	40 ml Glass V	Cool 4°C	14 days
8250, 8270	Semi-volatile organics	625, 1625	10 g	1 L	16 oz cwm	2-1/2 L jug	Cool 4°C	7 days until ex; 40 days after ex.
1310	EPTOX extraction		100 g	-	32 oz amber wm	No plastic	None	7 days until ex; 40 days after ex.

Solid Waste Method	Description	Water/ Wastewater Method	Sample Volume Required		Container Recommended by EPA		To Preserve	Holding Time
SW-846	Parameters	EPA-600/4-79-020	Solid	Liquid	Solid	Liquid		
1311	TCLP extraction		100 g	-	32 oz amber wm	No plastic	None	7 days until ex; 40 days after ex.
1010, 1110 9010, 9030	Hazardous waste characterization		200 g	-	8 oz cwm	No plastic	None	14 days

Appendix 7

Data Interpretation

Example 1

ABC LABORATORY
321 Your Street
Anytown, USA

Date:

To: (The Environmental Technician)

Job#:

P.O.#:

Subject: One drinking water sample for analysis

Sample Location: Kitchen sink

Date Sampled:

Time Sampled:

Sampled by:

Date Received at ABC Laboratory:

Parameter	ppm	EPA limits[1] (ppb)	Methods[2]
Arsenic	0.03	50	3114B
Barium	<0.05	100	3113B
Cadmium	<0.01	10	3113B
Chromium	0.03	50	3113B
Copper	0.05	1300	3113B
Lead	0.02	15	3113B
Mercury	<0.001	2	3112B
Selenium	<0.01	10	3114B
Silver	<0.05	50	3113B

1 Source: EPA "Safe Drinking Water Act of 7/8/87"; Copper and lead are action levels.
2 Source: American Public Health Association, "Standard Methods for the Examination of Water and Wastewater," 18th Edition, 1992.

________________ Supervisor

________________ Analyst

Example 2

ABC LABORATORY
321 Your Street
Anytown, USA

Date:

To: (The Environmental Technician)

Job#:

P.O.#:

Subject: Two air samples of welding booths for analysis

Sample Location:

Date Sampled:

Time Sampled:

Sampled by:

Date Received at ABC Laboratory:

Supervisor   Analyst

	Copper mg/m^3	Nickel mg/m^3	Lead mg/m^3	Chromium mg/m^3	Iron mg/m^3
John Smith	<0.01	<0.01	0.08	0.42	0.41
Sue Johnson	<0.01	0.03	0.11	0.36	0.47
OSHA TWA* Exposure Limits	1.0	1.0	0.05	0.50	10.0

*TWA = Time Weighted Average that must not be exceeded during any eight-hour work shift in a forty-hour work week.
Source: NIOSH, "Pocket Guide to Chemical Hazardous," US Government Printing Office, 1985.

Example 3

ABC LABORATORY
321 Your Street
Anytown, USA

Date:

To: (The Environmental Technician)

Job#:

P.O.#:

Subject: One soil sample for analysis

Sample Location:

Date Sampled:

Time Sampled:

Sampled by:

Date Received at ABC Laboratory:

Supervisor   Analyst

SAMPLE ID: Composite soil sample from waste pile NW at 2', 4' and 6' depths.

Parameter	Total Metals mg/kg	TCLP Metals mg/L	TCLP Limits mg/L
Arsenic	10.16	1.15	5.0
Barium	80.26	0.63	100.0
Cadmium	1.15	0.26	1.0
Chromium	142.10	3.36	5.0
Lead	1425.60	14.25	5.0
Mercury	<2.00	<0.05	0.2
Selenium	<2.00	<0.05	1.0
Silver	5.10	0.08	5.0

Source: "EPA Methods for the Examination of Solid Wastes," SW-846, 3rd Edition.

Glossary

Accuracy (in the laboratory) – The closeness of an individual measurement to the accepted range or true value. Accuracy is calculated as the percent recovery (%Rec) of the laboratory control or spiked sample.

Accuracy (of an instrument) – How close measurements are to true values. For direct reading instruments they are frequently expressed as a percent bias, or percent accuracy.

Aerosol – Liquids (mists and fogs) and solids (dusts, fumes and smoke) dispersed in the air.

Air Percussion Drilling – A type of drilling that uses air to power a down-hole percussion hammer to break the rock and carry the cuttings to the surface of the hole.

Aliquot – A part which is a definite fraction of a whole; as in aliquot samples for testing or analysis.

Ambient Atmosphere – Surrounding atmosphere.

Analyte – Any type of sample material to be analyzed.

Analytical Error – The difference between the reported concentration and the "true" concentration of a sample.

Anemometer – A meter designed to measure the speed (velocity) of the wind.

Annular Space – The space between the well riser and the borehole's wall.

Anthropogenic – Caused or produced by human activities.

Aquifer – A permeable geologic unit with the ability to store, transmit and yield fresh water in usable quantities.

Area Assessment – As used in the context of monitoring, it refers to sampling for the purpose of assessing not only a work site but also surrounding environmental air quality.

Artesian Well – A well drilled into a confined aquifer with a water level that stands at some height above the top of the aquifer, but not necessarily above the land surface.

Artifact – A physical object produced or shaped by human craft.

At-grade Well – A well that is almost flush with the ground.

Atomic Absorption (AA) Spectrophotometry – A laboratory process used to determine the metal ion concentration in a sample.

Auger – A tool for boring into earth, wood or ice.

Background – The normal or ambient concentration, outside the zone of contamination, of a substance under investigation.

Background Sample – Samples taken from media characteristic of the site being investigated, but outside the zone of contamination.

Backhoe – A tractor with attachments that can be used for excavating dirt or to perform other construction functions. Depending on the desired function, various tools are attached to the end of the hydraulically-powered and hinged arm. When used for excavating, it has a scoop or bucket attached to the arm that is drawn toward the tractor, then raised and dumped.

Bailer – A tube constructed of an inert material such as Teflon, stainless steel or polyvinylchloride (PVC).

Barometer – An instrument designed to measure atmospheric pressure.

Bedrock – The solid rock that underlies the loose material, such as soil, sand, clay or gravel.

Benchmark – An official, permanent marker placed by the USGS at the corner of each section (i.e., every mile) throughout the country expressing the specific location in longitude, latitude and elevation.

Bentonite – A clay derived from volcanic ash that swells when water is added.

Biased Sample – A sample collected to acquire information about particular places of interest at the site, such as areas of known or suspected contamination (also called judgmental sampling).

Bill of Lading – Shipping papers for common carriers, such as trucking, rail or vessel.

Biota – A collective term for all the animals and plants of an ecosystem.

Blank – A quality control sample taken to check some aspect of the sampling or analysis activities. See Equipment Blank, Rinsate Blank, Field Blank and Trip Blank.

Blind Sample – A sample in which its label does not reveal to lab personnel that it is a quality control sample.

BM – See Benchmark

Bomb Sampler – A liquid sampling device with two attached cords, one to retrieve the sampler and the other to activate the plunger that opens and closes the sampler at a desired depth.

Bonding – During container-to-container liquid transfer applications, a metal wire is clamped to each of the containers and a second wire connects from one of the containers to a pole in contact with the earth. This is called bonding. The purpose is to equalize any electrostatic charge differences between the two containers and the ground, thereby eliminating any chances for sparks (See also Grounding).

Breakthrough – Used in the context of sorbents, the condition that occurs when the sorbent becomes saturated and has exceeded its capacity to hold (retain) the sample.

Bulking – The mixing together (e.g., in a vacuum truck) of similar and compatible materials for shipment to treatment or disposal facilities.

Bung – Plug used to close a bunghole, which is the hole in a barrel through which the liquid is transferred.

Burette – A long graduated glass cylinder used for measuring volumes.

Cable-tool Drilling – A type of percussion drilling that uses a cable to raise a bit which is dropped to force the rock to break. It is a slow process, but it has the advantage of not requiring very much drilling fluid to carry the material out of the hole.

Calibration – The use of a standard with known concentration to adjust the true value readings of an instrument.

Canister – Evacuated metal container fitted with an inlet valve and used to collect gaseous grab samples. It may also be used to collect long-term gas samples (also called vacuum can).

Capillary Forces – The molecular attraction and surface tension between water and soil particles.

Carboy – A large (~ 5 gal) plastic or glass container designed to hold caustic materials that would react with metal.

Casing – See Riser

Casing Hammer – A percussion drilling method similar to the air percussion drilling method only it uses a second tool to carry rock cuttings to the surface.

CGI – see Combustible Gas Indicator

Chain of Custody – Documentation necessary to trace sample possession from the time of collection and throughout the time of analysis.

Chromatogram – The strip chart-paper product of a gas chromatograph. The position (retention time) and height of each peak is used to determine the components and concentrations of the sample.

CMS – See Corrective Measures Study

Coliform – A microbiological organism such as the bacterium Escherichia coli (E. coli).

COLIWASA – Acronym for Composite Liquid Waste Sampler, which is a tube made of inert material (such as Teflon) that is long enough to sample the depth of a 55-gallon drum.

Collocated Sample – Samples taken virtually from the same place and at the same time.

Colorimetric Indicator – A device designed to measure contaminant levels. It contains a chemical reagent, which reacts to the presence of a substance by undergoing a color change proportional to the amount of the substance.

Combustible Gas Indicator (CGI) – A direct reading instrument that measures the combined concentration of flammable gases or vapors in the air.

Comparability – A qualitative parameter that addresses the confidence with which one set of data can be compared to another. Its validity is contingent on the similarity of sampling techniques and analysis methods used for the two sets of data.

Completeness – As a quantitative aspect of quality assurance, completeness is calculated as the number of valid analyses reported divided by the total analyses undertaken. Valid analyses are dependent upon proper collection and handling of the samples both in the field and in the laboratory. Critical data point samples are also necessary to ensure completeness. Regulatory programs generally require an 80% to 90% factor of completeness.

Composite Sample – A sample obtained by mixing several samples collected from various locations, or from the same location but taken at different times (as opposed to grab sample).

Conductivity – The ability of a material (solid, liquid, or gas) to conduct or transmit electricity.

Conduit – Pathway for migration of contaminants from one environmental compartment to another.

Confidence Interval – A statistical range with a specified probability that a given parameter lies within the range.

Confining Unit – A sediment like clay, or a rock unit like shale, that does not readily permit water to move through it, and that restricts the movement of water either into or out of aquifers.

Continuous Sampler – A stainless steel tube that opens along the long axis into two halves. It is placed at the end of the drill rod and lowered to the bottom of a borehole through a hollow stem auger and pushed downward by the drill rig to obtain an undisturbed sample.

Contour Interval – The elevation difference between adjacent contour lines on a topographic map.

Contour Lines – The lines on a topographic map that join points of equal elevation on a land surface.

Contract Laboratory Program (CLP) – CLP laboratories are certified to perform EPA project analysis under rigorous QA/QC and documentation protocols. They are expected to provide the highest level of data quality with confirmed identification and quantitation of organic and inorganic analytes in air, soil, water and solid waste samples.

Control Water – Potable water, preferably from a municipal water system of known chemical composition, used in the decontaminating and rinsing of sampling equipment.

Corrective Measures Study (CMS) – The development and evaluation of remedial alternatives, so that decisions may be made regarding the limitation of the spread of contaminants and the reduction of their toxicity at a RCRA site (similar to a CERCLA Feasibility Study).

Corrosivity – One of the four RCRA characteristics that is used to determine if a waste is considered hazardous. If a sample has a pH of less than 2.0 or greater than 12.5, it is considered to be hazardous based on the characteristic of corrosivity.

Critical Data Points – Sample locations for which valid data must be obtained for the sampling event to be considered complete; for example, an upgradient well in a groundwater contamination study.

Critical Sample – A sample for which valid data must be obtained in order to satisfy the objectives of the sampling and analysis activity.

Data Interpretation – Comparison of project data with agency-set clean up objectives, toxicity data or hazardous waste criteria.

Data Qualifiers – Symbols included on lab reports that explain the uncertainty of the data and alert the data user to how much significance can be placed on a measurement; for example, a "B" may symbolize a data quality explanation such as "analyte found in associated blank as well as sample."

Data Quality Objectives (DQOs) – A list of statements in a Sampling Plan developed to ensure that the sampling and analysis data collected will be sufficient and of adequate quality for their intended uses. One of the main components of a sampling plan which specifies the methodology to be used during sampling to ensure that samples are representative and reliable.

Data Validation – The process of reviewing analytical data for errors, determining inconsistencies in the data, comparing the data against quality control samples and background data, deleting unusable data, and explaining limitations of project data.

Dedicated Equipment – Equipment used for only one purpose so that cross contamination is avoided.

Degassing – The loss of volatile organic compounds from a substance or material.

Detection Limit – The lowest measurement of analyte concentration that the instrument and analytical method can reliably achieve. Results that are reported as less than or equal to this limit should be taken to mean that if the analyte is present, it is present at some concentration less than this limit.

Detector Tube – A total vapor survey instrument composed of a glass tube containing a substance that is either specific or able to detect a family of substances. The tubes allow the versatility of being the only direct-reading indicators for some chemicals and able to measure a wide range of chemicals with a single pump.

Development Water – The water removed from a well until it is visually free of suspended materials or sediments and the temperature, conductivity and pH measurements remain relatively constant.

Digestion – A process in which a chemical substance, for example an acid, is used to dissolve and prepare a sample prior to analysis.

Dipper – A surface water sampling device composed of a cup at the end of a long pole.

Direct Push – A technique that pushes the tool into the ground to remove a soil sample without the use of drilling. One type of direct push sampling system is known as Geoprobe®.

Direct Reading Instrument (DRI) – A monitoring device that provides real-time (i.e., at the time of monitoring) data for almost immediate use.

Discrete Interval – Sampling at specified depths; for example, samples specified to be taken at precisely 10 feet deep at each target point.

Distillation – A separation process that makes use of the different boiling points of two or more liquid substances.

Dosimeter – See Self-reading Dosimeter

Dredge – A tool that grabs sediment below surface water.

DRI – See Direct Reading Instrument

Drum – A cylindrical, barrel-like container with flat ends. It is often made of metal with a 55-gallon capacity.

Drum Deheader – A manual tool that is used to remove the lid from a closed-head drum. It works much like a can opener, using the top chime or rim as a guide.

Duplicate – A quality control sample collected from the same place, in the same manner, using the same sampling tool, and placed in the same type of container. The two samples are analyzed separately, or sent to different labs, as a check against the analytical procedures.

Effluent – Material that flows out of a process; usually associated with wastewater.

EIS – See Environmental Impact Statement

Electron Volt (eV) – An extremely small unit of energy used to measure the amount of energy required to remove the outermost electron (ionization potential) from an atom or molecule of a substance.

Elevation – The numeric height of a point on a map above sea level.

Environmental Audit – A process analysis performed to determine compliance with applicable environmental laws and regulations and to identify opportunities for waste reduction, energy conservation, and pollution prevention.

Environmental Impact Statement (EIS) – The 1970 National Environmental Policy Act (NEPA) requires any federal branch or agency proposing a project to file an EIS which is meant to quantify its anticipated environmental impact, its potential benefits, as well as to describe and compare alternatives. The EIS is an integral part of all siting procedures.

Environmental Investigation – Also known as site characterization.

Environmental Sample – A representative portion of material collected from air, land, water, biota or waste to be analyzed for contamination.

EPA Level I (Screening) – Field screening or analysis using portable instruments such as a pH meters, conductivity meters, HNUs or filed test kits. It provides the fastest, least costly, and lowest level of data quality.

EPA Level II (Field Analysis) – Uses more sophisticated portable analytical instruments, including mobile onsite laboratories. It provides rapid, somewhat higher quality data than Level I. It is the level typically used for site characterization, engineering design and monitoring of remediation efforts.

EPA Level III (Engineering) – Performed off-site at a fixed analytical laboratory using EPA-approved analytical methods. Does not include stringent QA/QC or documentation requirements of a CLP. Data is widely used in environmental projects for risk assessment, health and safety issues, site characterization, establishing cleanup objectives, project design and monitoring during remediation efforts.

EPA Level IV (Confirmation) – Performed only at an EPA CLP certified laboratory and characterized by rigorous QA/QC protocols and documentation. It provides the highest level of data quality with confirmed identification and quantitation of organic and inorganic analytes.

EPA Level V (Nonstandard Analysis) – Analysis by nonstandard protocols. Used for one-of-a-kind analysis with no comparison data possible.

Equilibrate – A condition in which all acting influences are canceled by others, resulting in a stable, balanced or unchanging system.

Equipment Blank – see Rinsate Blank

eV – See Electron Volts

Evapotranspiration – The process of transferring moisture from the earth to the atmosphere by evaporation of water and transpiration from plants.

Exclusion Zone – An identified restricted area into which workers cannot enter without appropriate PPE and from which they cannot exit without going through decontamination procedures (also called hot zone).

Extraction – It may be part of an analysis scheme where a second substance, such as a liquid, is used to remove and/or concentrate a substance of interest from the sample matrix.

Fault – A fracture or rupture in the continuity of a rock formation caused by a shifting or dislodging of the earth's crust.

Feasibility Study (FS) – The development and evaluation of remedial alternatives, so that decisions may be made regarding the limitation of the spread of contaminants and the reduction of their toxicity at a CERCLA site (similar to a RCRA Corrective Measures Study).

FID – See Flame-ionization Detector

Field Analysis – Measurements made with test kits and hand-held instruments (pH meters, conductivity meters, and HNUs) which generally give direct readout of analytical results.

Field Blank – A sample composed of deionized water collected, processed and handled using the same procedures as the other samples. It is used to detect equipment or sampling procedure errors.

Filter Medium – Almost any porous material having a reasonable degree of rigidity can serve as a filter medium. Cotton, woven wire, nylon and glass cloth are sometimes used. Plastic membranes containing pore sizes down to 0.001 micron are also used. – (Note: A sorbent is different than a filter medium. A sorbent either adsorbs or absorbs a substance, whereas a filter restricts one substance but allows another to pass.)

First Draw Sample – A water sample collected after it has been sitting undisturbed in the plumbing system for a specified number of hours.

Fixed Laboratory Analysis – Performed in professionally staffed laboratories with the ability to perform a wide array of analysis on specialized instrumentation. These laboratories have the most sensitive instruments and can provide supporting documentation along with the analysis results.

Flame Atomic Absorption Spectrophotometer – In flame AA spectrophotometry a portion of a liquid sample is aspirated into an air/acetylene or nitrous oxide flame where the metal ions become atomized. The difference in intensity of a selected wavelength light before and after it passes through the flame is compared, and the concentration of metal in the sample calculated. Flame AAs have a sensitivity range from 0.1 to 100 mg/L (ppm).

Flame-ionization Detector (FID) – An instrument used to identify the amount of a substance by measuring the absorption of electrons resulting from its ionization as it passes through a hydrogen flame. The measurement readout is expressed as a ratio of parts of total contaminant per million parts air (ppm).

Flammable Range – The air and fuel concentration range between the Lower Explosive Limit (LEL) and the Upper Explosive Limit (UEL). It is the air and fuel concentration range of a material that will support a flame upon contact with an ignition source.

Flight Auger – A truck-mounted soil or sediment sampling device composed of a long hollow steel tube with a single blade (flight) that twists around and up the tube.Additional five-foot sections of tubing may be added for obtaining deeper samples.

Fracture – A cracked or broken zone in rock.

Free Product – Measurable hydrocarbon or non-aqueous phase liquid layers found in conjunction with groundwater.

FS – See Feasibility Study.

Furnace Atomic Absorption Spectrophotometer – In furnace AA spectrophotometry a portion of a liquid sample is injected into a graphite furnace where in Stage 1 it is heated driving off all liquids. In Stage 2 the heating is increased to char and drive off all organics. In Stage 3 strong heating atomizes the remaining metal ions present. The concentration of metal in the sample is calculated by comparing the difference in intensity of a selected wavelength light before and after it passes through the atomized metal ions. Furnace AAs have a sensitivity range from 0.1 to 1,000 mg/L (ppb).

Gas Chromatograph – An analytical instrument to separate and identify sample components. Used primarily to analyze the organic components present in a sample. The sample is introduced into a heated lined or packed chromatographic column and swept along by a carrier gas. Depending on its unique retention time with the packing material, each substance travels through the column at its own rate and is detected as it exits. The detection signal is sent to a chart recorder where it produces a series of peaks, called a chromatogram.

Gas Chromatograph/Mass Spectrometer (GC/MS) – The basic gas chromatograph function of separation of components remains the same (see gas chromatograph). However, as each substance exits the chromatograph, it enters the mass spectrometer where it is converted into charged particles (ions). Based on the unique pattern of each ion (along with its retention time) each ion fragment is identified; the compound is also identified by comparing it to the instrument's library.

Gas Sampling Bottle – Evacuated glass bottle fitted with an inlet valve and used to collect gaseous grab samples. It has the disadvantage of being breakable.

Geiger Counter – A monitor that uses a Geiger-Muller tube to detect the presence of ionizing radiation. Used to measure the degree of radioactivity or other ionizing radiation.

Geographic Information System (GIS) – A computer system with the capability to capture, store, display and analyze information of a geographic nature. It is designed to allow users to collect, manage and analyze large volumes of spatially referenced and associated attribute data. The major components of a GIS are: a user interface system; data base management capabilities; data base creation/data entry capacity; spatial data manipulation and analysis packages; and display/product generation functions.

Geotechnical Gauge – A visual aid used in field work. The references included are: soil color chips, a particle size classification chart, a particle percentage estimation chart, a particle shape illustration and the USCS system.

GIS – See Geographic Information System

Grab Sample – A single sample collected from a specific location at a given point in time (as opposed to composite sample).

Grab Sampler – A liquid sampling device that consists of a one-liter bottle at the end of a long pole.A spring-loaded plunger allows the sampling container to be opened and closed at the desired sampling depth.

Grain Sampler – A sampler for granular or powdered material; it consists of lengthwise slotted nested brass or stainless steel tubes. After inserting the sampler to the desired depth, the outer tube is rotated to open the slots and allow the sample to enter. The outer tube is again rotated to close the slots before withdrawing the sampler.

Gravimetric Analysis – The determination of the amount of a component in a sample by either directly weighing the analyte or by weighing a compound that contains a known portion of the analyte.

Gravity Corer – A sediment sampler composed of a metal tube attached to a retrieval cable. The tube is forcefully dropped, forcing the soft sediment to enter the sampler. The sampler is retrieved with the cable.

Grid – A pattern of regularly spaced lines forming squares, triangles or circles that determines how the sampling area is to be staked out and where the sampling points should be located.

Grounding – It is the use of a metal wire clamped on one container and the other end connected to a pole that is in contact with the earth during container-to-container liquid transfers. (See also Bonding.)

Groundwater – The water that lies beneath the Earth's land surface (i.e., in aquifers, which is an underground layer of water).

Hammer Blows – The number of hammer blows needed to reach the desired depth is counted and recorded in the logbook as a part of obtaining soil or sediment samples by the slide hammer method. The number of blows is valuable geotechnical information related to the strength and type of material being sampled.

Headspace – The void area above the soil and below the lid of a jar, where volatile gases can collect and potentially be lost from the analysis of the sample.

Heterogeneous – Referring to a substance that has variable characteristics within the container.

Holding Time – The maximum time allowable before a sample is analyzed, as specified by EPA in SW-846.

Hollow Stem Auger – A hollow pipe wrapped by a continuous spiral-shaped flight that looks like a corkscrew. Once a drill rig has added enough five-foot sections to reach the desired depth, the auger is used to prevent the walls from collapsing, while devices such as a split spoon sampler are lowered through the hollow pipe to obtain undisturbed soil samples below the hole.

Homogeneous – Referring to a substance that has the same characteristics throughout the container.

Horizon – The sequence of layers or soil profile in a vertical section of soil, subsoil and/or bedrock.

Hot Work – Activity that produces heat or a static electricity ignition source; for example; burning, welding, riveting, drilling, abrasive blasting and space heating.

Humidistat – An instrument for measuring and controlling humidity (moisture) in the air.

Humus – Decaying organic matter; that is, vegetation or animal matter.

Hydraulic – Operated or moved by means of a liquid under pressure. In the operation of a backhoe, for example, it is pressurized oil that is used to drive the hydraulic rams that move the backhoe's arm and attached tool.

Hygrometer – An instrument for measuring the water-vapor content of the atmosphere.

Ignitability – One of the four RCRA characteristics that is used to determine if a waste is considered hazardous. If a sample has a flash point of less than 140°F (60°C), it is considered to be hazardous based on the characteristic of ignitability.

Immiscible – The inability of one substance to be mixed or blended with another substance in the same state.

Immunoassay – A field test kit that uses specific chemicals to give a yes or no answer concerning the presence of a contaminant.

Impermeable – The inability of a substance to transmit a fluid.

Impinger – A collection bottle within a pumping system designed to remove a gas and/or allow its analysis.

Impoundment – A reservoir, lagoon or similar in-ground surface containment.

Inductively Coupled Plasma-Atomic Emission Spectrophotometer (ICP-AES) – A laboratory instrument that is used to quantitatively determine the metal ion concentration in a sample. A liquid sample is aspirated into and atomized by a plasma torch (6000° to 8000°C). The resultant radiant energy is measured and used to calculate the concentration of metals present. It can determine several metal concentrations simultaneously, with a sensitivity range from 0.01 to 1000 mg/L (ppm).

Influent – The material(s) that flow into a process.

Inherent Safety – A monitoring instrument's certified level of protective construction in which components (switches, batteries, etc.) do not trigger a spark that could ignite an explosive atmosphere while in operation.

Interface Meter – See Oil/Water Meter

Investigative Sample – The sample collected at a site for the purpose of analyzing environmental contamination (as opposed to quality control sample).

Ion Scan – The strip chart-recorder product of the mass spectrograph portion of a GC/MS. The compound can be identified by comparing the compound's ion scan to the instrument's computer library.

Ionization Potential (IP) – The amount of energy required to remove the outermost electron from an atom or molecule. The ionization potential for each substance is specific.

Ionizing Radiation – Extremely high frequency, short wavelength, gamma- and X-rays and energetic subatomic particles such as alpha particles, beta particles, electrons, protons and neutrons.

IP – See Ionization Potential

Isoconcentration Lines – On a map of a contaminated area, lines are drawn around the source of contamination in such a way that they connect monitoring points representing similar contamination levels. The somewhat circular lines depict a plume, which indicates the extent of movement of a contaminant.

Judgmental Sampling – see Biased Sampling

Lab Pack – The use of a large drum and packing material (vermiculite) to hold smaller containers of assorted chemicals that could react with one another.

Laboratory Certification – Laboratory certification provides assurance that a laboratory is truly capable of performing the analysis. There are no mandatory national certification programs; however, voluntary certifications can be obtained through the American Association for Laboratory Accreditation (A2LA) and the American Industrial Hygiene Association (AIHA).

Laboratory Control Sample – A quality control sample that is prepared by adding a known concentration of the analyte being determined to an aliquot of laboratory reagent water. The control sample then goes through each step of the preparation and analysis procedure. (also called Laboratory Standard)

Laboratory Reagent Water – Laboratory reagent water is a general term that includes both distilled and deionized water. EPA quality control (QC) documents list several grades of water purity. Grade I water, for example, may be suitable for use in wastewater analysis, but may not be pure enough for analysis of organic compounds in drinking water by GC/MS.

Laboratory Standard – See Laboratory Control Sample

Large Bore Sampler – A discrete interval sampler which obtains a sample 22 inches (55.88 cm) long by one inch (2.54 cm) wide. One such device is known as the Macro-Core®, which allows continuous soil cores to be recovered, with or without use of a liner.

Leachate – Any liquid that is collected and removed from under the specially designed liner in an EPA-regulated hazardous waste pile.

Lead Agency – The governmental agency with primary jurisdiction over the action being taken.

Legend – An explanation of the symbols that appear on a map indicating various natural and human-made features.

LEL – See Lower Explosive Limit

Liquid Limit – The amount of water required for soil to go from its plastic to a semiliquid state.

Litter – The organic matter (such as twigs, leaves, and other woody materials) that decays in the soil forming humus.

Lower Detection Limit (LDL) – The lowest concentration of a substance that can be measured by an instrument

Lower Explosive Limit (LEL) – The minimum concentration of an air and fuel mixture that will support a flame upon contact with an ignition source (also called Lower Flammable Limit).

Lower Flammable Limit (LFL) – See Lower Explosive Limit

Manway – An enclosed vault-type underground casing which allows human access.

Map Scale – The map scale describes the relationship between distance on a map and the corresponding distance on the ground. It is expressed as a ratio in which the first number represents measurement on a map and the second number represents the corresponding distance on the ground.

Matrix – The medium that carries or transmits the contaminant.

Matrix Spike – The addition of a known amount of the contaminant of interest to a quality control sample. Comparing the spiked sample with the nonspiked sample enables the analyst to determine if there are interferences from other matrix components in the samples.

Media – Elements of the environment (air, water, soil and biota) that carry or transmit the contaminant.

Media Sample – A media sample is taken from the environment: air, water, soil and biota (as opposed to waste sample).

Meniscus – The concave or convex upper surface of a non-turbulent liquid in a container.

Metal Oxide Semi-conductor – A solid-state sensor composed of a metal oxide film over a ceramic substrate, fused around a platinum wire coil. Monitors containing a metal oxide semi-conductor may be nonselective or selective depending on the metal oxide and operating temperature selected.

Method Blank – A blank consisting of an aliquot of laboratory reagent water processed through all steps of the preparation and analysis procedure for both inorganic and organic analysis. Its purpose is to indicate the degree of contamination that may have been introduced into a sample from a variety of sources in the laboratory–unclean glassware, vapors from another area of the laboratory or residue left inside a laboratory instrument from a previous analysis.

Mobile Laboratory Analysis – A laboratory contained in a van, trailer or boat. It provides improved data over field analysis; however, it may not be capable of providing the high quality results of a state-of-the-art fixed laboratory.

Monitoring – The use of a measuring device that provides real-time information about the presence and relative concentration of a substance.

MOS – See Metal Oxide Semi-conductor

Munsell Soil Color Chart – A color comparison chart used to describe soil colors.

National Priorities List (NPL) – Currently a list of more than 1000 sites that have been evaluated and prioritized for the level of risk they present to the public health and the environment. EPA can only take remedial actions at hazardous waste sites that are on the NPL.

Nonaqueous Phase Liquids (NAPLs) – An immiscible hydrocarbon layer found either at the bottom or floating on the top of a column of water.

Nonattainment Area – A geographic area in which the level of a criteria air pollutant is higher than the level allowed by the federal standards. Criteria air pollutants are very common air pollutants regulated by EPA on the basis of information on health and/or environmental effects of pollution.

Nonpoint Source – A general source of pollution without a specific point of origin, such as rain water carrying topsoil and chemicals from a field (as opposed to Point Source).

NPDES Permit – National Pollutant Discharge Elimination System regulatory permit under the Clean Water Act.

NPL – See National Priorities List

Oil/Water Meter – A meter capable of measuring hydrocarbon layers on water. Oil/water meters typically have two signals, one to indicate when the hydrocarbon layer is encountered and a second to indicate when it contacts the water layer (also called Interface Meter).

Overland Flow – Water that flows directly over the land surface, e.g., a stream or river.

Overpack – A polyethylene or steel container that is large enough to enclose the damaged drum.

Parameters – A set of measurable conditions, such as temperature and pH, that define a system and determine its behavior.

Percent Recovery (%Rec) – Used to report accuracy, percent recovery is the absolute difference between the analyte concentration in the spiked and the unspiked samples divided by the amount of spike added and multiplied by 100 to change it to a percent.

Percussion Drilling – Any drilling method that uses rapping or a succession of mechanical hammer blows to force the sampler into the soil or sediment, or to break the rock.

Performance Audit – Samples of unknown concentration that are analyzed by the laboratory and then sent to an independent agent for verification.

Performance Sample – A sample that has values known to the client (i.e., data user) and that is sent along for analysis with the investigative samples as a check of the laboratory's performance. The laboratory is not made aware of the inclusion of a performance sample within the group of investigative samples.

Permeability – The ability of a substance to transmit a fluid.

Personal Assessment – As used in the context of monitoring, it refers to sampling the environment of individual workers who are exposed to potentially hazardous workplace materials.

Personal Protective Equipment (PPE) – Includes special clothing, devices, or equipment that protects the worker from a hazardous environment.

Petrology – The systematic classification of rocks.

pH Meter – A meter consisting of a solid state voltmeter connected to a pH and reference electrode. Hydrogen ions, H^+, pass through the porous glass electrode in the pH electrode and change the solution's conductivity, which is converted into a meter reading.

Phase I Environmental Site Assessment – A historical study of local governmental records, aerial maps, interviews with involved parties, and other data to establish a site's former and current uses.

Phase II Environmental Site Assessment – A field investigation that includes collecting samples and laboratory analysis to determine possible contamination.

Phase III Environmental Site Assessment – A process that fully characterizes the vertical and lateral extent of soil and groundwater contamination. Phase III also identifies potential cleanup alternatives, their feasibility, and implementation of appropriate cleanup activities.

Photoionization Detector (PID) – An instrument used to identify the amount of a substance by measuring the absorption of electrons resulting from its ionization by a powerful ultraviolet light source. The ultraviolet light source lamps may be changed, but must provide enough energy to reach the substance's ionization potential (IP). The measurement readout is expressed as a ratio of parts of total contaminant per million parts air (ppm).

PID – See Photoionization Detector

Pin Set – A permanent marker placed in one corner of a site as a surveying reference point.The pin's size and material is specified as 5/8 inch steel, so that it may be picked up by a metal detector.

Plasticity – The "stickiness" of fine-grained soil. Plasticity is a function of the water content of a soil which allows it to be molded.

Plume – A body of contamination originating from a specific source. In site characterization it is graphically depicted with isoconcentration lines.

Pneumatic – Operated or moved by means of air pressure.

Point Source – A source of pollution with a specific point of origin, such as a wastestream from a pipe (as opposed to Nonpoint Source).

Pollution Prevention Assessment – Determines the opportunities for waste reduction, energy conservation and pollution prevention.

Pond Sampler – A surface water sampler (see Dipper).

Poorly-graded – Soil that consists mostly of particles with similar sizes. Also known as uniformly graded soil.

Porosity – The ratio of the volume of void spaces to the total volume of the rock or sediment.

Potable Water – Water that is fit or suitable for drinking, preferably from a municipal water system of known chemical composition.

Potentiometric Surface – The elevation or level that groundwater will rise to in wells that tap an aquifer (also called Piezometric Surface).

PPE – See Personal Protective Equipment

Precision (in the laboratory) – The reproducibility–or the agreement–of individual measurements made under a given set of conditions. The precision of an analytical procedure is commonly determined through analysis of laboratory duplicate samples and reported as Relative Percent Difference (%RPD) or as Relative Standard Deviation (%RSD).

Precision (of an instrument) – The ability of an instrument to repeatedly yield the same measurement.

Preservation – Methods used to decrease the likelihood of sample deterioration during storage and transport to the laboratory; i.e., refrigeration and/or the addition of certain chemicals in the sample container.

Pretreatment – Treatment on an effluent to meet discharge standards prior to releasing it to a POTW or the environment.

Protozoans – Single-celled organisms such as Giardia lamblia, which causes intestinal infections in humans.

Pumping Test – Used to measure the hydraulic characteristics of an aquifer as well as the rate at which a well will yield water. The test is performed by pumping a well for a long period of time and observing the change in the water level in the well and/or other surrounding wells.

Purging – Removing impurities, i.e., removing stale well water before obtaining a sample that is representative of the groundwater.

QA/QC – See Quality Assurance and Quality Control

Quadrangle Map – An area of land shown on a United States Geological Survey atlas sheet. The measurement of a quadrangle represents an area between two latitude lines and two longitude lines and is expressed as 7-1/2 minutes by 7-1/2 minutes. Also known as a 7-1/2 minute map.

Quality Assurance (QA) – The procedures that ensure that a sampling event meets the data quality objectives, such as checking field calculations to make sure mathematical results are correct, documenting all field and sampling activities (such as the testing and calibration of equipment), making sure all signatures are obtained where necessary on field notes and chain-of-custody forms, and using a sample designation system to identify each sample uniquely.

Quality Assurance Project Plan (QAPjP) – A list of directives for the collection, handling, storing, packaging and shipping of quality control (QC) samples. One of the main components of the Sampling Plan which specifies the methodology to be used during sample collection in order to ensure that samples are representative and reliable.

Quality Control (QC) – The procedures specified in the Sampling Plan ensuring that the resulting analytical data can be trusted, such as collecting quality control samples (i.e., blank and duplicate samples).

Radionuclide – A radionuclide is a radioactive type of atom; e.g., carbon-14 is a radionuclide of carbon.

Random Error – An error due to such things as inherent limitations of an instrument (instrument noise) or lack of care in making the measurements (weighing uncertainty or aliquot variability).

Random Sample – see Unbiased Sample

RCRA Closure – A Resource Conservation and Recovery Act closure is an EPA enforcement action against a company that has uncontrolled hazardous wastes on its site.

RCRA Facility Investigation (RFI) – Data-gathering activities undertaken to determine the presence, degree and extent of contamination at a RCRA site.

RCRA Hazardous Waste – A waste is considered hazardous if it is listed in 40 CFR 261.31-261.33 or if it exhibits ignitability, corrosivity, reactivity or toxicity characteristics.

Reactivity – One of the four RCRA characteristics that is used to determine if a waste is considered hazardous. If a sample is unstable, if it reacts violently with water, if it is sufficiently cyanide- or sulfide-bearing to produce toxic gas, or if it is capable of detonation, it is considered to be hazardous based on the characteristic of reactivity.

Real-time Use – The use of monitoring devices that provide current information about the presence and relative concentration of a substance.

Regulatory Inspection – An inspection conducted by an agency inspector to determine compliance with federal, state or local requirements.

Regulatory Threshold – The threshold level of concentration of a contaminant determined by a regulatory agency to be hazardous and therefore not to be exceeded.

Relative Percent Difference (%RPD) – Used for the calculation of precision, %RPD is the difference between the sample and duplicate values divided by one-half the sum of the sample and duplicate values, multiplied by 100 to change it into a percent.

Relative Response – It is usually expressed as a ratio or percentage conversion factor. This known variability is used to calculate the actual concentration of a substance during monitoring, even though the instrument was standardized to another.

Remedial Investigation (RI) – A hazardous waste site characterization undertaken to determine the degree and extent of contamination according to EPA's Superfund or CERCLA regulations.

Representativeness – The degree to which the sample data accurately represent the site conditions. It is a qualitative parameter contingent on both a proper site sampling program and proper laboratory protocols.

Retention Time – The unique time interval that it takes for each component of a sample to be swept through the column of a gas chromatograph by the carrier gas.

RFI – See RCRA Facility Investigation.

RI – See Remedial Investigation.

Rinsate Blank – A quality control sample in which the last deionized water used to rinse or decontaminate sampling equipment is collected to check the effectiveness of the decontamination methods (also called equipment blank).

Riser – In a sampling well it is the outer piece of solid pipe that is connected to a screen and sealed off at the bottom with an end cap. The riser and screen may be constructed of several different types of materials, including stainless steel, rigid polyvinyl chloride (PVC) pipe or Teflon® (also called Casing).

Sample Collection Bag – A specially designed bag constructed of polyethylene, Saran®, Mylar®, Tedlar® or Teflon® used to collect gas samples.

Sample Contacting Equipment – Equipment used in the sampling activities that directly contacts the sample. ASTM's standard D5088-90 indicates that this equipment is subjected to a different decontamination protocol than sample non-contacting equipment.

Sample Custodian – A lab employee authorized to receive samples.

Sample Non-contacting Equipment – Equipment used in the sampling activities that does not directly contact the sample itself. ASTM's standard D5088-90 indicates that this equipment may be subjected to a different decontamination protocol than sample contacting equipment.

Sampling – The collection of representative portions of gas, liquid or solid environmental media and its preparation for transport to an analytical laboratory.

Sampling Chart – A chart that summarizes EPA 600 or SW846 recommended sample holding times, volumes requirements, preservatives, containers, etc. for a sampling event.

Sampling Points – see Sampling Targets

Sampling Protocol – A written description of a detailed procedure to be followed for sample collecting, preserving, labeling, packaging, storing, transporting and documenting samples as well as decontaminating the equipment. One of the components of the sampling plan which specifies the methodology to be used during sampling to ensure that samples are representative and reliable.

Sampling Pump – A hand-operated, battery-operated or AC-current-powered device used for drawing gas samples through a sampling device. The pumps must have calibrated rates of flow ranging from 10 cc/min to high volume (1,130 liters/min) pumps.

Sampling Targets – All biased and unbiased sampling points identified on the sampling grid.

Sampling Train – A sequence of sampling equipment connections ending in a gas collection bag or a gas monitor.

Sanborn Fire Insurance Map – Detailed city map which provides detailed information for risk assessment on urban property.

Saturation Concentration – See Upper Detection Limit (UDL)

Sediment – Geologic material that settles below flowing or standing surface waters.

Selectivity – An instrument's ability to detect and measure a specific chemical or group of similar chemicals.

Self-reading Dosimeter – A small wearable monitor that enables an employee to read a real-time exposure measurement to a substance.

Sensitivity – The ability of an instrument to detect small changes in concentration, for example in the parts per million or parts per billion ranges.

Septum – A partition that allows a syringe to be inserted (e.g., into a collection bag) to draw out a sample for analysis.

Simple Random Sampling – Collection of single samples from anywhere within the site. If the sample is taken from a homogeneous waste, then this sample should yield accurate and precise information about the nature of the waste.

Site – As it pertains to an investigation for a contaminant release, a site may be a specific property having specific legal boundaries or multiple properties; furthermore, a single property may have more than one site to investigate for contaminants.

Site Health and Safety Program – A written site-specific plan to protect the general health and safety of workers.The program must meet the requirements of 29 CFR 1910.120(b) Worker Protection Standard developed jointly by EPA and OSHA.

Slide Hammer – A manual tool for collecting relatively shallow soil samples. The tool is composed of a stainless steel barrel that can have extensions added and a hammer that is used to drive the barrel into the ground. The barrel accommodates stainless steel liners or plastic liners for ease in removing the soil sample from the device.

Sling Psychrometer – An instrument that twirls two thermometers around its center handle. By comparing the readings of the wet and the dry bulb thermometers the atmospheric humidity can be determined.

Sludge Judge – A sludge sampling device composed of a 3/4 inch tube in 5-foot sections with a check valve on the bottom. One-foot intervals are marked to determine the sludge depth.

Slug Test – A test used to determine the hydraulic conductivity of an aquifer by removing a known volume or "slug" of water from a well and measuring its rate of recovery.

Slurry – A thin watery mixture containing a finely divided material.

Soap Bubble Meter – A device used to calibrate the flow of gas. Its basic operation depends on the measurement of the amount of time required for a soap bubble to sweep through a given volume. It functions electronically, but it is also available in the form of basic chemistry equipment (burette, squeeze-bulb and beaker).

Sole Source Aquifer – Aquifers (identified by the SDWA) that are especially valuable because they are the only source of drinking water in an area.

Sorbent – An absorbing or adsorbing substance in solid or liquid form that can hold (adsorb) gas or vapors on its surface; for example, activated charcoal and silica gel.

Specific Chemical Instrument – As used in the context of monitoring, it refers to instruments that are made to measure the concentrations of only one specific chemical.

Specific Ion Analyzer – Similar to a pH meter; however, the pH electrode is replaced by a specific ion electrode that contains an ion-selective, thin membrane separating it from the inner solution-filled electrode. Changes in the conductivity of this solution convert to a direct reading.

Spectrophotometric Analysis – An analysis that relies on the change in the amount of light passing through a colored solution. The greater the color, the greater the absorbency of light. The amount of light being absorbed by the sample is compared to a calibration curve to determine the analyte concentration.

Split Sample – A quality control sample in which one sample is divided into two or more portions and then analyzed separately, either at the same lab or at different labs (also called replicate sample).

Split Spoon Sampler – A hand-held or truck-mounted sampling device that allows for the collection of an undisturbed core of soil. After the two-foot tube is bored to the desired depth in undisturbed soil, it is withdrawn and split open lengthwise to remove the sample.

Stabilize – A pumped monitoring well is considered to be stabilized when its water is visually free of suspended materials, and when temperature, conductivity, and pH measurements remain relatively constant.

Stage – To assemble drums in an upright and orderly manner, with enough space between the containers to allow for technicians to maneuver and take samples.

Stratified Random Sampling – The collection of samples from the various layers of a heterogeneous material.

Subsoil – The material beneath the soil but above bedrock. Composed of shattered or weathered bedrock material, which has not broken down into soil.

Superfund – The remediation fund operated under the legislative authority of CERCLA, the 1980 Comprehensive Environmental Response, Compensation and Liability Act, which requires EPA to provide a structure for cleaning up old, dangerous and hazardous waste sites.

Surface Waters – Natural waters that lie on the earth's surface in either streams, rivers, lakes or ponds and in human-made surface impoundments and lagoons.

Surrogate Standard – An organic compound that is similar in chemical composition to the analyte being determined, but not normally found in environmental samples.

SW-846 – See Test Methods for Evaluating Solid Waste, Physical/Chemical Methods

SW-846 Formula – SW-846 is the Environmental Protection Agency's (EPA's) standard set of analytical procedures. The formula provides a statistical method for determining the number of samples to be collected at a site.

System Audit – The review of a laboratory's total data production process, including an examination of its operations, instruments and facilities.

Systematic Error – A consistent error that generally biases the measurement in one direction. Examples of systematic errors include instrument calibration error, uncorrected instrument drift and operational errors.

Systematic Random Sampling – Sampling in which the first sample is randomly selected, but all subsequent samples are taken at fixed space or time intervals.

Tank – A container varying in sizes and ranging from small storage tanks or process containers less than five feet deep to giant chemical storage units that hold tens of thousands of gallons. A tank can also be a cargo vessel (on a truck or train), a process tank (a vat) containing chemical inputs or outputs for some manufacturing operation, or a storage structure for chemicals or wastes.

Tedlar® Bag – A type of sampling storage bag that gases cannot penetrate.

Ternary Diagram – An equidimensional triangular graph that can be used to determine a soil's texture, for example, by plotting its percentages of sand, silt and clay-sized grains of soil.

Test Methods for Evaluating Solid Waste, Physical/Chemical Methods – An EPA publication specifying the analytical laboratory methods that must be used to determine the nature of pollutants. Also known as SW-846 Procedures.

Texture – As used in a soil classification system, it refers to the percentage of each grain size in a particular soil.

Thermoset – A polymer that is permanently solidified through the application of heat.

Thief – A sampling device for liquids composed of a tube made from an inert material and open at both ends. The tube is lowered to the desired depth, the upper end stoppered, and the tube and sample withdrawn.

Titration – The use of a measured amount of a standard reagent (known concentration) to determine the amount of a second substance present. Typically the end of the reaction is determined by the use of a pH meter or an indicator substance that changes color.

Top of Casing (TOC) – A surveying mark on the top of the monitoring well. It is used as a starting point for measuring to the water and to the bottom of the well.

Topographic Map – A map showing the surface features of a region, usually by means of contour lines.

Total Vapor Survey Instrument – An direct reading monitoring instrument that provides information about the total amount of flammable vapors or mixture of vapors present.

Toxic Characteristics Leaching Procedure (TCLP) – An EPA test for solid wastes intended to simulate leaching action in a municipal landfill. At the end of the 18-hour test the leachate is analyzed for any of the listed eight metals, eight herbicides/pesticides and 25 organic compounds. If any of the TCLP extraction results exceed the EPA's established TCLP limits, the waste is considered a hazardous waste.

Toxicity – One of the four RCRA characteristics that is used to determine if a waste is considered hazardous. If the analyte resulting from TCLP contains any of the regulated contaminants at or above the regulatory level, it is considered to be hazardous based on the characteristic of toxicity.

Transpiration – The loss of water from the pores of plants (i.e., stomata) into the air.

Trier – A long pointed tubular scoop designed to sample granular or powdered materials.

Trip Blank – A sample bottle in which deionized water is poured before going into the field. This sample is kept with the investigative samples and subjected to the same handling, packaging for shipment, and analysis as the other samples. It is used to identify container contamination, transportation or handling errors.

True Value – In analytical chemistry, the "true" value refers to an acceptable range of values.

Turbid – The condition of having sediment or suspended material present in an otherwise clear solution.

Turnaround Time – The time required for the laboratory to complete and report the analysis of a sample.

UEL – See Upper Explosive Limit

Unbiased Sample – A sample obtained by a sampling strategy in which samples are collected from statistically selected targets to predict an average or overall picture of a site (also called random sample).

Unconsolidated Material – Loose, uncemented material, such as sediments present in streams, creeks and rivers.

Uncontrolled Hazardous Waste Site – A site where an accumulation of hazardous substances, often from illegally or poorly managed waste disposal, creates a threat to the health and safety of individuals or to the environment or both.

Upper Detection Limit (UDL) – The highest concentration of substance that can be measured by an instrument. Also known as the Saturation Concentration.

Upper Explosive Limit (UEL) – The maximum concentration of an air and fuel mixture that will support a flame upon contact with an ignition source (also called Upper Flammable Limit).

Upper Flammable Limit (UFL) – See Upper Explosive Limit

Vacuum Can – See Canister

Vadose Zone – The unsaturated zone in the earth's crust; it is composed of the soil (root) zone, the intermediate zone and the capillary fringe.

Visual-Manual Method – An ASTM method used to describe soils, based only on physical observations and senses.

Waste Pile – A pile of material generated by an industrial process is not a waste pile until its owner/manager determines that it is a waste with no other use. The pile may be controlled and planned or it may be uncontrolled and composed of illegally dumped hazardous wastes.

Waste Sample – A waste sample is one taken from a waste disposal container or area (as opposed to media sample).

Water Table – The top of the zone of saturation.

Water-table Well – A well placed in such a way that the well screen extends a few feet into the saturated zone of an unconfined aquifer. The water level in this type of well indicates the position of the water table in the surrounding aquifer.

Weathering – The various natural processes caused by water, wind, etc., that physically and chemically decompose rock.

Well-graded – A soil sample is considered to be well-graded if it has a wide range of particle sizes with substantial amounts of intermediate-sized particles.

Acknowledgments

Chapter 1

Figure 1-1: "The care of the Earth is our most ancient and, after all, our most pleasing responsibility. To cherish what remains of it, and to foster its renewal, is our only legitimate hope." Photography by John Livzey, © 1995.

Figure 1-2: A technician packs a sampling jar with soil from a bucket auger. HMTRI Photo.

Figure 1-3: Aerial photograph of Kearns, Utah. Photo courtesy of the USDA-FSA Aerial Photography Field Office.

Figure 1-4: Students analyzing an aerial photograph and a USGS map. HMTRI Photo.

Technical Corner: Map. Courtesy of Department of the Interior U.S. Geological Survey National Mapping Division.

Technical Corner: Chart. Courtesy of Department of the Interior U.S. Geological Survey National Mapping Division.

Figure 1-5: The oily staining and discoloration of the doors and concrete indicate there is reason to suspect that the doors, concrete, soil and groundwater may be contaminated. HMTRI Photo.

Figure 1-6: The decontamination process is demonstrated at a training event. HMTRI Photo.

Figure 1-7: Students and local Hazmat team members practice donning personal protective equipment. HMTRI Photo.

Chapter 2

Figure 2-1: The environmental media air, water, soil and biota are the carriers of the contaminants. Illustration by Doug Millhoff, Zogmorph Designs.

Figure 2-2: Uncontrolled hazardous waste site simulated for training purposes. HMTRI Photo.

Point of View: David Foss. HMTRI Photo.

Figure 2-7: A technician stakes a grid at a sampling site. HMTRI Photo.

Table 2-3: Guidelines for minimum QA/QC samples for field sampling programs. Source: EPA - Data Quality Objectives for Remedial Response Activities, Appendix C-13.

Chapter 3

Figure 3-1: An array of field sampling equipment. Photo courtesy of Versar Inc., Springfield, VA.

Table 3-1: Checklist of items required in field sampling. Sources: Appendix C, Samplers and Sampling Procedures for Hazardous Waste Streams and How to Sample Ground Water and Soils, by Peter Garnett.

Figure 3-2: Environmental technology class practices taking a soil sample with a hand auger. Turning the auger can be a two-person task! HMTRI Photo.

Figure 3-3: Student with air monitoring equipment. HMTRI Photo.

Figure 3-4: A hand auger with a sample. HMTRI Photo.

Figure 3-5: Truck-mounted hollow-core flight auger. HMTRI Photo.

Figure 3-9: Students using a pond sampler with telescoping pole are careful not to disturb the sediment as they take a surface water sample. HMTRI Photo.

Figure 3-15: Refrigerant sealed between layers of plastic film provides greater thermal efficiency than wet ice without mess or damage to samples; it also protects the samples against breakage. Photo courtesy of Polyfoam Packers, Wheeling, IL.

Figure 3-16: Filling out a label on a rounded surface may be challenging! HMTRI Photo.

Figure 3-17: Example of chain-of-custody record. Courtesy of California Department of Health Hazardous Materials Laboratory.

Chapter 4

Figure 4-1: Technicians in Level A equipment monitor a drum. HMTRI Photo.

Figure 4-2: The problem of pollution is not limited to the big cities. Here we have a photograph taken at the Great Smoky Mountains National Park on a clear day (left) and one of the same location on a day when the air was quite polluted (right). The visibility impairment is primarily due to sulfates, which come from combustion of fossil fuels. Photo courtesy of U.S. National Park Service.

Figure 4-4: This technician routinely monitors storage tanks at the paint manufacturing company where he is employed. HMTRI Photo.

Table 4-1: Sample protective ensemble. Source: NIOSH/OSHA/USCG/EPA "Occupational Safety and Health Guidance Manual for Hazardous Waste Site Activities," U.S. Government Printing Office, Washington DC, 1985.

Point of View: Sally Gaines. HMTRI Photo.

Technical Corner: Source: EPA, Air Surveillance for Hazardous Materials, Environmental Response Team, 1990. MSA detector tube inserted in Deluxe Kwik-Draw™ pump. Photo courtesy of Lab Safety Supply, Inc., Janesville, WI.

Table 4-3: NEC hazardous atmospheres. Source: "Classification of Gases, Vapors and Dusts for Electrical Equipment in Hazardous (classified) Locations" Nat'l. Fire Prot. Assoc., 1986. ANSI/ NFPA 497M.

Figure 4-11: The personal sampler pump may be worn on a belt, while the sorbent portion of the unit is clipped onto the shirt within the wearer's breathing zone. Photo courtesy of Lab Safety Supply, Inc. Janesville, WI.

Figure 4-12: Parts of a sorbent sampler. Illustration source: EPA.

Chapter 5

Figure 5-1: Sequence of tilted strata along Interstate 70, Jefferson County, Colorado. Photo courtesy of U.S. Geological Survey/W.R. Hansen, 187.

Figure 5-2: Ternary diagram for USDA, Soil Conservation Service Soil Classification System. Source: U.S. Department of Agriculture Textural Chart.

Table 5-1: Plasticity description. Source: ASTM Unified Soil Classification System.

Table 5-2: U.S. Department of Agriculture's Textural Soil Classification of Particle Sizes.

Box 5-2: Soil Porosity. Source: ASTM.

Table 5-3: Rock permeability. Source: ASTM.

Figure 5-4: The ASTM Laboratory Method is based in part on a sieve test. Photo courtesy of Ben Meadows Co., Atlanta, GA.

Table 5-4: Particle sizes. Source: ASTM.

Figure 5-5: The Geotechnical Gauge is an excellent visual aid for field work. Reprinted with permission of W.F. McCollough, Beltsville, MD.

Table 5-5: Criteria for describing consistency in soil/sediments. Source: ASTM.

Table 5-6: Criteria for describing structures in soil/sediments. Source: ASTM.

Table 5-7: Identification of inorganic fine-grained soils. Source: ASTM.

Table 5-8: Soil classification chart. Source: ASTM.

Figure 5-7: Three steps to classify soil by the SCS system. Illustration. Source: U.S. Department of Agriculture Textural Chart.

Figure 5-8: Different auger buckets are used for different types of soil. Photos courtesy of JMC/Clements Associates, Newton, IA.

Figure 5-9: Slide hammer. Photo courtesy of JMC/Clements Associates, Newton, IA.

Figure 5-10: Since the Geoprobe® is pushed directly into the soil, it does not produce soil cuttings; this reduces the necessity of handling investigation-derived wastes. Photo courtesy of Geoprobe® Systems, Salina, KS.

Figure 5-11: Phases of Macro-Core® Closed-Piston soil sampling. Illustration of diagram from Geoprobe Macro-Core® Soil Sampler SOP Technical Bulletin 95-8500, Sept. 1996 rev.

Table 5-9: Examples of samples contacting and non-contacting equipment. Source: ASTM.

Figure 5-13: Each sprayer should be dedicated to use with only one type of decontamination solution. Photo courtesy of Forestry Suppliers, Inc., Jackson, MS.

Box 5-4: Utility ground marker. HMTRI Photo.

Chapter 6

Figure 6-1: Water is one of our most precious resources. We protect it by setting up and implementing regulatory standards for groundwater and surface waters. Photography by John Livzey © 1997.

Figure 6-4: The screen area at the bottom of a groundwater monitoring well is built in such a way as to allow the water from the aquifer to flow through it and into the well. Photo courtesy of Cook Screen Technologies, Inc., Cincinnati, OH.

Figure 6-12b: Close-up of submersible pump. Photo courtesy of Ben Meadows Co., Atlanta, GA.

Figure 6-13: Bomb samplers. Photo courtesy of Ben Meadows Co., Atlanta, GA.

Figure 6-14: Kemmerer bottle. Photo courtesy of Ben Meadows Co., Atlanta, GA.

Figure 6-15: Subsurface grab sampler. Photo courtesy of Ben Meadows Co., Atlanta, GA.

Figure 6-16: Dip sampler. Photo courtesy of Ben Meadows Co., Atlanta, GA.

Figure 6-17: Dissolved oxygen meter. Photo courtesy of Corning Inc., Corning, NY.

Figure 6-18: Filters for dissolved metals. Photo courtesy of Gelman Sciences, Ann Arbor, MI.

Point of View: Cheryl Stith. Photo courtesy of Glenn Black.

Chapter 7

Figure 7-1: Valley of the Drums, Brooks, Kentucky. Photo courtesy of Steve Delaney/ EPA.

Box 7-1: Lab packs may be used to transport any unmarked containers of potentially hazardous materials. For example, materials from a school chemistry lab that has not been properly maintained through the years would be transported in lab packs to a hazardous waste facility. Photo courtesy of Lab Safety Supply, Inc., Janesville, WI.

Figure 7-3: The technicians use the HNU photoionization detector (PID) to check for air contaminants around the drums. Photo Courtesy of Versar Inc., Springfield, VA.

Figure 7-4: Drum opening using a manual drum deheader. Photo courtesy of Lab Safety Supply, Inc., Janesville, WI.

Box 7-2: Drum data sheet. Source: EPA SOP #2009.

Figure 7-6: The COLIWASA is long enough to take a composite sample from the entire depth of a 55-gallon drum. Photo courtesy of Lab Safety Supply, Inc., Janesville, WI.

Technical Corner: Industrial plants or shops that store flammable materials or hazardous wastes will keep those materials in a clearly marked storage shed. Each such storage drum is grounded by attaching a bonding wire from the drum to a metal pole that is in contact with the earth. Note that the technician in the photograph above has also used a bonding wire from the drum to the portable container. Photos courtesy of Lab Safety Supply, Inc., Janesville, WI.

Box 7-4: The vacuum truck operator removes from the sump an oil-water mixture, which is bulked in the truck with similar materials from other sites. Photo courtesy of Safety-Kleen Corp., Elgin, IL.

Box 7-5: Bruin Lagoon, Bruin Borough, Pennsylvania. The wastes are primarily from the white oil manufacturing industry. They contain sulfuric acid, heavy metals and other materials, which have produced contamination of air, surface water and groundwater. Photo courtesy of Steve Delaney/ EPA.

Point of View: David Foss. HMTRI Photo.

Figure 7-9: Among the items of PPE worn by these technicians are full-facepiece canister-type respirators and hard hats. This indicates Level C, which is appropriate if the atmospheric contaminants and liquid splashes at the site would not adversely affect skin, and if the air contaminants have been identified so that the proper canisters are used. Photo courtesy of Steve Delaney/EPA.

Chapter 8

Figure 8-1: The extraction method can analyze the solubility of various wastes in solvent. Photo courtesy of National Environmental Testing Inc., Bartlett, IL.

Figure 8-2: The technician performs an immunoassay test to detect the presence of PCBs in soil. Photo courtesy of Millipore Corp., Bedford, MA.

Point of View: Mike Steinmaus.

Figure 8-3: A technician performs a field screening for volatile organics using a mobile gas chromatograph. Photo courtesy of Versar Inc., Springfield, VA.

Table 8-2: Excerpt from 40 CFR 136.3 listing the various methods which the EPA has approved inorganic test procedures.

Figure 8-4: AA instrument. Photo courtesy of National Environmental Testing Inc., Bartlett, IL.

Figure 8-5: The ICP instrument analyzes the concentration of metal ions in a sample. Photo courtesy of National Environmental Testing Inc., Bartlett, IL.

Table 8-3: AA concentration ranges. Source: EPA Methods for Chemical Analysis of Water and Wastes, 600/4-79-020.

Figure 8-6a: The GC indicates the likely compounds in a sample as peaks on a graph. The computer matches the peaks against those generated by pure forms of numerous compounds. The MS further refines the analysis by determining the molecular weight of each component of the sample. Photo courtesy of National Environmental Testing Inc., Bartlett, IL.

Figure 8-7: Environmental instructors demonstrate the use of titration in a "sheen screen" lab exercise at a professional development workshop. HMTRI Photo.

Figure 8-8: Hazardous waste sticker. Source: EPA

Table 8-5: TCLP regulated contaminants. Source: 40 CFR 261.24.

Chapter 9

Figure 9-1: The Total Organic Carbon (TOC) Analyzer converts all carbon-based components of a sample to carbon dioxide to determine the sample's "carbon load" (amount of carbon). Photo courtesy of National Environmental Testing, Inc., Bartlett, IL.

Figure 9-2: Environmental Manager Ricky Peer interprets air quality lab reports for a U.S. Army manufacturer and compares the data to Illinois EPA permit limitations. HMTRI Photo.

Figure 9-3: To generate high quality data, the laboratory technician must follow standard methodology designated by EPA while meeting the lab's quality assurance guidelines. Photo courtesy of National Environmental Testing, Inc., Bartlett, IL.

Figure 9-4: The sampling technician must perform every field duty with diligence. The expression "Garbage in? Garbage out!" reminds us that the laboratory analysis is a costly waste of time and money if the samples have not been properly collected and stored. An excellent field technician is a great asset to the site characterization team. Photo courtesy of National Environmental Testing, Inc., Bartlett, IL.

Technical Corner: Underground storage tank removal. HMTRI Photo.

Bibliography

American Association of State Highway and Transportation Officials (AASHTO), Designation M 145-87, "Recommended Practice for The Classification of Soils and Soil-Aggregate Mixtures for Highway Construction Purposes," Washington, DC, 1987.

American Society for Testing and Materials (ASTM), "Standard Practice for Sampling Chain-of-Custody Procedures," D4840-88. West Conshohocken, PA: ASTM, Reapproved 1993.

American Society for Testing and Materials (ASTM), Standard D 2487-92, "Standard Classification of Soils for Engineering Purposes (Unified Soil Classification System)," West Conshohocken, PA: ASTM, 1992.

American Society for Testing and Materials (ASTM), Standard D 2487-93, "Standard Practice for Description and Identification of Soils (Visual-Manual Procedure)," West Conshohocken, PA: ASTM, 1993.

American Society for Testing and Materials (ASTM), Standard D 3370-82, "Standard Practices for Sampling Water," West Conshohocken, PA: ASTM, 1982.

American Society for Testing and Materials (ASTM), Standard D 4448-85a, "Standard Guide for Sampling Groundwater Monitoring Wells," West Conshohocken, PA: ASTM, Reapproved 1992.

American Society for Testing and Materials (ASTM), Standard D 5088-90 "Standard Practice for Decontamination of Field Equipment Used at Nonradioactive Waste Sites," West Conshohocken, PA: ASTM, 1990.

American Society for Testing and Materials (ASTM), Standard D 5092-90, "Standard Practice for Design and Installation of Ground Water Monitoring Wells in Aquifers," West Conshohocken, PA: ASTM, 1990.

American Water Works Association, Water Pollution Control Association, and American Public Health Association, *Standard Methods for the Examination of Water and Wastewater*, 17th edition, Washington, D.C., American Public Health Association, 1989.

Berry, Wendell, *The Unsettling of America: Culture and Agriculture*, San Francisco, Sierra Club Books, 1977.

Boulding, J. Russell, *Description and Sampling of Contaminated Soils*, 2nd ed. Boca Raton, CRC Press Inc., 1994.

Brady, James and Humiston, Gerard, *General Chemistry Principles and Structure*, 2nd ed., John Wiley and Sons, 1978.

Byrnes, David E., *Field Sampling Methods for Remedial Investigations*, Boca Raton, CRC Press Inc., 1994.

Chiras, Daniel D., *Lessons from Nature*, Washington, D.C., Island Press, 1992.

Csuros, Maria, *Environmental Sampling and Analysis for Technicians*, Boca Raton, CRC Press Inc., 1994.

Driscoll, Fletcher G., *Groundwater and Wells*, 2nd ed., St. Paul, Johnson Filtration Systems Inc., 1986.

Feil, Doug et al., *Hazardous Waste Site Health and Safety Guide*, Raleigh, Hazardous Materials Training and Research Institute, 1993.

Fetter, C.W., *Applied Hydrology*, 2nd ed. Columbus, Merrill Publishing Company, 1988.

Freeze, R. Allan and Cherry, John A., *Groundwater*, Englewood Cliffs, Prentice-Hall Inc., 1979.

Friedman, Gerald M. and Sanders, John E., *Principles of Sedimentology*, New York, John Wiley & Sons, 1978.

Garrett, Peter, *How to Sample Ground Water and Soils*, Dublin, National Water Well Association, 1988.

Geoprobe Systems, "Geoprobe Large Bore Soil Sampler Standard Operating Procedure: Technical Bulletin 93-660," Salina, Kejr Engineering Inc., 1996.

Geoprobe Systems, "Geoprobe Macro-Core Soil Sampler Standard Operating Procedure: Technical Bulletin 95-8500," Salina, Kejr Engineering Inc., 1996.

Geoprobe Systems, "Geoprobe Tools and Equipment Catalog," Salina, Kejr Engineering Inc., 1995.

Gere, Diane and Walker, Christine, *Living and Writing Responsibly*, Raleigh, Hazardous Materials Training and Research Institute, 1992.

Hazardous Materials Training and Research Institute, *Sampling and Monitoring Procedures (HMT270)*, Raleigh, HMTRI, 1993.

Heath, Ralph C., *Basic Ground-Water Hydrology*, United States Geological Survey Water-Supply Paper 2220, U.S. Government Printing Office, Washington, DC, 1983.

Hoyer, Bernard E. et al., *Iowa Groundwater Protection Strategy*, prepared for the Environmental Protection Commission, Iowa Department of Natural Resources, 1987.

Keith, Lawrence H. ed., *Principles of Environmental Sampling*, Washington, D.C., American Chemical Society, 1996.

National Technical Information Service (NTIS), *Samplers and Sampling Procedures for Hazardous Waste Streams*, Springfield, U.S. Department of Commerce National Technical Information Service, 1980 (EPA/600/2-80-018).

Nielsen, David M., ed., *Practical Handbook of Ground-Water Monitoring*, Chelsea, Lewis Publishers Inc., 1991.

NIOSH/OSHA/USCG/EPA, *Occupational Safety and Health Guidance Manual for Hazardous Waste Site Activities*, Washington D.C., Government Printing Office, 1985.

Powers, M.C., "A New Roundness Scale For Sedimentary Particles," *Journal of Sedimentary Petrology*, 23 (1953): p. 117-119.

Ritter, Dale F., *Process Geomorphology*, Dubuque, Wm. C. Brown Company, 1978.

Root, Cynthia, *Operator Training Guide on the Sampling Procedures for the Safe Drinking Water Act*, National Environmental Training Association and the Coalition of Environmental Training Centers, Phoenix, AZ, 1995.

Sanborn Map Company, 629 Fifth Ave., Pelham, NY 10803

Soil Conservation Service, "Soil Classification, A Comprehensive System - 7th Approximation," Washington, D.C., U.S. Department of Agriculture, 1960.

Title 29, Code of Federal Regulations, Part 1910, Section 1910.120.

Title 29, Code of Federal Regulations, Part 1910, Section 1910.120.

Title 40, Code of Federal Regulations, Part 264.

Title 49, Code of Federal Regulations, Part 172, Section 172.101.

Title 49, Code of Federal Regulations, Part 178, Subpart L.

U.S. Environmental Protection Agency, *A Compendium of Superfund Field Operations Methods*, Washington, D.C., Office of Emergency and Remedial Response and Office of Waste Programs Enforcement, 1987, (EPA/540/P-87/001).

U.S. Environmental Protection Agency, *Air Monitoring for Hazardous Materials*, Washington, D.C., Office of Emergency and Remedial Response, Environmental Response Team, 1993.

U.S. Environmental Protection Agency, *Data Quality Objectives for Remedial Response Activities*, Washington D.C., Office of Emergency and Remedial Response and Office of Waste Programs Enforcement, Environmental Response Team, 1987.

U.S. Environmental Protection Agency, *Data Quality Objectives Process for Superfund: Interim Final Guidance*, Washington, D.C., General Printing Office, 1993, (EPA 540/G-93/071).

U.S. Environmental Protection Agency, *Ground Water*, Washington D.C., Government Printing Office, 1987, (EPA/625/6-87/016).

U.S. Environmental Protection Agency, *Guidance for Conducting Remedial Investigations and Feasibility Studies Under CERCLA*, Office of Emergency and Remedial Response, Washington, D.C., 1988, (EPA/540/G-89/004)

U.S. Environmental Protection Agency, *Methods for Chemical Analysis of Water and Wastes*, Washington, D.C., General Printing Office, 1983, (EPA 600/4-79-020).

U.S. Environmental Protection Agency, *Methods for Determination of Organic Compounds in Drinking Water*, Washington, D.C., General Printing Office, 1988, (EPA 600/4-88-039).

U.S. Environmental Protection Agency, *Methods for the Determination of Metals in Environmental Samples*, Washington, D.C., General Printing Office, 1993, (EPA 600/4-91-010).

U.S. Environmental Protection Agency, *Sampling for Hazardous Materials*, Washington D.C., Office of Emergency and Remedial Response–Hazardous Response Support Division, Environmental Response Team, 1987.

U.S. Environmental Protection Agency, *Standard Operating Safety Guides*, Office of Emergency and Remedial Response, Washington, D.C. 20460, 1992. (Pub 9285.1-03)

U.S. Environmental Protection Agency, *Test Methods for Evaluating Solid Waste, Physical/Chemical Methods*, (SW-846), Washington, D.C., General Printing Office, 1991.

Wilson, Neal, *Soil Water and Ground Water Sampling*, Boca Raton, CRC Press Inc., 1995.

Winegardner, Duane L., *An Introduction to Soils for Environmental Professionals*, Boca Raton, CRC Press Inc., 1996.

Index

A

G

H

I

J

K

L

M

N

O

P

T

U

V

W